TOBY WILLIAMSON

Advanced Strength and Applied Elasticity

The SI Version

Advanced Strength and Applied Elasticity

The SI Version

A.C. Ugural
Fairleigh Dickinson University

S. K. Fenster
New Jersey Institute of Technology

ELSEVIER
New York • Oxford

Elsevier North Holland
52 Vanderbilt Avenue, New York, New York 10017

Distributors outside the United States and Canada:
Edward Arnold (Publishers), Ltd.
41 Bedford Square, London
WC1B 3DQ England

Library of Congress Cataloging in Publication Data

Ugural, A.C.
 Advanced strength and applied elasticity: The SI version.

 Includes bibliographical references and index.
 1. Strength materials. 2. Elasticity. I. Fenster,
 Saul K., 1933– joint author. II. Title.
TA405.U42 1981   620.1'12   80-26638
ISBN 0-444-00428-9

Desk Editor Louise Calabro Schreiber
Design Edmée Froment
Art Editor Glen Burris
Cover Design Paul Agule Design
Production Manager Joanne Jay
Compositor Science Typographers, Inc.
Printer Haddon Craftsmen

Manufactured in the United States of America

Contents

Preface to the SI Version

The new edition of this text seeks to preserve the general character of the original version. While the change to the System of International Units provided the immediate impetus for this work, a major effort has been made to provide a more complete and current text through the inclusion of new material dealing with the following topics: polynomial solutions, metal fatigue, failure criteria for combined fluctuating loads, torsion of curved bars, thermal stresses in cylinders, elastic-plastic stresses in rotating disks, and the inelastic behavior of structural members. An appendix concerned with the practical solution of the stress cubic equation and direction cosines has been added.

The authors have resisted the temptation to increase greatly the material covered. However, it was considered desirable to add a number of new illustrative examples and problems of practical importance. It is hoped that clarity of presentation has been maintained, as well as simplicity as permitted by the nature of the subject, unpretentious depth, an effort to encourage intuitive understanding, and a shunning of the irrelevant.

The subject matter of this text is usually covered in one-semester senior and one-semester graduate level courses in *Strength of Solids* and *Applied Elasticity*. However, inasmuch as sufficient material is presented for a full year of study, the book may simulate the development of courses in *Structural Analysis* and *Stress Analysis*. Topics selected and methods of presentation are also directed to the needs of practicing design and research engineers.

Because of extensive subdivision into a variety of topics and employment of more than one method of analysis, the text should provide flexibility in choice of assignments to cover courses of different length and content. Chapters 1 and 2 address the analysis of stress and strain and should be studied first. The treatment of two-dimensional problems in elasticity (Chapter 3) is illustrated repeatedly throughout and compared with the mechanics of materials approach. The remaining chapters may be studied in any order.

We are indebted to the many readers who have contributed numerous constructive suggestions. Our thanks also to our colleagues who have found the book useful through the years, and who have given encouragement to the preparation of the SI version.

A. C. Ugural

S. K. Fenster

Teaneck, New Jersey
January, 1981

Preface to the First Edition

This text is a development of classroom notes prepared in connection with advanced undergraduate and first year graduate courses in elasticity and the mechanics of solids. It is designed to satisfy the requirements of courses subsequent to an elementary treatment of the strength of materials. In addition to its applicability to aeronautical, civil, and mechanical engineering and to engineering mechanics curricula, the authors have endeavored to make the text useful to practicing engineers. Emphasis is given to *numerical techniques* (which lend themselves to computerization) in the solution of problems resisting *analytical treatment*. The stress placed upon numerical solutions is not intended to deny the value of classical analysis, which is given a rather full treatment. It instead attempts to fill what the authors believe to be a void in the world of textbooks.

An effort has been made to present a balance between the theory necessary to gain insight into the mechanics, but which can offer no more than crude approximations to real problems because of simplifications related to geometry and conditions of loading, and numerical solutions, which are so useful in presenting stress analysis in a more realistic setting. The authors have thus attempted to emphasize those aspects of theory and application which prepare a student for more advanced study or for professional practice in design and analysis.

The theory of elasticity plays three important roles in the text: it provides exact solutions where the configurations of loading and boundary are relatively simple; it provides a check upon the limitations of the strength of materials approach; it serves as the basis of approximate solutions employing numerical analysis.

To make the text as clear as possible, attention is given to the presentation of the fundamentals of the strength of materials. The physical significance of the solutions and practical applications are given emphasis. The authors have made a special effort to illustrate important principles and applications with numerical examples. Consistent with announced national policy, included in the text are problems in which the physical quantities are expressed in the International System of Units (SI).

It is a particular pleasure, upon the completion of a project of this nature, to acknowledge the contributions of those who assisted the authors in the evolution

xiii

of the text. Thanks are, of course, due to the many students who have made constructive suggestions throughout the several years when drafts of this work were used as a text. To Professor F. Freudenstein of Columbia University and Professor R. A. Scott of the University of Michigan, we express our appreciation for their helpful recommendations and valuable perspectives in connection with their review of the manuscript. And, as has always been the case, Mrs. Helen Stanek has provided intelligent editorial and typing assistance throughout the several drafts of this work; to her, the authors express their special thanks and appreciation.

A. C. Ugural
S. K. Fenster

Teaneck, New Jersey
May, 1975

List of Symbols

A	area, constant
a,b	constants, dimensions, distances
C	carry-over factor, torsional rigidity
c	distance from neutral axis to outer fiber
D	distribution factor, flexural rigidity of plate
$[D]$	elasticity matrix
d	diameter, distance
E	modulus of elasticity in tension or compression
E_s	modulus of plasticity or secant-modulus
E_t	tangent modulus
e	dilatation, displacement, distance, eccentricity
F	body force per unit volume, concentrated force
f_s	form factor for shear
G	modulus of elasticity in shear
g	acceleration of gravity (≈ 9.81 m/s²)
h	depth of beam, height, membrane deflection, mesh width
I	moment of inertia of area, stress invariant
J	polar moment of inertia of area
K	bulk modulus, spring constant of an elastic support, stiffness factor, thermal conductivity, fatigue factor
$[K]$	stiffness matrix of whole structure
k	constant, modulus of elastic foundation, spring constant, strength coefficient, stress concentration factor
$[k]$	stiffness matrix of finite element
L	length, span
M	moment
M_t, M_{xy}	twisting moment or torque
m	area property, moment caused by unit load

N	factor of safety, fatigue life (cycles), force, newton
n	number, strain hardening index
l,m,n	direction cosines
P	concentrated force
p	distributed load per unit length or area, pressure
Q	first moment of area, heat flow per unit length, shearing force
$\{Q\}$	nodal force matrix of finite element
R	radius, reaction
r	radius, radius of gyration
r, θ	polar coordinates
s	distance along a line or a curve, second
T	surface load per unit area or stress resultant, temperature
t	thickness
U	strain energy
U_0	strain energy per unit volume
$U*$	complementary energy
V	shearing force, velocity, volume
W	weight, work
u,v,w	components of displacement
Z	section modulus
x,y,z	rectangular coordinates
α	angle, coefficient of thermal expansion
β	numerical factor
γ	shear strain, weight per unit volume or specific weight
δ	deflection, finite difference operator, variational symbol
$\{\delta\}$	nodal displacement matrix of finite element
Δ	change of a function
ε	normal strain
θ	angle, angle of twist per unit length
ν	Poisson's ratio
κ	curvature
λ	axial load factor, Lamé constant
Π	potential energy
ρ	density (mass per unit volume), radius
σ	normal stress
τ	shear stress
ϕ	angle, stress function
φ	total angle of twist
ω	angular velocity

**Advanced
Strength
and Applied
Elasticity**

The SI Version

Chapter 1

Analysis of Stress

1.1 Introduction

The basic structure of matter is characterized by nonuniformity and discontinuity attributable to its various subdivisions: molecules, atoms, and subatomic particles. Our concern in this text is not with the particulate level of matter, however, and it is to our advantage to replace the actual system of particles with a *continuous* distribution of matter. There is the clear implication in such an approach that any small volumes with which we may deal are large enough to contain a great many particles. Random fluctuations in the properties of the material are thus of no consequence. Of the states of matter, we are here concerned only with the solid, with its ability to maintain its shape without the need of a container, and to resist continuous shear and tension.

In contrast with rigid body statics and dynamics, which treat the external behavior of bodies (i.e., the equilibrium and motion of bodies without regard to small deformations associated with the application of load), the mechanics of solids is concerned with the relationships of external effect (forces and moments) to internal stresses and strains.

External forces acting on a body may be classified as *surface forces* and *body forces*. A surface force is of the *concentrated* type when it acts at a point; a surface force may also be distributed *uniformly* or *nonuniformly* over a finite area. Body forces act on volumetric elements rather than surfaces, and are attributable to fields such as gravity and magnetism.

The principal topics under the general heading *mechanics of solids* may be summarized as follows:

a. Analysis of the stresses and deformations within a body subject to a prescribed system of forces. This is accomplished by solving the governing equations which describe the stress and strain fields (theoretical stress analysis). It is often advantageous, where the shape of the structure or conditions of loading preclude a theoretical solution or where verification is required, to apply the laboratory techniques of experimental stress analysis.

   b. Determination by theoretical analysis or by experiment of the limiting values of load which a structural element can sustain without suffering damage, failure, or compromise of function.

   c. Determination of the body shape and selection of those materials which are most efficient for resisting a prescribed system of forces under specified conditions of operation such as temperature, humidity, vibration, and ambient pressure. This is the *design* function and more particularly that of *optimum design*. Efficiency may be gaged by such criteria as minimum weight or volume, minimum cost, or any criterion deemed appropriate.

The design function, item (c) above, clearly relies upon the performance of the theoretical analyses cited under (a) and (b), and it is these to which this text is directed. The role of analysis in design is observed in examining the following steps comprising the systematic design of a load-carrying member:

   1. Evaluation of the most likely modes of failure under anticipated conditions of service.

   2. Determination of expressions relating external influences such as force and torque to such effects as stress, strain, and deformation. Often, the member under consideration and the conditions of loading are so significant or so amenable to solution as to have been the subject of prior analysis. For these situations textbooks, handbooks, journal articles, and technical papers are good sources of information. Where the situation is unique, a mathematical derivation specific to the case at hand is required.

   3. Determination of the maximum or allowable value of a significant quantity such as stress, strain, or energy, either by reference to compilations of material properties or by experimental means such as a simple tension test. This value is used in connection with the relationship derived in (2).

   4. Selection of a *design factor of safety*, usually referred to simply as the factor of safety, to account for uncertainties in a number of aspects of the design, including those related to the actual service loads, material properties, or environmental factors. An important area of uncertainty is connected with the assumptions made in the analysis of stress and deformation. Also, one is not likely to have a secure knowledge of the stresses which may be introduced during machining, assembly, and shipment of the element. The design factor of safety also reflects the consequences of failure, e.g., the possibility that failure will result in loss of human life or injury, and the possibility that failure will result in costly repairs or danger to other components of the overall system. For the above-mentioned reasons the design factor of safety is also sometimes called the *factor of ignorance*. The uncertainties encountered during the design phase

may be of such magnitude as to lead to a design-carrying extreme weight, volume, or cost penalties. It may then be advantageous to perform thorough tests or more exacting analysis rather than to rely upon overly large design factors of safety. The so-called *true factor of safety* can only be determined after the member is constructed and tested. This factor is the ratio of the maximum load the member *can sustain* under severe testing without damage to the maximum load *actually* carried under normal service conditions.

The foregoing procedure is not always conducted in as formal a fashion as may be implied. In some design procedures, one or more steps may be regarded as unnecessary or obvious on the basis of previous experience.

We conclude this section with an appeal for the reader to exercise a degree of skepticism with regard to the application of formulas for which there is uncertainty as to the limitations of use or the areas of applicability. The relatively simple form of many formulas usually results from rather severe restrictions in their derivation. These relate to simplified boundary conditions and shapes, limitations upon stress and strain, and the neglect of certain complicating factors. Designer and stress analysts must be aware of such restrictions lest their work be of no value, or worse, lead to dangerous inadequacies.

In this chapter, we are concerned with the state of *stress at a point* and the *variation of stress* throughout an elastic body. The latter is dealt with in Secs. 1.4 and 1.12, and the former in the balance of the chapter.

1.2 Definition of Stress

Consider a body in equilibrium, subject to the system of forces shown in Fig. 1.1a. An element of area ΔA, located on an exterior or interior surface (the latter as in Fig. 1.1b), is acted on by force $\Delta \mathbf{F}$. Let n, s_1, s_2 constitute a set of orthogonal axes, origin placed at the point P, with n normal and s_1, s_2 tangent to ΔA. In general $\Delta \mathbf{F}$ does not lie along n, s_1, or s_2. Decomposition of $\Delta \mathbf{F}$ into components parallel to n, s_1, and s_2 (Fig. 1.1c) leads to the following definitions of the normal stress σ_n and the shear stresses τ_s:

$$\sigma_n = \lim_{\Delta A \to 0} \frac{\Delta F_n}{\Delta A}$$

$$\tau_{s_1} = \lim_{\Delta A \to 0} \frac{\Delta F_{s_1}}{\Delta A}, \qquad \tau_{s_2} = \lim_{\Delta A \to 0} \frac{\Delta F_{s_2}}{\Delta A} \tag{1.1}$$

These expressions provide the stress components at a point P to which the area ΔA is reduced in the limit. Clearly, the expression $\Delta A \to 0$ depends upon the idealization discussed in Sec. 1.1. In the International System of Units (SI), stress is measured in *newtons per square meter* or *pascals* (Pa). The SI system replaces the English system of units, which has long been

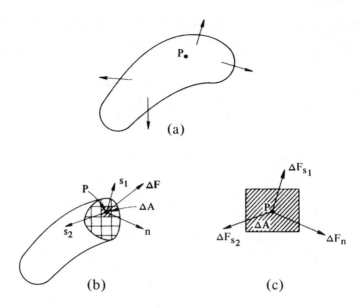

Figure 1.1

used by engineers in this country. Table 1.1 compares the two systems. The common prefixes in SI usage are μ(micro), m(milli), c(centi), d(deci), k(kilo), M(mega), and G(giga) which represent the multiplication factors 10^{-6}, 10^{-3}, 10^{-2}, 10^{-1}, 10^3, 10^6, and 10^9, respectively.

The values obtained in the limiting processes of Eq. (1.1) differ from point to point on the surface as ΔF varies. The stress components depend not only upon ΔF, however, but also upon the orientation of the plane on which it acts at point P. Even at a given point, therefore, the stresses will differ as different planes are considered. The complete description of stress at a point thus requires the specification of the stress on all planes passing through the point.

Table 1.1

Quantity	SI Units	U.S. Units
Length	meter (m)	inch
Force	newton (N)	pound force
Time	second (s)	second
Mass	kilogram (kg)	pound mass, slug

Some conversion factors

1 m = 39.37 in.	1 N = 0.2248 lbf
1 MPa = 145 psi	1 kg = 2.2046 lbm
1 kN/m = 68.53 lb/ft	1 N·m = 0.7376 lbf·ft

1.3 The Stress Tensor

It is verified in Sec. 1.8 that in order to enable the determination of the stresses on an infinite number of planes passing through a point P, thus defining the stress at that point, one need only specify the stress components on three mutually perpendicular planes passing through the point. These three planes, perpendicular to the coordinate axes, contain three sides of an infinitesimal parallelepiped (Fig. 1.2). A three-dimensional state of stress is shown in the figure. Consider the stresses to be identical at points P and P', and uniformly distributed on each face, represented by a single vector acting at the center of each face. In accordance with the foregoing, a total of nine scalar stress components define the state of stress at a point. The stress components can be assembled in the following *matrix form*, wherein each row represents the group of stresses acting on a plane passing through $P(x, y, z)$:

$$\begin{bmatrix} \tau_{xx} & \tau_{xy} & \tau_{xz} \\ \tau_{yx} & \tau_{yy} & \tau_{yz} \\ \tau_{zx} & \tau_{zy} & \tau_{zz} \end{bmatrix} = \begin{bmatrix} \sigma_x & \tau_{xy} & \tau_{xz} \\ \tau_{yx} & \sigma_y & \tau_{yz} \\ \tau_{zx} & \tau_{zy} & \sigma_z \end{bmatrix} \tag{1.2}$$

The above array represents a tensor of second rank (refer to Sec. 1.8), requiring two indices to identify its elements or components. A vector is a tensor of first rank; a scalar is of zero rank.

The double subscript notation is interpreted as follows: the first subscript indicates the direction of a normal to the plane or face on which the stress component acts; the second subscript relates to the direction of the stress itself. Repetitive subscripts will be avoided in this text, so that the normal stresses τ_{xx}, τ_{yy}, and τ_{zz} will be designated σ_x, σ_y, and σ_z, as indicated in Eq. (1.2). *A face or plane is usually identified by the axis normal to it*, e.g., the x faces are perpendicular to the x axis.

Referring again to Fig. 1.2, we observe that *both* stresses labeled τ_{yx} tend to twist the element in a clockwise direction. It would be convenient,

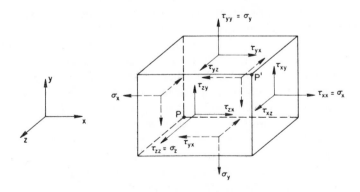

Figure 1.2

therefore, if a sign convention were adopted under which these stresses carried the same sign. Applying a convention relying solely upon the coordinate direction of the stresses would clearly not produce the desired result, inasmuch as the τ_{yx} stress acting on the upper surface is directed in the positive x direction, while τ_{yx} acting on the lower surface is directed in the negative x direction. The following sign convention, which applies to both normal and shear stresses, is related to the deformational influence of a stress, and is based upon the relationship between the direction of an outward normal drawn to a particular surface, and the directions of the stress components on the same surface. When *both* the outer normal and the stress component face in a positive direction relative to the coordinate axes, the stress is positive. When *both* the outer normal and the stress component face in a negative direction relative to the coordinate axes, the stress is positive. When the normal points in a positive direction while the stress points in a negative direction (or vice versa), the stress is negative. In accordance with this sign convention, tensile stresses are always positive and compressive stresses always negative. Figure 1.2 depicts a system of positive normal and shear stresses.

Many of the equations of elasticity become quite unwieldy when written in full, unabbreviated form; see, for example, Eq. (1.17). As the complexity of the situation described increases, so does that of the formulations, tending to obscure the fundamentals in a mass of symbols. For this reason the more compact *indicial* or *tensor notation* described in Appendix A is sometimes found in technical publications. A stress tensor is written in indicial notation as τ_{ij}, where i and j each assume the values x, y, and z as required by Eq. (1.2). Generally, such notation is not employed in this text.

1.4 Variation of Stress Within a Body

As pointed out in Sec. 1.2, the components of stress generally vary from point to point in a stressed body. These variations are governed by the conditions of equilibrium of *statics*. Fulfillment of these conditions establishes certain relationships, known as the *differential equations of equilibrium*, which involve the derivatives of the stress components.

Consider a thin element of sides dx and dy (Fig. 1.3), and assume that σ_x, σ_y, τ_{xy}, τ_{yx} are functions of x, y but do not vary throughout the thickness (are independent of z) and that the other stress components are zero. Also assume that the x and y components of the body forces per unit volume, F_x and F_y, are independent of z, and that the z component of the body force $F_z = 0$. This combination of stresses, satisfying the conditions described, is termed *plane stress*. Note that as the element is very small, for the sake of simplicity, the stress components may be considered to be distributed uniformly over each face. In the figure they are shown by a single vector, representing the mean values applied at the center of each face.

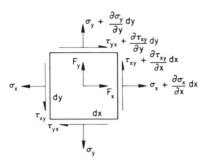

Figure 1.3

As we move from one point to another—as, for example, from the lower left corner to the upper right corner of the element—one of the stress components, say σ_x, acting on the negative x face, changes in value on the positive x face. The stresses σ_y, τ_{xy}, and τ_{yx} similarly change. The variation of stress with position may be expressed by a truncated Taylor's expansion:

$$\sigma_x + \frac{\partial \sigma_x}{\partial x} dx \tag{a}$$

The partial derivative is used because σ_x is a function of x and y. Treating all the components similarly, the state of stress shown in Fig. 1.3 is obtained.

We consider now the equilibrium of an element of unit thickness, taking moments of force about the lower left corner:

$$\left(\frac{\partial \sigma_y}{\partial y} dx\, dy \right) \frac{dx}{2} - \left(\frac{\partial \sigma_x}{\partial x} dx\, dy \right) \frac{dy}{2} + \left(\tau_{xy} + \frac{\partial \tau_{xy}}{\partial x} dx \right) dx\, dy$$

$$- \left(\tau_{yx} + \frac{\partial \tau_{yx}}{\partial y} dy \right) dx\, dy + F_y\, dx\, dy\, \frac{dx}{2} - F_x\, dx\, dy\, \frac{dy}{2} = 0$$

Neglecting the triple products involving dx and dy, the above reduces to

$$\tau_{xy} = \tau_{yx} \tag{1.3}$$

In a like manner, it may be shown that

$$\tau_{xz} = \tau_{zx} \quad \text{and} \quad \tau_{yz} = \tau_{zy}$$

The shearing forces are therefore characterized by subscripts that are commutative, and the stress tensor is symmetric. This symmetry means that each pair of equal shear stresses acts on mutually perpendicular planes. Because of this, we shall hereafter not distinguish between τ_{xy} and τ_{yx}, τ_{xz} and τ_{zx}, or τ_{yz} and τ_{zy}.

From the equilibrium of x forces we obtain

$$\left(\sigma_x + \frac{\partial \sigma_x}{\partial x} dx\right) dy - \sigma_x\, dy + \left(\tau_{xy} + \frac{\partial \tau_{xy}}{\partial y} dy\right) dx - \tau_{xy}\, dx + F_x\, dx\, dy = 0 \quad \text{(b)}$$

Upon simplification, Eq. (b) becomes

$$\left(\frac{\partial \sigma_x}{\partial x} + \frac{\partial \tau_{xy}}{\partial y} + F_x\right) dx\, dy = 0 \quad \text{(c)}$$

Inasmuch as $dx\, dy$ is nonzero, the quantity in the parentheses must vanish. A similar expression is written to describe the equilibrium of y forces. The x and y equations yield the following differential equations of equilibrium:

$$\frac{\partial \sigma_x}{\partial x} + \frac{\partial \tau_{xy}}{\partial y} + F_x = 0$$

$$\frac{\partial \sigma_y}{\partial y} + \frac{\partial \tau_{xy}}{\partial x} + F_y = 0$$

(1.4)

The differential equations of equilibrium for the case of three-dimensional stress may be generalized from the above expressions as follows:

$$\frac{\partial \sigma_x}{\partial x} + \frac{\partial \tau_{xy}}{\partial y} + \frac{\partial \tau_{xz}}{\partial z} + F_x = 0$$

$$\frac{\partial \sigma_y}{\partial y} + \frac{\partial \tau_{xy}}{\partial x} + \frac{\partial \tau_{yz}}{\partial z} + F_y = 0$$

$$\frac{\partial \sigma_z}{\partial z} + \frac{\partial \tau_{xz}}{\partial x} + \frac{\partial \tau_{yz}}{\partial y} + F_z = 0$$

(1.5)

A succinct representation of the above expressions is given Eq. (A.1).

1.5   Two-Dimensional Stress at a Point

A two-dimensional state of stress exists when the stresses and body forces are independent of one of the coordinates, here taken as z. Such a state is described by stresses σ_x, σ_y, and τ_{xy} and the x and y body forces. Two-dimensional problems are of two classes: *plane stress* and *plane strain*. In the case of plane stress, as described in the previous article, the stresses σ_z, τ_{xz}, τ_{yz} and the z directed body forces are assumed to be zero. In the case of plane strain, the stresses τ_{xz}, τ_{yz} and the body force F_z are likewise taken to be zero, but σ_z does not vanish,* and can be determined from stresses σ_x and σ_y.

---

*More details and illustrations of these assumptions are given in Chapter 3.

We shall now determine the equations for transformation of the stress components σ_x, σ_y, τ_{xy} at any point of a body represented by an *infinitesimal* element (Fig. 1.4a). The z-directed normal stress σ_z, even if it is nonzero, need not be considered here.

Consider an infinitesimal wedge cut from the loaded body shown in Fig. 1.4a, b. It is required to determine the stresses $\sigma_{x'}$ and $\tau_{x'y'}$, which refer to axes x', y' making an angle θ with axes x, y as shown in the figure. Let side BC be normal to the x' axis. Note that in accordance with the sign convention, $\sigma_{x'}$ and $\tau_{x'y'}$ are positive stresses as shown in the figure. If the area of side BC is taken as unity, then sides AB and AC have area $\cos\theta$ and $\sin\theta$, respectively.

Equilibrium of forces in the x and y directions requires that

$$T_x = \sigma_x \cos\theta + \tau_{xy}\sin\theta$$
$$T_y = \tau_{xy}\cos\theta + \sigma_y \sin\theta \tag{1.6}$$

where T_x and T_y are the *components of stress resultant* acting on BC in the x and y directions, respectively. The normal and shear stresses on the x' plane (BC plane) are obtained by projecting T_x and T_y in the x' and y' directions:

$$\sigma_{x'} = T_x \cos\theta + T_y \sin\theta$$
$$\tau_{x'y'} = T_y \cos\theta - T_x \sin\theta \tag{a}$$

Upon substitution of the stress resultants from Eq. (1.6), Eqs. (a) become

$$\sigma_{x'} = \sigma_x \cos^2\theta + \sigma_y \sin^2\theta + 2\tau_{xy}\sin\theta\cos\theta$$
$$\tau_{x'y'} = \tau_{xy}(\cos^2\theta - \sin^2\theta) + (\sigma_y - \sigma_x)\sin\theta\cos\theta \tag{b}$$

The stress $\sigma_{y'}$ may readily be obtained by substituting $\theta + \pi/2$ for θ in the

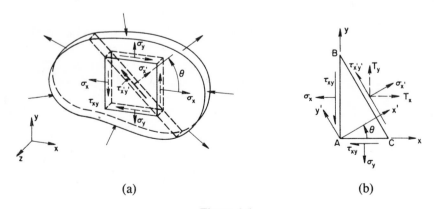

(a)   (b)

Figure 1.4

expression for $\sigma_{x'}$. By means of the trigonometric identities

$$\cos^2\theta = \tfrac{1}{2}(1 + \cos 2\theta), \qquad \sin\theta\cos\theta = \tfrac{1}{2}\sin 2\theta,$$

$$\sin^2\theta = \tfrac{1}{2}(1 - \cos 2\theta)$$

the transformation equations for stress are now written in the following form:

$$\sigma_{x'} = \tfrac{1}{2}(\sigma_x + \sigma_y) + \tfrac{1}{2}(\sigma_x - \sigma_y)\cos 2\theta + \tau_{xy}\sin 2\theta \qquad (1.7a)$$

$$\tau_{x'y'} = -\tfrac{1}{2}(\sigma_x - \sigma_y)\sin 2\theta + \tau_{xy}\cos 2\theta \qquad (1.7b)$$

$$\sigma_{y'} = \tfrac{1}{2}(\sigma_x + \sigma_y) - \tfrac{1}{2}(\sigma_x - \sigma_y)\cos 2\theta - \tau_{xy}\sin 2\theta \qquad (1.7c)$$

The foregoing expressions permit the computation of stresses acting on all possible planes BC—the *state of stress* at a point—provided that three stress components on a set of orthogonal faces are known.

Consider, for example, the possible states of stress corresponding to $\sigma_x = 14$ MPa, $\sigma_y = 4$ MPa, and $\tau_{xy} = 10$ MPa. Substituting these values into Eq. (1.7) and permitting θ to vary from $0°$ to $360°$ yields the data upon which the curves shown in Fig. 1.5 are based. The plots shown, called *stress trajectories,* are polar representations: $\sigma_{x'}$ vs θ (Fig. 1.5a) and $\tau_{x'y'}$ vs θ (Fig. 1.5b). It is observed that the direction of each maximum shear stress bisects the angle between the maximum and minimum normal stresses. Note that the normal stress is either a maximum or a minimum on planes at $\theta = 31.66°$ and $\theta = 31.66° + 90°$, respectively, for which the shearing stress is zero. The conclusions drawn from this example are valid for any two-dimensional (or three-dimensional) state of stress and will be observed in the sections to follow.

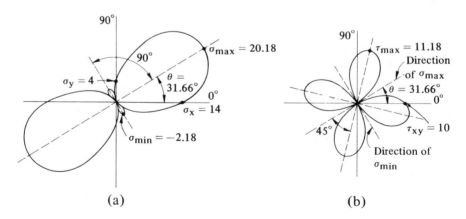

(a)   (b)

Figure 1.5

1.6 Principal Stresses in Two Dimensions

In order to ascertain the orientation of $x'y'$ corresponding to maximum or minimum $\sigma_{x'}$, the necessary condition $d\sigma_{x'}/d\theta=0$ is applied to Eq. (1.7a), yielding

$$-(\sigma_x-\sigma_y)\sin 2\theta+2\tau_{xy}\cos 2\theta=0 \qquad\text{(a)}$$

or

$$\tan 2\theta_p=\frac{2\tau_{xy}}{\sigma_x-\sigma_y} \qquad (1.8)$$

Inasmuch as $\tan 2\theta=\tan(\pi+2\theta)$, two directions, mutually perpendicular, are found to satisfy Eq. (1.8). These are the *principal directions*, along which the principal or maximum and minimum normal stresses act. When Eq. (1.7b) is compared with Eq. (a), it becomes clear that $\tau_{x'y'}=0$ on a principal plane. A principal plane is thus a plane of zero shear. The principal stresses are determined by substituting Eq. (1.8) into Eq. (1.7a):

$$\sigma_{1,2}=\frac{\sigma_x+\sigma_y}{2}\pm\sqrt{\left(\frac{\sigma_x-\sigma_y}{2}\right)^2+\tau_{xy}^2} \qquad (1.9)$$

Note that the *algebraically* larger stress given above is the maximum principal stress, denoted by σ_1. The minimum principal stress is represented by σ_2. Similarly, by using the above approach and employing Eq. (1.7b), an expression for the maximum shear stress may also be derived.

1.7 Mohr's Circle for Two-Dimensional Stress

A graphical technique, predicated upon Eq. (1.7), permits the rapid transformation of stress from one plane to another, and leads also to the determination of the maximum normal and shear stresses. In this approach Eqs. (1.7) are depicted by a stress circle, called Mohr's circle.* In the Mohr representation, the normal stresses obey the sign convention of Sec. 1.3. However, for the purposes only of *constructing and reading values of stress from Mohr's circle*, the sign convention for shear stress is as follows: if the shearing stresses on opposite faces of an element would produce shearing forces that result in a *clockwise* couple, these stresses are regarded as *positive*. Accordingly, the shearing stresses on the y faces of the element in Fig. 1.4a are taken as positive (as before), but those on the x faces are now negative. Given σ_x, σ_y, and τ_{xy} with algebraic sign in accordance with the foregoing sign convention, the procedure for obtaining Mohr's circle (Fig.

---

*After Otto Mohr (1835–1918), Professor at Dresden Polytechnic. For further details refer to any text dealing with the elementary mechanics of solids.

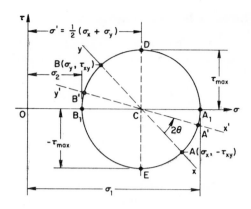

Figure 1.6

1.6) is as follows:

1. Establish a rectangular coordinate system, indicating $+\tau$ and $+\sigma$. Both stress scales must be identical.
2. Locate the center C of the circle on the horizontal axis a distance $\frac{1}{2}(\sigma_x + \sigma_y)$ from the origin.
3. Locate point A by coordinates σ_x, $-\tau_{xy}$. These stresses may correspond to any face of an element such as in Fig. 1.4a. It is usual to specify the stresses on the positive x face, however.
4. Draw a circle with center at C and of radius equal to CA.
5. Draw line AB through C.

The angles on the circle are measured in the same direction as θ is measured in Fig. 1.4. An angle of 2θ on the circle corresponds to an angle of θ on the element. The state of stress associated with the original x and y planes corresponds to points A and B on the circle, respectively. Points lying on diameters other than AB, such as A' and B', define states of stress with respect to any other set of x' and y' planes rotated relative to the original set through an angle θ. It is clear that points A_1 and B_1 on the circle locate the principal stresses and provide their magnitudes as defined by Eqs. (1.8) and (1.9), while D and E represent the maximum shearing stresses. The maximum values of shear stress (regardless of algebraic sign) will be denoted by τ_{max}, and are given by

$$\tau_{max} = \pm \frac{1}{2}(\sigma_1 - \sigma_2) = \pm \sqrt{\left(\frac{\sigma_x - \sigma_y}{2}\right)^2 + \tau_{xy}^2} \qquad (1.10)$$

Mohr's circle shows that the planes of maximum shear are always located at 45° from planes of principal stress, as already indicated in Fig. 1.5.

Note that a diagonal of an infinitesimal stress element along which the algebraically larger principal stress acts is referred to as the *positive shear diagonal.* The maximum shearing stresses *act toward* the positive shear diagonal. The normal stresses $\sigma_{x'}$ and $\sigma_{y'}$ associated with τ_{max} are equal to one another and given by

$$\sigma' = \tfrac{1}{2}(\sigma_1 + \sigma_2) = \tfrac{1}{2}(\sigma_x + \sigma_y) \tag{1.11}$$

The use of Mohr's circle is illustrated in the following examples.

Example 1.1. At a point in the structural member, the stresses (in megapascals) are represented as in Fig. 1.7a. Employ Mohr's circle to determine (a) the magnitude and orientation of the principal stresses and (b) the magnitude and orientation of the maximum shearing stresses and associated normal stresses. In each case show the results on a properly oriented element; represent the stress tensor in matrix form.

SOLUTION. Mohr's circle, constructed in accordance with the procedure outlined, is shown in Fig. 1.7b. The center of the circle is at $(40+80)/2 = 60$ MPa on the σ axis.

(a) The principal stresses are represented by points A_1 and B_1. Hence the maximum and minimum principal stresses, referring to the circle, are

$$\sigma_{1,2} = 60 \pm \sqrt{\tfrac{1}{4}(80-40)^2 + (30)^2}$$

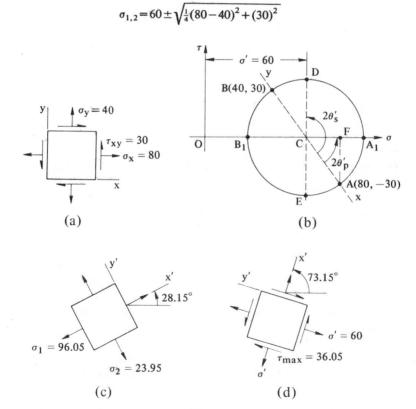

Figure 1.7

or

$$\sigma_1 = 96.05 \text{ MPa} \quad \text{and} \quad \sigma_2 = 23.95 \text{ MPa}$$

The planes on which the principal stresses act are given by

$$2\theta'_p = \tan^{-1} \frac{30}{20} = 56.30° \quad \text{and} \quad 2\theta''_p = 56.30° + 180° = 236.30°$$

Hence

$$\theta'_p = 28.15° \quad \text{and} \quad \theta''_p = 118.15°$$

Mohr's circle clearly indicates that θ'_p locates the σ_1 plane. The results may readily be checked by substituting the two values of θ_p into Eq. (1.7a). The state of principal stress is shown in Fig. 1.7c.

(b) The maximum shearing stresses are given by points D and E. Thus,

$$\tau_{\max} = \pm \sqrt{\tfrac{1}{4}(80-40)^2 + (30)^2} = \pm 36.05 \text{ MPa}$$

It is seen that $(\sigma_1 - \sigma_2)/2$ yields the same result. The planes on which these stresses act are represented by

$$\theta'_s = 28.15° + 45° = 73.15° \quad \text{and} \quad \theta''_s = 163.15°$$

As Mohr's circle indicates, the positive maximum shearing stress acts on a plane whose normal x' makes an angle θ'_s with the normal to the original plane (x plane). Thus, $+\tau_{\max}$ on two opposite x' faces of the element will be directed so that a clockwise couple results. The normal stresses acting on maximum shear planes are represented by OC, $\sigma' = 60$ MPa on each face. The state of maximum shearing stress is shown in Fig. 1.7d. The direction of the $\tau_{\max}$'s may also be readily predicted by recalling that they act toward the positive shear diagonal. We note that according to the general sign convention (Sec. 1.3), the shearing stress acting on the x' plane in Fig. 1.7d is negative. As a check, if $2\theta'_s = 146.30°$, and the given initial data are substituted into Eq. (1.7b), we obtain $\tau_{x'y'} = -36.05$ MPa, as already found.

We may now describe the state of stress at the point in the following matrix forms:

$$\begin{bmatrix} 80 & 30 \\ 30 & 40 \end{bmatrix}, \quad \begin{bmatrix} 96.05 & 0 \\ 0 & 23.95 \end{bmatrix}, \quad \begin{bmatrix} 60 & -36.05 \\ -36.05 & 60 \end{bmatrix}$$

These three representations, associated with the $\theta = 0°$, $\theta = 28.15°$, and $\theta = 73.15°$ planes passing through the point, are equivalent.

Note that if we assume $\sigma_z = 0$ in this example, a much *higher* shearing stress is obtained in the planes bisecting the x' and z planes (Problem 1.11). Thus, three-dimensional analysis, Sec. 1.11, should be considered for determining the *true maximum shearing stress* at a point.

Example 1.2. The stresses (in MPa) acting on an element of a loaded body are shown in Fig. 1.8a. apply Mohr's circle to determine the normal and shear stresses acting on a plane defined by $\theta = 30°$.

SOLUTION. Mohr's circle of Fig. 1.8b describes the state of stress given in Fig. 1.8a. Points A_1 and B_1 represent the stress components on the x and y faces, respectively.

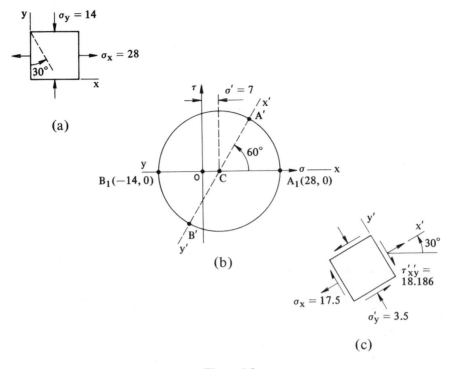

Figure 1.8

The radius of the circle is $(14+28)/2=21$. Corresponding to the 30° plane within the element, it is necessary to rotate through 60° counterclockwise on the circle to locate point A'. A 240° counterclockwise rotation locates point B'. Referring to the circle,

$$\sigma_{x'}=7+21\cos 60°=17.5 \text{ MPa}$$
$$\sigma_{y'}=-3.5 \text{ MPa}$$

and

$$\tau_{x'y'}=\pm 21\sin 60°=\pm 18.186 \text{ MPa}$$

Figure 1.8c indicates the orientation of the stresses. The results can be checked by applying Eq. (1.7), using the initial data.

1.8 Three-Dimensional Stress at a Point

Equations governing the transformation of stress in the three-dimensional case may be obtained by the use of a similar approach to that used for the two-dimensional state of stress.

Consider a small tetrahedron isolated from a continuous medium (Fig. 1.9), subject to a general state of stress. The body forces are taken to be negligible. In the figure T_x, T_y, T_z are the Cartesian components of stress

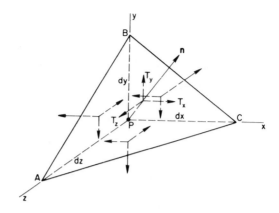

Figure 1.9

resultant **T**, acting on oblique plane ABC. It is required to relate the stresses on the perpendicular planes intersecting at the origin to the normal and shear stresses acting on ABC.

The orientation of plane ABC may be defined in terms of the angles between a unit normal **n** to the plane and the x, y, z directions. The direction cosines associated with these angles are

$$\cos(n, x) = l$$
$$\cos(n, y) = m \quad\quad (1.12)$$
$$\cos(n, z) = n$$

The three direction cosines for the **n** direction are related by

$$l^2 + m^2 + n^2 = 1 \quad\quad (1.13)$$

The area of the perpendicular plane PAB, PAC, PBC may now be expressed in terms of A, the area of ABC, and the direction cosines:

$$A_{PAB} = A_x = \mathbf{A} \cdot \mathbf{i} = A(l\mathbf{i} + m\mathbf{j} + n\mathbf{k}) \cdot \mathbf{i} = Al$$

The other two areas are similarly obtained. In so doing, we have altogether

$$A_{PAB} = Al, \quad\quad A_{PAC} = Am, \quad\quad A_{PBC} = An \quad\quad (1.14)$$

Here **i**, **j**, and **k** are unit vectors in the x, y, and z directions, respectively.

Next, from the equilibrium of x, y, z-directed forces together with Eq. (1.14), we obtain after canceling A,

$$T_x = \sigma_x l + \tau_{xy} m + \tau_{xz} n$$
$$T_y = \tau_{xy} l + \sigma_y m + \tau_{yz} n \quad\quad (1.15)$$
$$T_z = \tau_{xz} l + \tau_{yz} m + \sigma_z n$$

The stress resultant on A is thus determined on the basis of known stresses $\sigma_x, \sigma_y, \sigma_z, \tau_{xy}, \tau_{xz}, \tau_{yz}$ and a knowledge of the orientation of A. In the limit as the sides of the tetrahedron approach zero, plane A contains point P. It is thus demonstrated that the stress resultant at a point is specified. This in turn gives the stress components acting on any three mutually perpendicular planes passing through P as shown below. While perpendicular planes have been used above for convenience, these planes need not be perpendicular to define the stress at a point.

Consider now a Cartesian coordinate system x', y', z', wherein x' coincides with $\mathbf{n}$ and y', z' lie on an oblique plane. The $x'y'z'$ and xyz systems are related by the direction cosines: $l_1 = \cos(x', x)$, $m_1 = \cos(x', y)$, etc. (The notation corresponding to a complete set of direction cosines is shown in Table 1.2.) The normal stress $\sigma_{x'}$ is found by projecting T_x, T_y, and T_z in the x' direction and adding:

$$\sigma_{x'} = T_x l_1 + T_y m_1 + T_z n_1 \qquad (1.16)$$

Equations (1.15) and (1.16) are combined to yield

$$\sigma_{x'} = \sigma_x l_1^2 + \sigma_y m_1^2 + \sigma_z n_1^2 + 2(\tau_{xy} l_1 m_1 + \tau_{yz} m_1 n_1 + \tau_{xz} l_1 n_1) \qquad (1.17a)$$

Similarly, by projecting T_x, T_y, T_z in the y' and z' directions we obtain, respectively,

$$\tau_{x'y'} = \sigma_x l_1 l_2 + \sigma_y m_1 m_2 + \sigma_z n_1 n_2 + \tau_{xy}(l_1 m_2 + m_1 l_2)$$

$$+ \tau_{yz}(m_1 n_2 + n_1 m_2) + \tau_{xz}(n_1 l_2 + l_1 n_2) \qquad (1.17b)$$

$$\tau_{x'z'} = \sigma_x l_1 l_3 + \sigma_y m_1 m_3 + \sigma_z n_1 n_3 + \tau_{xy}(l_1 m_3 + m_1 l_3)$$

$$+ \tau_{yz}(m_1 n_3 + n_1 m_3) + \tau_{xz}(n_1 l_3 + l_1 n_3) \qquad (1.17c)$$

Recalling that the stresses on three mutually perpendicular planes are required to specify the stress at a point (one of these planes being the oblique plane in question), the remaining components are found by considering those planes perpendicular to the oblique plane. For one such plane $\mathbf{n}$ would now coincide with the y' direction, and expressions for the stresses $\sigma_{y'}$, $\tau_{y'x'}$, $\tau_{y'z'}$ would be derived. In a similar manner, the stresses $\sigma_{z'}$, $\tau_{z'x'}$, $\tau_{z'y'}$ are determined when $\mathbf{n}$ coincides with the z' direction. Owing to the symmetry of the stress tensor, only six of the nine stress components thus

Table 1.2

	x	y	z
x'	l_1	m_1	n_1
y'	l_2	m_2	n_2
z'	l_3	m_3	n_3

developed are unique. The remaining stress components are as follows:

$$\sigma_{y'} = \sigma_x l_2^2 + \sigma_y m_2^2 + \sigma_z n_2^2 + 2(\tau_{xy} l_2 m_2 + \tau_{yz} m_2 n_2 + \tau_{xz} l_2 n_2) \quad (1.17d)$$

$$\sigma_{z'} = \sigma_x l_3^2 + \sigma_y m_3^2 + \sigma_z n_3^2 + 2(\tau_{xy} l_3 m_3 + \tau_{yz} m_3 n_3 + \tau_{xz} l_3 n_3) \quad (1.17e)$$

$$\tau_{y'z'} = \sigma_x l_2 l_3 + \sigma_y m_2 m_3 + \sigma_z n_2 n_3 + \tau_{xy} (m_2 l_3 + l_2 m_3)$$
$$+ \tau_{yz} (n_2 m_3 + m_2 n_3) + \tau_{xz} (l_2 n_3 + n_2 l_3) \quad (1.17f)$$

Equations (1.17) represent expressions transforming the quantities σ_x, σ_y, σ_z, τ_{xy}, τ_{xz}, τ_{yz}, which, as we have noted, completely define the state of stress. Quantities such as stress (and mass moment of inertia) which are subject to such transformations are tensors of second rank. Mohr's circle is thus a *graphical representation* of a tensor transformation. Equations (1.17) are succinctly expressed by Eq. (A.3).

It is interesting to note that, because x', y', and z' are orthogonal, the nine direction cosines must satisfy trigonometric relations of the following form:

$$l_i^2 + m_i^2 + n_i^2 = 1 \qquad (i = 1, 2, 3) \tag{a}$$

and

$$l_1 l_2 + m_1 m_2 + n_1 n_2 = 0$$
$$l_2 l_3 + m_2 m_3 + n_2 n_3 = 0 \tag{b}$$
$$l_1 l_3 + m_1 m_3 + n_1 n_3 = 0$$

From Table 1.2 it is observed that the equations (a) are the sums of the squares of the cosines in each row, and the equations (b) are the sums of the products of the adjacent cosines in any two rows.

1.9   Principal Stress in Three Dimensions

For the three-dimensional case, it is now demonstrated that three planes of zero shear stress exist, that these planes are mutually perpendicular, and that on these planes the normal stresses have maximum or minimum values. As has been discussed, these normal stresses are referred to as *principal stresses*, usually denoted σ_1, σ_2, and σ_3. The algebraically largest stress is represented by σ_1, and the smallest by σ_3.

We begin by again considering an oblique x' plane. The normal stress acting on this plane is given by Eq. (1.17a):

$$\sigma_{x'} = \sigma_x l^2 + \sigma_y m^2 + \sigma_z n^2 + 2(\tau_{xy} lm + \tau_{yz} mn + \tau_{xz} ln) \tag{a}$$

The problem at hand is the determination of *extreme* or *stationary values* of $\sigma_{x'}$. To accomplish this, we examine the variation of $\sigma_{x'}$ relative to the direction cosines. Inasmuch as l, m, and n are not independent, but connected by $l^2 + m^2 + n^2 = 1$, only l and m may be regarded as independent variables. Thus,

$$\frac{\partial \sigma_{x'}}{\partial l} = 0, \qquad \frac{\partial \sigma_{x'}}{\partial m} = 0 \tag{b}$$

Differentiating Eq. (a) as indicated above in terms of the quantities in Eq. (1.15), we obtain

$$T_x + T_z \frac{\partial n}{\partial l} = 0, \qquad T_y + T_z \frac{\partial n}{\partial m} = 0 \tag{c}$$

From $n^2 = 1 - l^2 - m^2$, we have $\partial n / \partial l = -l/n$ and $\partial n / \partial m = -m/n$. Introducing these into Eq. (c), the following relationships between the components of T and n are determined:

$$\frac{T_x}{l} = \frac{T_y}{m} = \frac{T_z}{n} \tag{d}$$

These proportionalities indicate that the stress resultant must be *parallel* to the unit normal and therefore contains no shear component. It is concluded that on a plane for which $\sigma_{x'}$ has an extreme or principal value, a principal plane, the shearing stress vanishes.

It is now shown that three principal stresses and three principal planes exist. Denoting the principal stresses by σ_p, Eq. (d) may be written as

$$T_x = \sigma_p l, \qquad T_y = \sigma_p m, \qquad T_z = \sigma_p n \tag{e}$$

These expressions together with Eq. (1.15) lead to

$$(\sigma_x - \sigma_p)l + \tau_{xy}m + \tau_{xz}n = 0$$
$$\tau_{xy}l + (\sigma_y - \sigma_p)m + \tau_{yz}n = 0 \tag{1.18}$$
$$\tau_{xz}l + \tau_{yz}m + (\sigma_z - \sigma_p)n = 0$$

A nontrivial solution for the direction cosines requires that the characteristic determinant vanish:

$$\begin{vmatrix} \sigma_x - \sigma_p & \tau_{xy} & \tau_{xz} \\ \tau_{xy} & \sigma_y - \sigma_p & \tau_{yz} \\ \tau_{xz} & \tau_{yz} & \sigma_z - \sigma_p \end{vmatrix} = 0 \tag{1.19}$$

Expanding Eq. (1.19) leads to

$$\sigma_p^3 - I_1\sigma_p^2 + I_2\sigma_p - I_3 = 0 \tag{1.20}$$

where

$$I_1 = \sigma_x + \sigma_y + \sigma_z$$
$$I_2 = \sigma_x\sigma_y + \sigma_x\sigma_z + \sigma_y\sigma_z - \tau_{xy}^2 - \tau_{yz}^2 - \tau_{xz}^2 \tag{1.21}$$
$$I_3 = \begin{vmatrix} \sigma_x & \tau_{xy} & \tau_{xz} \\ \tau_{xy} & \sigma_y & \tau_{yz} \\ \tau_{xz} & \tau_{yz} & \sigma_z \end{vmatrix}$$

The three roots of the *stress cubic equation* (1.20) are the principal stresses, corresponding to which are three sets of direction cosines, which establish the relationship of the principal planes to the origin of the nonprincipal axes. The principal stresses are the characteristic values or *eigenvalues** of the stress tensor τ_{ij}. Since the stress tensor is a symmetric tensor whose elements are all real, it has real eigenvalues. That is, the three principal stresses are *real*. The direction cosines l, m, n are the *eigenvectors* of τ_{ij}.

It is clear that the principal stresses are independent of the orientation of the original coordinate system. It follows from Eq. (1.20) that the coefficients I_1, I_2, and I_3 must likewise be independent of x, y, z, since otherwise the principal stresses would change. For example, one may demonstrate that adding the expressions for $\sigma_{x'}$, $\sigma_{y'}$, $\sigma_{z'}$ given by Eq. (1.17) and making use of Eqs. (a) and (b) of the previous section leads to $I_1 = \sigma_{x'} + \sigma_{y'} + \sigma_{z'} = \sigma_x + \sigma_y + \sigma_z$. Thus, the coefficients I_1, I_2 and I_3 represent three *invariants* of the stress tensor.

If now one of the principal stresses, say σ_1 [obtained from Eq. (1.20)], is substituted into Eq. (1.18), the resulting expressions, together with $l^2 + m^2 + n^2 = 1$, provide enough information to solve for the direction cosines, thus specifying the orientation of σ_1 relative to the xyz system. The direction cosines of σ_2 and σ_3 are similarly obtained. In the event that $\sigma_1 = \sigma_2 = \sigma_3$, it is a simple matter to show that *all* planes within the continuum are principal planes. This situation exists in an ideal fluid, in which there are no shearing stresses, and in the case of hydrostatic stress. In the latter, fluid elements experience no relative motion and consequently viscous effects are absent.

Example 1.3. A steel shaft is to be force fitted into a fixed-ended cast iron hub. The shaft is subjected to a bending moment M, a torque M_t, and a vertical force P, Fig. 1.10a. Suppose that at a point A in the hub, the stress field is as shown in Fig. 1.10b, represented by the matrix

$$\begin{bmatrix} -19 & -4.7 & 6.45 \\ -4.7 & 4.6 & 11.8 \\ 6.45 & 11.8 & -8.3 \end{bmatrix} \text{MPa}$$

Determine the principal stresses and their orientation with respect to the original coordinate system.

SOLUTION. Substituting the given stresses into Eq. (1.20) we obtain from Eqs. (B.2)

$$\sigma_1 = 11.618 \text{ MPa}, \qquad \sigma_2 = -9.002 \text{ MPa}, \qquad \sigma_3 = -25.316 \text{ MPa}$$

Successive introduction of these values into Eq. (1.18), together with Eq. (1.13), or application of Eqs. (B.6), yields the direction cosines that define the orientation of

---

*For details, see L. A. Pipes, *Matrix Methods for Engineering*, Englewood Cliffs, NJ: Prentice-Hall, 1963, Chapter 5.

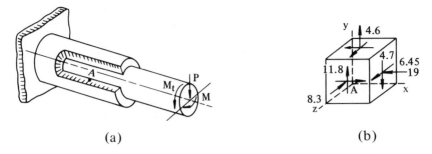

(a)   (b)

Figure 1.10

the planes on which σ_1, σ_2, and σ_3 act:

$$l_1 = 0.0266, \qquad l_2 = -0.6209, \qquad l_3 = 0.7834$$
$$m_1 = -0.8638, \qquad m_2 = 0.3802, \qquad m_3 = 0.3306$$
$$n_1 = -0.5031, \qquad n_2 = -0.6855, \qquad n_3 = -0.5262$$

It should be noted that the directions of the principal stresses are seldom required for purposes of predicting the behavior of structural components.

Example 1.4. The stress tensor at a point in a machine element with respect to a Cartesian coordinate system is given by the following array:

$$\begin{bmatrix} 50 & 10 & 0 \\ 10 & 20 & 40 \\ 0 & 40 & 30 \end{bmatrix} \text{MPa} \qquad (f)$$

Determine the state of stress and I_1, I_2, I_3 for an x', y', z' coordinate system defined by rotating x, y through an angle of $45°$ counterclockwise about the z axis.

SOLUTION. The direction cosines $l_1 = m_1 = m_2 = 1/\sqrt{2}$, $l_2 = -1/\sqrt{2}$, $n_3 = 1$, and $n_1 = n_2 = l_3 = m_3 = 0$ correspond to the prescribed rotation of axes. Through the use of Eq. (1.17) we obtain

$$\begin{bmatrix} 45 & -15 & 28.28 \\ -15 & 25 & 28.28 \\ 28.28 & 28.28 & 30 \end{bmatrix} \text{MPa} \qquad (g)$$

It is seen that the arrays (f) and (g), when substituted into Eq. (1.21), both yield $I_1 = 100$, $I_2 = 1400$, $I_3 = -53,000$; and the invariance of I_1, I_2, I_3 under the orthogonal transformation is confirmed.

1.10   Stresses on an Oblique Plane in Terms of Principal Stresses

It is sometimes required to determine the shear and normal stresses acting on an arbitrary oblique plane as in Fig. 1.11a, given in the principal stresses acting on perpendicular planes. In the figure the x, y, z axes are parallel to the principal axes. Denoting the direction cosines of plane ABC by l, m, n, the equations (1.15) with $\sigma_x = \sigma_1$, $\tau_{xy} = \tau_{xz} = 0$, etc. reduce to

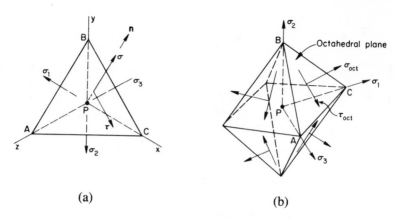

Figure 1.11

$T_x = \sigma_1 l$, $T_y = \sigma_2 m$, $T_z = \sigma_3 n$. The resultant stress on the oblique plane is thus

$$T^2 = \sigma_1^2 l^2 + \sigma_2^2 m^2 + \sigma_3^2 n^2 = \sigma^2 + \tau^2 \qquad (a)$$

The normal stress on this plane, from Eq. (1.17a), is found as

$$\sigma = \sigma_1 l^2 + \sigma_2 m^2 + \sigma_3 n^2 \qquad (1.22)$$

When this expression is substituted into Eq. (a), the result is

$$\tau^2 = \sigma_1^2 l^2 + \sigma_2^2 m^2 + \sigma_3^2 n^2 - \left(\sigma_1 l^2 + \sigma_2 m^2 + \sigma_3 n^2 \right)^2 \qquad (1.23)$$

Expanding the above, and using the expressions $1 - l^2 = m^2 + n^2$, $1 - n^2 = l^2 + m^2$, etc., the following result is obtained:

$$\tau = \left[(\sigma_1 - \sigma_2)^2 l^2 m^2 + (\sigma_2 - \sigma_3)^2 m^2 n^2 + (\sigma_3 - \sigma_1)^2 n^2 l^2 \right]^{1/2} \qquad (1.24)$$

This again clearly indicates that if the principal stresses are all equal, the shear stress vanishes, regardless of the choices of the direction cosines.

Example 1.5. Determine the shear stresses acting on a plane of particular importance in failure theory, represented by face ABC in Fig. 1.11a with $PA = PB = PC$.

SOLUTION. The normal to this oblique face thus has equal direction cosines relative to the principal axes. Since $l^2 + m^2 + n^2 = 1$, we have

$$l = m = n = \frac{1}{\sqrt{3}} \qquad (b)$$

Plane ABC is clearly one of eight such faces of a regular octahedron (Fig. 1.11b). Equations (1.23) and (b) are now applied to provide an expression for the *octahedral shearing stress* which may be rearranged to the form

$$\tau_{oct} = \frac{1}{3} \left[(\sigma_1 - \sigma_2)^2 + (\sigma_2 - \sigma_3)^2 + (\sigma_3 - \sigma_1)^2 \right]^{1/2} \qquad (1.25)$$

Through the use of Eqs. (1.22) and (b) we obtain

$$\sigma_{oct} = \frac{1}{3} (\sigma_1 + \sigma_2 + \sigma_3) \qquad (1.26)$$

The normal stress acting on an octahedral plane is thus the average of the principal stresses, the *mean stress*. The orientations of σ_{oct} and τ_{oct} are indicated in Fig. 1.11b. Another useful form of Eq. (1.25) is developed in Sec. 2.9.

1.11   Mohr's Circle for Three-Dimensional Stress

It has been demonstrated that given the magnitudes and direction cosines of the principal stresses, the stresses on any oblique plane may be ascertained through the application of Eqs. (1.22) and (1.24). This may also be accomplished by means of a graphical technique due to Mohr, in which the aforementioned equations are represented by three circles of stress.

Consider the element shown in Fig. 1.12a, resulting from the cutting of a small cube by an oblique plane. The element is subjected to principal stresses σ_1, σ_2, and σ_3 represented as coordinate axes with the origin at P. We are interested in determining the normal and shear stresses acting at point Q on the slant face (plane *abcd*). This plane is oriented so as to be tangent at Q to a quadrant of a spherical surface inscribed within a cubic element as shown. Note that PQ, running from the origin of the principal axis system to point Q, is the line of intersection of the shaded planes (Fig. 1.12a). the inclination of plane PA_2QB_3 relative to the σ_1 axis is given by the angle θ (measured in the σ_1, σ_3 plane), and that of plane PA_3QB_1, by the angle ϕ (measured in the σ_1, σ_2 plane). Circular arcs $A_1B_1A_2$ and $A_1B_3A_3$ are located on the cube faces. It is clear that angles θ and ϕ unambiguously define the orientation of PQ with respect to the principal axes.

To determine σ and τ given σ_1, σ_2, σ_3, the following procedure is applied (refer to Fig. 1.12b):

1. Establish a Cartesian coordinate system, indicating $+\sigma$ and $+\tau$ as shown. Lay off the principal stresses along the σ axis, with $\sigma_1 > \sigma_2 > \sigma_3$ (algebraically).
2. Draw three Mohr semicircles centered at C_1, C_2, C_3 (referred to as circles c_1, c_2, c_3) with diameters A_1A_3, A_2A_3, and A_1A_3.
3. At point C_1, draw line C_1B_1 at angle 2ϕ; at C_3, draw C_3B_3 at angle 2θ. These lines cut circles c_1 and c_3 at points B_1 and B_3, respectively.
4. By trial and error, draw arcs through points A_3 and B_1 and through A_2 and B_3, with their centers on the σ axis. The intersection of these arcs locates point Q on the σ, τ plane.

In connection with the foregoing construction, several points are of particular interest:

a. Point Q will be located within the shaded area or along the circumference of circles c_1, c_2, or c_3, for all combinations of θ and ϕ.
b. For the particular case $\theta = \phi = 0$, Q coincides with A_1 in Fig. 1.12a and b.

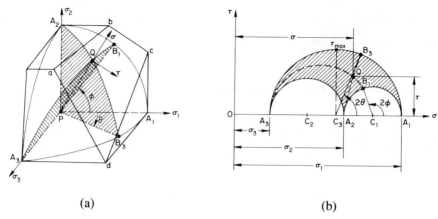

(a)   (b)

Figure 1.12

c. For the case $\theta = 45°$, $\phi = 0°$, the shearing stress is a maximum located as the highest point on circle c_3 ($2\theta = 90°$). The value of the maximum shearing stress is therefore

$$\tau_{max} = \tfrac{1}{2}(\sigma_1 - \sigma_3) \tag{1.27}$$

acting on the planes bisecting the planes of maximum and minimum principal stress, as indicated in Fig. 1.13. It is noted that the planes of maximum shear may also be ascertained by substituting $n^2 = 1 - l^2 - m^2$ into Eq. (1.23), differentiating with respect to l and m, and equating the resulting expressions to zero (Problem 1.23).

d. For $\theta = \phi = 45°$, line PQ will make equal angles with the principal axes. The oblique plane is, in this case, an octahedral plane, and the stresses acting on the plane, the octahedral stresses. Recall that Eqs. (1.25) and (1.26) provide algebraic expressions for these stresses.

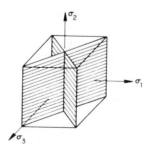

Figure 1.13

1.12   Variation of Stress at the Boundary of a Body

We now consider the relationship between the stress components and the given surface forces acting on the boundary of a body. The equations of equilibrium which must be satisfied within a body are derived in Sec. 1.4. The distribution of stress in a body must also be such as to accommodate the conditions of equilibrium with respect to externally applied forces. The external forces may thus be regarded a continuation of the internal stress distribution. Consider the equilibrium of the forces acting on the tetrahedron shown in Fig. 1.9, and assume that oblique face *ABC* is coincident with the surface of the body. The components of the stress resultant **T** are thus now the *surface forces* per unit area or so-called *surface tractions*, T_x, T_y, T_z. The equations of equilibrium for this element, representing *boundary conditions* are, from Eqs. (1.15),

$$T_x = \sigma_x l + \tau_{xy} m + \tau_{xz} n$$
$$T_y = \tau_{xy} l + \sigma_y m + \tau_{yz} n \qquad (1.28)$$
$$T_z = \tau_{xz} l + \tau_{yz} m + \sigma_z n$$

For example, if the boundary is a plane with an x directed surface normal, Eqs. (1.28) give $T_x = \sigma_x$, $T_y = \tau_{xy}$, and $T_z = \tau_{xz}$; under these circumstances, the applied surface force components T_x, T_y, and T_z are balanced by σ_x, τ_{xy}, and τ_{xz}, respectively.

It is of interest to note that instead of prescribing the distribution of surface forces on the boundary, the boundary conditions of a body may also be given in terms of displacement components. Furthermore, we may be given boundary conditions which prescribe surface forces on one part of the boundary and displacements on another. When displacement boundary conditions are given, the equations of equilibrium express the situation in terms of strain, through the use of Hooke's law and subsequently in terms of the displacements by means of strain-displacement relations (Sec. 2.2). It is usual, in engineering problems, however, to specify the boundary conditions in terms of surface forces as in Eq. (1.28), rather than surface displacements. This practice is adhered to in this text.

Chapter 1—Problems

Secs. 1.1 to 1.7

1.1.   Determine whether the following stress fields are possible within an elastic structural member:

$$\text{(a)} \quad \begin{bmatrix} c_1 x + c_2 y & c_5 x - c_1 y \\ c_5 x - c_1 y & c_3 x + c_4 \end{bmatrix}, \qquad \text{(b)} \quad \begin{bmatrix} -\frac{3}{2} x^2 y^2 & xy^3 \\ xy^3 & -\frac{1}{4} y^4 \end{bmatrix}$$

The c's are constant, and it is assumed that the body forces are negligible.

1.2. For what body forces will the following stress field describe a state of equilibrium?

$$\sigma_x = -2x^2 + 3y^2 - 5z, \qquad \tau_{xy} = z + 4xy - 7$$
$$\sigma_y = -2y^2, \qquad \tau_{xz} = -3x + y + 1$$
$$\sigma_z = 3x + y + 3z - 5, \qquad \tau_{yz} = 0$$

1.3. The states of stress at two points in a loaded beam are represented in Fig. P1.3a and b. All stresses are in megapascals. Determine the following for each point: (a) The magnitude of the maximum and minimum principal stresses and the maximum shearing stress; use Mohr's circle. (b) The orientation of the principal and maximum shear planes; use Mohr's circle. (c) Sketch the results on properly oriented elements. Check the values found in (a) and (b) by applying the appropriate equations.

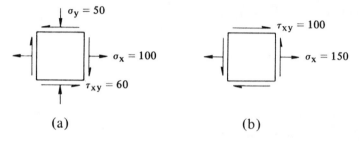

(a)   (b)

Figure P1.3

1.4. By means of Mohr's circle, show that for any two-dimensional stress field, the following quantities are invariant: $\sigma_{x'} + \sigma_{y'}, \sigma_{x'} \cdot \sigma_{y'} - \tau_{x'y'}^2$.

1.5. Given the stress acting uniformly over the sides of a thin flat plate, Fig. P1.5, determine (a) the stresses on planes inclined at 20° to the horizontal and (b) the principal stresses and their orientations.

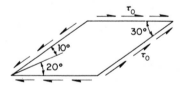

Figure P1.5

1.6. A steel shaft of radius $r = 75$ mm. is subjected to an axial compression $P = 81$ kN, a twisting couple $M_t = 15.6$ kN·m, and a bending moment $M = 13$ kN·m at both ends. Calculate the magnitude of the principal stresses, the maximum shear stress, and the planes on which they act in the shaft. (The elementary stress formulas are found in Secs. 5.2 and 6.1.)

1.7. A structural member is subjected to a set of forces and moments. Each separately produces the stress conditions (megapascals) at a point shown in Fig. P1.7. Determine the principal stresses and their orientations at the point under the effect of combined loading.

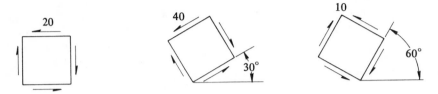

Figure P1.7

1.8. Redo Problem 1.7 for the case shown in Fig. P1.8.

Figure P1.8

1.9. A square prismatic bar of 0.0013 m^2 cross-sectional area is composed of the pieces of wood glued together along plane **n**, which lies in a longitudinal plane parallel to a side face and makes an angle θ with the axial direction. The normal and shearing stresses acting simultaneously on the joint plane are limited to 20 MPa and 10 MPa, respectively, and on the bar itself, to 56 MPa, and 28 MPa, respectively. Find the maximum allowable load that the bar can carry and the corresponding value of the angle θ.

1.10. The state of stress (MPa) at a point in a machine member is shown in Fig. P1.10. The allowable compression stress at the point is 14 MPa. Determine (a) the tensile stress σ_x and (b) the maximum principal and maximum shearing stresses in the member. Sketch the results on property oriented elements.

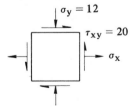

Figure P1.10

1.11. In Example 1.1, taking $\sigma_z = 0$, investigate the maximum shearing stresses on all possible (three-dimensional) planes.

1.12. A long, thin-walled pipe is subjected to an internal pressure p and uniform axial tensile load P. The radius and the thickness of the pipe are $r = 0.45$ m, and $t = 5$ mm. The axial and tangential stresses under internal pressure are $\sigma_a = pr/2t$ and $\sigma_\theta = 2\sigma_a$ (Sec. 8.2). The normal stresses at a point A on the surface of the cylinder are restricted to $\sigma_{x'} = 84$ MPa, $\sigma_{y'} = 56$ MPa, while shear stress $\tau_{x'y'}$ is not specified. Determine the values of P and p. Note that x and y are in the axial and radial directions, respectively. The x', y' axes make an angle of $30°$ with the x, y axes

1.13. A thin-walled cylindrical pressure vessel of 0.3-m radius and 6-mm wall thickness has a welded spiral seam at an angle of $30°$ with the axial direction. The vessel is subjected to an internal gage pressure of p Pa and an axial compressive load of 9π kN applied through rigid end plates. Find the allowable value of p if the normal and shearing stresses acting simultaneously in the plane of welding are limited to 21 MPa and 7 MPa, respectively.

Secs. 1.8 and 1.9

1.14. At a specified point in a member, the state of stress with respect to a Cartesian coordinate system is given by

$$\begin{bmatrix} 12 & 6 & 9 \\ 6 & 10 & 3 \\ 9 & 3 & 14 \end{bmatrix} \text{MPa}$$

Calculate the magnitude and direction of the maximum principal stress.

1.15. The stress at a point, with respect to an x, y, z coordinate system, is described by

$$\sigma_x = x^2 + y, \qquad \sigma_z = -x + 6y + z$$

$$\sigma_y = y^2 - 5, \qquad \tau_{xy} = \tau_{xz} = \tau_{yz} = 0$$

At point $(3, 1, 5)$ determine (a) the stress components with respect to x', y', z' if

$$l_1 = 1, \qquad m_2 = \frac{1}{2}, \qquad n_2 = \frac{\sqrt{3}}{2}, \qquad n_3 = \frac{1}{2}, \qquad m_3 = -\frac{\sqrt{3}}{2}$$

and (b) the stress components with respect to x'', y'', z'' if $l_1 = 2/\sqrt{5}$, $m_1 = -1/\sqrt{5}$, and $n_3 = 1$. Show that the quantities given by Eq. (1.21) are invariant under the transformations (a) and (b).

1.16. Determine the stresses with respect to the x', y', z' axes in the element of Problem 1.14 if

$$l_1 = \frac{1}{2}, \qquad l_2 = -\frac{\sqrt{3}}{2}, \qquad l_3 = 0$$

$$m_1 = \frac{\sqrt{3}}{2}, \qquad m_2 = \frac{1}{2}, \qquad m_3 = 0$$

$$n_1 = 0, \qquad n_2 = 0, \qquad n_3 = 1$$

1.17. Obtain the principal stresses and the related direction cosines for the following cases.

(a) $\begin{bmatrix} 3 & 4 & 6 \\ 4 & 2 & 5 \\ 6 & 5 & 1 \end{bmatrix}$ MPa, (b) $\begin{bmatrix} 14.32 & 0.8 & 1.55 \\ 0.8 & 6.97 & 5.2 \\ 1.55 & 5.2 & 16.3 \end{bmatrix}$ MPa

1.18. If x, y, z represent the directions of the principal axes, show by the use of Eq. (1.15) that

$$\left(\frac{T_x}{\sigma_x}\right)^2 + \left(\frac{T_y}{\sigma_y}\right)^2 + \left(\frac{T_z}{\sigma_z}\right)^2 = 1$$

Sketch the *stress ellipsoid* curve represented by this equation. What do the semiaxes indicate?

Secs. 1.10 to 1.12

1.19. Determine the magnitude and direction of the maximum shearing stress for the cases given in Problem 1.17.

1.20. Given the principal stresses σ_1, σ_2, σ_3 at a point in an elastic solid, prove that the maximum shearing stress at the point always exceeds the octahedral shearing stress.

1.21. For the case of plane stress determine the octahedral stress invariants.

1.22. Determine the value of the octahedral stresses of Problem 1.14.

1.23. By using Eq. (1.23) verify that the planes of maximum shearing stress in three dimensions bisect the planes of maximum and minimum principal stresses. Also find the normal stresses associated with the shearing plane by applying Eq. (1.22).

1.24. The principal stresses at a point in a solid are $\sigma_1 = 56$ MPa, $\sigma_2 = 35$ MPa, and $\sigma_3 = 14$ MPa. Obtain, by use of Mohr's circle, (a) the maximum shearing stress and (b) the normal and shearing stresses on the octahedral planes.

1.25. Rework Problem 1.24 for $\sigma_1 = 35$ MPa, $\sigma_2 = 14$ MPa, $\sigma_3 = -7$ MPa.

1.26. Employ Mohr's circle to find the normal and shearing stresses on an oblique plane defined by $\theta = 60°$, $\phi = 30°$ (see Fig. 1.12). The principal stresses are $\sigma_1 = 35$ MPa, $\sigma_2 = -14$ MPa, and $\sigma_3 = -28$ MPa. Apply Eqs. (1.22) and (1.24) to check the values thus determined. If this plane is on the boundary of a structural member, what should be the values of surface forces T_x, T_y, T_z on the plane?

Chapter 2

Strain and Stress–Strain Relations

2.1 Introduction

In the preceding chapter, our concern was with the stress field within the continuum. We now turn to the deformation field. Let us consider a three-dimensional body subjected to external loading such that point A is displaced to A', B to B', and so on until all the points in the body are displaced to new positions (Fig. 2.1). The displacements of *any two* points such as A and B are simply AA' and BB', respectively, and may be a consequence of deformation (straining), rigid body motion (translation and rotation), or some combination. The body has experienced straining if the *relative positions* of points in the body are altered. If now straining has taken place, displacements AA' and BB' are attributable to rigid body motion. In the latter case, the distance between A and B remains fixed; such displacements are not discussed in this chapter.

In order to describe the magnitude and direction of the displacements, points within the body are located with respect to an appropriate coordinate reference as, for example, the xyz system of Fig. 2.1. The components of displacement at a point, occurring in the x, y, and z directions, are denoted by u, v, and w, respectively. The displacement at *every* point within the body constitutes the *displacement field*, $u = u(x, y, z)$, $v = v(x, y, z)$, $w = w(x, y, z)$. In this text, only *small displacements* are considered, a simplification consistent with the magnitude of deformation commonly found in engineering structures. The strains produced by small deformations are small compared with unity, and their products (higher order terms) are neglected.

2.2 Strain Defined

For purposes of defining normal strain, refer to Fig. 2.2, where line AB of an axially loaded member has suffered deformation to become $A'B'$. The length of AB is Δx (Fig. 2.2a). As shown in Fig. 2.2b, points A and B have each been displaced: A an amount u, and B, $u + \Delta u$. Stated differently,

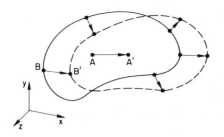

Figure 2.1

point B has been displaced by an amount Δu in addition to displacement of point A, and the length Δx has been increased by Δu. Normal strain, the unit change in length, is defined

$$\varepsilon_x = \lim_{\Delta x \to 0} \frac{\Delta u}{\Delta x} = \frac{du}{dx} \tag{2.1}$$

In view of the limiting process, the above represents the strain at a point, the point to which Δx shrinks.

If the deformation is distributed *uniformly* over the original length, the normal strain may be written

$$\varepsilon_0 = \frac{L - L_0}{L_0} = \frac{\delta}{L_0} \tag{2.2}$$

where L, L_0, and δ are the final length, the original length, and the change of length of the member, respectively. When uniform deformation does not occur, the aforementioned is the average strain.

We now investigate the case of two-dimensional or *plane strain*, wherein all points in the body, before and after application of load, remain in the same plane.

Referring to Fig. 2.3, consider an element with dimensions dx, dy and and of unit thickness. The total deformation may be regarded as possess-

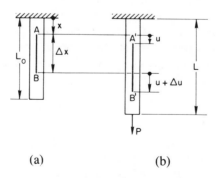

(a)   (b)

Figure 2.2

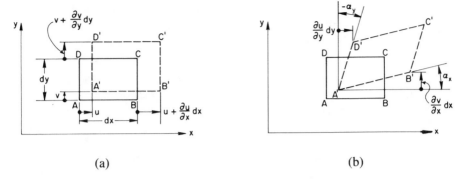

(a)   (b)

Figure 2.3

ing the following features: a change in length experienced by the sides (Fig. 2.3a), and a relative rotation without accompanying changes of length (Fig. 2.3b).

Recalling the basis of Eq. (2.1), two normal or longitudinal strains are apparent upon examination of Fig. 2.3a:

$$\varepsilon_x = \frac{\partial u}{\partial x}, \qquad \varepsilon_y = \frac{\partial v}{\partial y}$$

A *positive* sign applies to *elongation*; a *negative* sign, to *contraction*.

Now consider the change experienced by right angle DAB, Fig. 2.3b. We shall assume the angle α_x between AB and $A'B'$ to be so small as to permit the approximation $\alpha_x \approx \tan \alpha_x$. Also in view of the smallness of α_x, the normal strain is small, so that $AB \approx A'B'$. As a consequence of the above-mentioned considerations $\alpha_x \approx \partial v / \partial x$, where the *counterclockwise rotation* is defined as *positive*. Similar analysis leads to $-\alpha_y \approx \partial u / \partial y$. The total angular change of angle DAB, i.e., the angular change between lines in the x and y directions, is defined as the shearing strain, and denoted by γ_{xy}:

$$\gamma_{xy} = \alpha_x - \alpha_y = \frac{\partial u}{\partial y} + \frac{\partial v}{\partial x}$$

The shear strain is *positive* when the *right angle* between two positive (or negative) axes *decreases*. That is, if the angle between $+x$ and $+y$ or $-x$ and $-y$ decreases, we have positive γ_{xy}; otherwise the shear strain is negative.

In the case of a three-dimensional element, a rectangular prism with sides dx, dy, dz, an essentially identical analysis leads to the following normal and shearing strains:

$$\varepsilon_x = \frac{\partial u}{\partial x}, \qquad \varepsilon_y = \frac{\partial v}{\partial y}, \qquad \varepsilon_z = \frac{\partial w}{\partial z}$$

$$\gamma_{xy} = \frac{\partial u}{\partial y} + \frac{\partial v}{\partial x}, \qquad \gamma_{yz} = \frac{\partial v}{\partial z} + \frac{\partial w}{\partial y}, \qquad \gamma_{zx} = \frac{\partial w}{\partial x} + \frac{\partial u}{\partial z}$$

(2.3)

Clearly, the angular change is not different if it is said to occur between the x and y directions, or between the y and x directions; $\gamma_{xy} = \gamma_{yx}$. The remaining components of shearing strain are similarly related:

$$\gamma_{xy} = \gamma_{yx}, \qquad \gamma_{yz} = \gamma_{zy}, \qquad \gamma_{zx} = \gamma_{xz}$$

The symmetry of shearing strains may also be deduced from an examination of Eq. (2.3). The expressions (2.3) are the *strain-displacement* relations of continuum mechanics. They are also referred to as the *kinematic* relations, treating the *geometry* of strain rather than the matter of cause and effect.

A succinct statement of Eq. (2.3) is made possible by tensor notation:

$$\varepsilon_{ij} = \frac{1}{2}\left(\frac{\partial u_i}{\partial x_j} + \frac{\partial u_j}{\partial x_i} \right) \qquad (i, j = x, y, z) \tag{2.4}$$

where $u_x = u$, $u_y = v$, $x_x = x$, etc. The factor $\frac{1}{2}$ in Eq. (2.4) facilitates the representation of the strain transformation equations in indicial notation. The longitudinal strains are obtained when $i = j$; the shearing strains are found when $i \neq j$ and $\varepsilon_{ij} = \varepsilon_{ji}$. It is apparent from Eqs. (2.3) and (2.4) that

$$\varepsilon_{xy} = \tfrac{1}{2}\gamma_{xy}, \qquad \varepsilon_{yz} = \tfrac{1}{2}\gamma_{yz}, \qquad \varepsilon_{xz} = \tfrac{1}{2}\gamma_{xz} \tag{2.5}$$

Just as the state of stress at a point is described by a nine-term array, so Eq. (2.4) represents nine strains composing the symmetric strain tensor $(\varepsilon_{ij} = \varepsilon_{ji})$

$$\varepsilon_{ij} = \begin{bmatrix} \varepsilon_x & \tfrac{1}{2}\gamma_{xy} & \tfrac{1}{2}\gamma_{xz} \\ \tfrac{1}{2}\gamma_{yx} & \varepsilon_y & \tfrac{1}{2}\gamma_{yz} \\ \tfrac{1}{2}\gamma_{zx} & \tfrac{1}{2}\gamma_{zy} & \varepsilon_z \end{bmatrix} \tag{2.6}$$

It is interesting to observe that the Cartesian coordinate systems of Chapters 1 and 2 are not identical. In Chapter 1, the equations of statics pertain to the deformed state, and the coordinate set is thus established in a deformed body; xyz is, in this instance, a so-called *Eulerian coordinate system*. In discussing the kinematics of deformation in this chapter, recall that the xyz set is established in the undeformed body. In this case, xyz is referred to as a *Lagrangian coordinate system*. Though these systems are clearly not the same, the assumption of small deformation permits us to regard x, y, z, the coordinates in the undeformed body, as applicable to equations of stress *or* strain. Choice of the Lagrangian system should lead to no errors of consequence unless applications in finite elasticity or large deformation theory are attempted. Under such circumstances, the approximation discussed is not valid, and the resulting equations are more difficult to formulate.

Throughout the text, strains are indicated as dimensionless quantities. They are also frequently described in terms of units such as millimeters per millimeter, meters per meter, or micrometers per meter.

2.3 Equations of Compatibility

Expressions of compatibility have both mathematical and physical significance. From a mathematical point of view, they assert that the displacements u, v, w are single-valued and continuous functions.* Physically, this
means that the body must be pieced together.

The kinematic relations, Eq. (2.3), connect six components of strain to
only three components of displacement. One cannot therefore arbitrarily
specify all of the strains as functions of x, y, z. As the strains are evidently
not independent of one another, in what way are they related? In two-
dimensional strain, differentiation of ε_x twice with respect to y, ε_y twice
with respect to x, and γ_{xy} with respect to x and y results in

$$\frac{\partial^2 \varepsilon_x}{\partial y^2} = \frac{\partial^3 u}{\partial x \, \partial y^2}, \qquad \frac{\partial^2 \varepsilon_y}{\partial x^2} = \frac{\partial^3 v}{\partial x^2 \, \partial y}, \qquad \frac{\partial^2 \gamma_{xy}}{\partial x \, \partial y} = \frac{\partial^3 u}{\partial x \, \partial y^2} + \frac{\partial^3 v}{\partial x^2 \, \partial y}$$

or

$$\frac{\partial^2 \varepsilon_x}{\partial y^2} + \frac{\partial^2 \varepsilon_y}{\partial x^2} = \frac{\partial^2 \gamma_{xy}}{\partial x \, \partial y} \qquad (2.7)$$

This is the *condition of compatibility* of the two-dimensional problem,
expressed in terms of strain. The three-dimensional *equations of compatibility* are derived in a like manner:

$$\frac{\partial^2 \varepsilon_x}{\partial y^2} + \frac{\partial^2 \varepsilon_y}{\partial x^2} = \frac{\partial^2 \gamma_{xy}}{\partial x \, \partial y}, \qquad 2\frac{\partial^2 \varepsilon_x}{\partial y \, \partial z} = \frac{\partial}{\partial x}\left(-\frac{\partial \gamma_{yz}}{\partial x} + \frac{\partial \gamma_{xz}}{\partial x} + \frac{\partial \gamma_{xy}}{\partial z}\right)$$

$$\frac{\partial^2 \varepsilon_y}{\partial z^2} + \frac{\partial^2 \varepsilon_z}{\partial y^2} = \frac{\partial^2 \gamma_{yz}}{\partial y \, \partial z}, \qquad 2\frac{\partial^2 \varepsilon_y}{\partial z \, \partial x} = \frac{\partial}{\partial y}\left(\frac{\partial \gamma_{yz}}{\partial x} - \frac{\partial \gamma_{xz}}{\partial y} + \frac{\partial \gamma_{xy}}{\partial z}\right) \qquad (2.8)$$

$$\frac{\partial^2 \varepsilon_z}{\partial x^2} + \frac{\partial^2 \varepsilon_x}{\partial z^2} = \frac{\partial^2 \gamma_{xz}}{\partial z \, \partial x}, \qquad 2\frac{\partial^2 \varepsilon_z}{\partial x \, \partial y} = \frac{\partial}{\partial z}\left(\frac{\partial \gamma_{yz}}{\partial x} + \frac{\partial \gamma_{xz}}{\partial y} - \frac{\partial \gamma_{xy}}{\partial z}\right)$$

To gain further insight into the meaning of compatibility, imagine an
elastic body subdivided into a number of small cubic elements prior to
deformation. These cubes may, upon loading be deformed into a system of
parallelepipeds. The deformed system will, in general, be impossible to
arrange in such a way as to compose a continuous body unless the
components of strain satisfy the equations of compatibility.

---

*See I. S. Sokolnikoff, *Mathematical Theory of Elasticity*, New York: McGraw-Hill, 1956,
pp. 25–59.

2.4   State of Strain at a Point

Recall from Chapter 1 that given the components of stress at a point, it is possible to determine the stresses on any plane passing through the point. A similar operation pertains to the strains at a point.

Consider a small linear element AB of length ds in an unstrained body (Fig. 2.4a). The projections of the element on the coordinate axes are dx and dy. After straining, AB is displaced to position $A'B'$ and is now ds' long. The x and y displacements of point A are u and v, respectively, while for B, the displacements are $u + du$ and $v + dv$, respectively. The variation with position of the displacement is expressed by a truncated Taylor's expansion as follows:

$$du = \frac{\partial u}{\partial x} dx + \frac{\partial u}{\partial y} dy, \qquad dv = \frac{\partial v}{\partial x} dx + \frac{\partial v}{\partial y} dy \tag{a}$$

Figure 2.4b shows the relative displacement of B with respect to A, i.e., the straining of AB. It is observed that AB has been translated so that A coincides with A'; it is now in the position $A'B''$. Here $B''D = du$ and $DB' = dv$ are the components of displacement. We now choose a new coordinate system, $x'y'$ as shown in the figure, and examine the components of strain with respect to it: $\varepsilon_{x'}, \varepsilon_{y'}, \gamma_{x'y'}$. First we shall determine the unit elongation of ds', $\varepsilon_{x'}$. The projections of du and dv upon the x' axis, after taking $EB' \cos \alpha = EB'(1)$ by virtue of the small angle approximation, lead to the approximation (Fig. 2.4b)

$$EB' = du \cos \theta + dv \sin \theta \tag{b}$$

By definition, $\varepsilon_{x'}$ is found from EB'/ds. Thus, applying Eq. (b) together with Eqs. (a), one obtains

$$\varepsilon_{x'} = \left(\frac{\partial u}{\partial x} \frac{dx}{ds} + \frac{\partial u}{\partial y} \frac{dy}{ds} \right) \cos \theta + \left(\frac{\partial v}{\partial y} \frac{dy}{ds} + \frac{\partial v}{\partial x} \frac{dx}{ds} \right) \sin \theta$$

Substituting $\cos \theta$ for dx/ds, $\sin \theta$ for dy/ds, and Eq. (2.3) into the above, we have

$$\varepsilon_{x'} = \varepsilon_x \cos^2 \theta + \varepsilon_y \sin^2 \theta + \gamma_{xy} \sin \theta \cos \theta \tag{c}$$

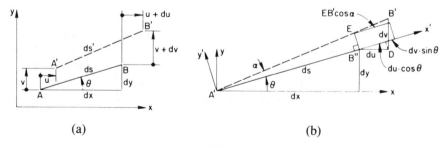

(a)   (b)

Figure 2.4

The above represents the transformation equation for the x-directed normal strain which, through the use of trigonometric identities, may be converted to the form

$$\varepsilon_{x'} = \frac{\varepsilon_x + \varepsilon_y}{2} + \frac{\varepsilon_x - \varepsilon_y}{2}\cos 2\theta + \frac{\gamma_{xy}}{2}\sin 2\theta \qquad (2.9a)$$

The normal strain $\varepsilon_{y'}$ is determined by replacing θ by $\theta + \pi/2$ in the above expression.

To derive an expression for the shearing strain $\gamma_{x'y'}$, we first determine the angle α through which AB (the x' axis) is rotated. Referring again to Fig. 2.4b, $\tan\alpha = B''E/ds$, where $B''E = dv\cos\theta - du\sin\theta - EB'\sin\alpha$. By letting $\sin\alpha = \tan\alpha = \alpha$, we have $EB'\sin\alpha = \varepsilon_{x'}ds\,\alpha = 0$. The latter is a consequence of the smallness of both $\varepsilon_{x'}$ and α. Substituting Eqs. (a) and (2.3) into $B''E$, $\alpha = B''E/ds$ may be written as follows:

$$\alpha = -(\varepsilon_x - \varepsilon_y)\sin\theta\cos\theta + \frac{\partial v}{\partial x}\cos^2\theta - \frac{\partial u}{\partial y}\sin^2\theta \qquad (d)$$

Next, the angular displacement of y' is readily derived by replacing θ by $\theta + \pi/2$ in Eq. (d):

$$\alpha_{\theta+\pi/2} = -(\varepsilon_y - \varepsilon_x)\sin\theta\cos\theta + \frac{\partial v}{\partial x}\sin^2\theta - \frac{\partial u}{\partial y}\cos^2\theta$$

Now, taking counterclockwise rotations to be positive (see Fig. 2.3b), it is necessary, in finding the shear strain $\gamma_{x'y'}$, to add α and $-\alpha_{\theta+\pi/2}$:

$$\gamma_{x'y'} = 2(\varepsilon_y - \varepsilon_x)\sin\theta\cos\theta + \left(\frac{\partial v}{\partial x} + \frac{\partial u}{\partial y}\right)(\cos^2\theta - \sin^2\theta)$$

Through the use of trigonometric identities, the above expression for the transformation of the shear strain becomes

$$\gamma_{x'y'} = -(\varepsilon_x - \varepsilon_y)\sin 2\theta + \gamma_{xy}\cos 2\theta \qquad (2.9b)$$

Comparison of Eqs. (1.7) with Eqs. (2.9), the two-dimensional transformation equations of strain, reveals an identity of form. It is observed that transformation expressions for stress are converted into strain relationships by *replacing σ with ε and τ with $\gamma/2$*. By analogy with stress, the principal strain directions (where $\gamma_{x'y'} = 0$) are found from Eq. (1.8):

$$\tan 2\theta_p = \frac{\gamma_{xy}}{\varepsilon_x - \varepsilon_y} \qquad (2.10)$$

Similarly, the magnitudes of the principal strains are

$$\varepsilon_{1,2} = \frac{\varepsilon_x + \varepsilon_y}{2} \pm \sqrt{\left(\frac{\varepsilon_x - \varepsilon_y}{2}\right)^2 + \left(\frac{\gamma_{xy}}{2}\right)^2} \qquad (2.11)$$

The transformation of three-dimensional strain proceeds from Eq. (1.17), inasmuch as we have concluded that the transformation properties of stress and strain are identical.

It is now apparent that a *Mohr's circle for strain* may be drawn and that the construction technique does not differ from that of Mohr's circle for stress. In Mohr's circle for strain, the normal strains are plotted on the horizontal axis, positive to the right. When the shear strain is positive, the point representing the x axis strains is plotted a distance $\gamma/2$ *below* the ε line, and the y axis point a distance $\gamma/2$ *above* the ε line; and vice versa when the shear strain is negative. Note that this convention for shear strain, used *only* in constructing and reading values from Mohr's circle, agrees with the convention employed for stress in Sec. 1.7.

An illustration of the use of Mohr's circle of strain is given in the solution of the following numerical problem.

Example 2.1. The state of strain at a point on a steel plate is given by $\varepsilon_x = 510 \times 10^{-6}$, $\varepsilon_y = 120 \times 10^{-6}$, and $\gamma_{xy} = 260 \times 10^{-6}$. Determine, using Mohr's circle of strain: (a) the state of strain associated with axes x', y' which make an angle $\theta = 30°$ with the axes x, y (Fig. 2.5a); (b) the principal strains and directions of the principal axes; (c) the maximum shear strains and associated normal strains.

SOLUTION. A sketch of Mohr's circle of strain is shown in Fig. 2.5b, constructed by determining the position of point C at $\frac{1}{2}(\varepsilon_x + \varepsilon_y)$ and A at $(\varepsilon_x, \frac{1}{2}\gamma_{xy})$ from the origin O. Note that $\gamma_{xy}/2$ is positive, so that point A, representing x axis strains, is plotted below the ε axis (or B above). Carrying out calculations similar to that for Mohr's circle of stress (Sec. 1.7), the required quantities are determined. The radius of the circle is $r = (195^2 + 130^2)^{1/2} \times 10^{-6} = 234 \times 10^{-6}$, and the angle $2\theta'_p = \tan^{-1}(130/195) = 33.7°$.

(a) At a position 60° counterclockwise from the x axis lies the x' axis on Mohr's circle, corresponding to twice the angle on the plate. The angle $A'CA_1$ is 60° − 33.7°

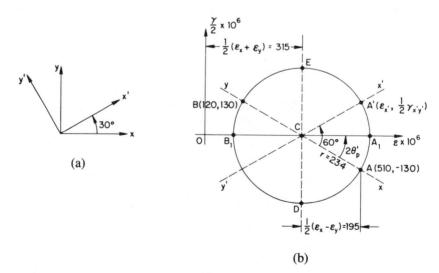

(b)

Figure 2.5

$= 26.3°$. The strain components associated with $x'y'$ are therefore

$$\varepsilon_{x'} = 315 \times 10^{-6} + 234 \times 10^{-6}\cos 26.3° = 525 \times 10^{-6}$$

$$\varepsilon_{y'} = 315 \times 10^{-6} - 234 \times 10^{-6}\cos 26.3° = 105 \times 10^{-6}$$

$$\gamma_{x'y'} = -2(234 \times 10^{-6}\sin 26.3°) = -207 \times 10^{-6}$$

The shear strain is taken as negative because the point representing the x axis strains, A', is above the ε axis. The negative sign indicates that the angle between the element faces x' and y' at the origin increases (Sec. 2.2). As a check, Eq. (2.9b) is applied with the given data to obtain -207×10^{-6}, as above.

(b) The principal strains, represented by points A_1 and B_1 on the circle, are found to be

$$\varepsilon_1 = 315 \times 10^{-6} + 234 \times 10^{-6} = 549 \times 10^{-6}$$

$$\varepsilon_2 = 315 \times 10^{-6} - 234 \times 10^{-6} = 81 \times 10^{-6}$$

The axes of ε_1 and ε_2 are directed at 16.85° and 106.85° from the x axis, respectively.

(c) Points D and E represent the maximum shear strains. Thus,

$$\gamma_{max} = \pm 468 \times 10^{-6}$$

It is observed from the circle that the axes of maximum shear strain make an angle of 45° with respect to the principal axes. The normal strains associated with the axes of γ_{max} are equal, represented by OC on the circle: 315×10^{-6}.

2.5 Engineering Materials

The equations of equilibrium derived in Chapter 1 and the kinematic relations of this chapter together represent nine equations involving 15 unknowns (six stresses, six strains, and three displacements). The insufficiency noted in the number of available equations is made up for by a set of material dependent relationships, discussed in the next section, which connect stress with strain. We first define some important characteristics of engineering materials, e.g., those in widespread commercial usage, including a variety of metals, plastics, and concretes. Following this, the tension test is discussed, this procedure providing information basic to material behavior.

An *elastic material* is one that returns to its original (unloaded) shape upon the removal of applied forces. Elastic behavior thus precludes permanent deformation. In many cases, the *elastic range* includes a region throughout which stress and strain bear a linear relationship. This portion of the stress–strain variation ends at a point termed the *proportional limit*. Such materials are linearly elastic. It is not necessary for a material to possess such linearity for it to be elastic. In a *viscoelastic* material, the state of stress is a function not only of the strains, but of the time rates of change of stress and strain as well. Viscoelastic solids return to their original state when unloaded. A *plastically* deformed solid, on the other hand, does not return to its original shape when the load is removed; there is some *permanent* deformation. With the exception of Chapter 12, our considerations will be limited to the behavior of elastic materials.

It is also assumed in this text that the material is homogeneous and isotropic. A *homogeneous* material displays identical properties throughout. If the properties are identical in all directions at a point, the material is termed *isotropic*. A nonisotropic or *anisotropic* solid such as wood displays direction-dependent properties, e.g., greater strength in a direction parallel to the grain than perpendicular to the grain. Single crystals also display pronounced anisotropy, manifesting different properties along the various crystallographic directions. Materials comprised of many crystals (poly-crystalline aggregates) may exhibit either isotropy or anisotropy. Isotropy results when the crystal size is small relative to the size of the sample, provided that nothing has acted to disturb the random distribution of crystal orientations within the aggregate. Mechanical processing operations such as cold rolling may contribute to minor anisotropy, which in practice is often disregarded. These processes may also result in high internal stress, termed *residual stress*. In the cases treated in this volume, materials are assumed initially *entirely free* of such stress.

Let us now discuss briefly the nature of the typical static tensile test. In such a test, a specimen is inserted in the jaws of a machine which permits tensile straining at a relatively low rate (since material strength is strain rate dependent). Normally, the stress–strain curve resulting from a tensile test is predicated upon *engineering* (conventional) *stress* as the ordinate, and *engineering* (conventional) *strain* as the abscissa. The latter is defined by Eq. (2.2). The former is the load or tensile force (P) divided by the original cross-sectional area (A_0) of the specimen, and as such, is simply a measure of load (force divided by a constant) rather than *true stress*. True stress is the load divided by the *actual instantaneous* or *current* area (A) of the specimen.

In Fig. 2.6a are shown two stress–strain plots, one (indicated by solid line) based upon engineering stress, the other upon true stress. The material tested is a relatively ductile, polycrystalline metal such as steel. A ductile metal is capable of substantial elongation prior to failure, as in a drawing process. The converse applies to brittle materials. Note that beyond the point labeled "proportional limit" is a point labeled "*yield point*" (for most cases these two points are taken as one). At the yield point, a great deal of deformation occurs while the applied loading remains essentially constant. The engineering stress curve for the material when strained beyond the yield point shows a characteristic maximum termed the *ultimate* tensile stress, and a lower value, the *rupture* stress, at which failure occurs. Bearing in mind the definition of engineering stress, this decrease is indicative of a decreased load-carrying capacity of the specimen with continued straining beyond the ultimate tensile stress.

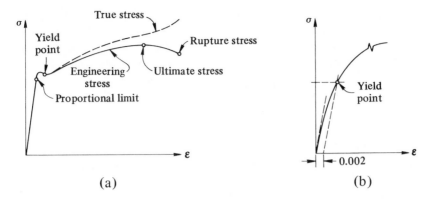

Figure 2.6

The large disparity between the engineering stress and true stress curves in the region of a large strain is attributable to the significant localized decrease in area (necking down) prior to fracture. In the area of large strain, particularly that occurring in the *plastic range*, the engineering strain, based upon small deformation, is clearly inadequate. It is thus convenient to introduce so-called true or logarithmic strain. The *true strain*, denoted by ε, is defined by

$$\varepsilon = \int_{L_0}^{L} \frac{dL}{L} = \ln\frac{L}{L_0} = \ln(1 + \varepsilon_0) \tag{2.12}$$

This strain is observed to represent the sum of the increments of deformation divided by the length L corresponding to a particular increment of length, dL. Here L_0 is the original length and ε_0 is the engineering strain. For small strains, Eqs. (2.2) and (2.13) yield approximately the same results. Note that the curve of true stress versus true strain is more informative in examining plastic behavior and will be discussed in detail in Chapter 12. In the plastic range, the material is assumed to be *incompressible* and the *volume constant* (Sec. 2.6). Hence,

$$A_0 L_0 = AL \tag{a}$$

where the left and right sides of the above equation represent the original and the current volume, respectively. If P is the current load, then

$$\sigma = \frac{P}{A} = \frac{P}{A_0}\frac{L}{L_0} = \sigma_0 \frac{L}{L_0}$$

But, from Eq. (2.2), we have $L/L_0 = 1 + \varepsilon_0$. The *true stress* is thus defined by

$$\sigma = \sigma_0(1 + \varepsilon_0) \tag{2.13}$$

That is, the *true stress is equal to the engineering stress multiplied by one plus the engineering strain.*

For materials that do not exhibit a distinctive yield point it is usual to employ a quasiyield point. According to the so-called 0.2-*percent offset* method, a line is drawn through a strain of 0.002, parallel to the initial straight line portion of the curve (Fig. 2.6b). The intersection of this line with the stress–strain curve defines the yield point, as shown.

Referring to a stress–strain diagram, the *modulus of elasticity E* is equal to the slope of the linearly elastic portion,

$$E = \frac{\sigma}{\varepsilon} \tag{2.14}$$

This modulus is clearly a measure of stiffness. Two additional moduli of interest are the *modulus of resilience* and the *modulus of toughness*. The former refers to the area under a stress–strain curve up to the proportional limit, while the latter refers to the entire area.

It was stated above that the tensile load (longitudinal) induces reduction in the cross-sectional area of the specimen (lateral contraction). Similarly, a contraction due to a compressive load is accompanied by lateral extension. In the linearly elastic range, it is found experimentally that lateral strains, say in the y and z directions, are related to the longitudinal strain, in the x direction, by a constant of proportionality, ν:

$$\varepsilon_y = \varepsilon_z = -\nu \frac{\sigma_x}{E}$$

where ν is called Poisson's ratio.

2.6 Generalized Hooke's Law

In the case of uniaxial loading, stress is related to strain by Eq. (2.14), applicable within the linear-elastic range. For x-directed loading,

$$\sigma_x = E \varepsilon_x$$

which is Hooke's law.

For the three-dimensional state of stress, each of six stress components is expressed as a linear function of six components of strain within the linear elastic range, and vice versa. This is the generalization of Hooke's law for any homogeneous elastic material. For example

$$\varepsilon_x = c_1 \sigma_x + c_2 \sigma_y + c_3 \sigma_z + c_4 \tau_{xy} + c_5 \tau_{yz} + c_6 \tau_{xz} \tag{a}$$

The c's are the material-dependent *elastic constants*. Similar expressions for ε_y, ε_z, γ_{xy}, γ_{yz}, and γ_{xz} may also be written, involving different constants. In a homogeneous body each of these constants has the same value at all points. For a homogeneous isotropic material, the constants must be identical in *all* directions at *any* point. It will be shown below that the number of constants, for an isotropic material, reduces to 2 from 36.

In the following derivation, we rely upon certain experimental evidence: a normal stress (σ_x) creates no shear strain whatever, and a shear stress

(τ_{xy}) creates only a shear strain (γ_{xy}). Also, according to the small deformation assumption, the principle of superposition applies under multiaxial stressing. Consider now a two-dimensional homogeneous isotropic rectangular element of unit thickness, subjected to a biaxial state of stress (Fig. 2.7a). Were σ_x to act, not only would the direct strain σ_x/E take place, but a y contraction as well, $-\nu\sigma_x/E$. Application of σ_y alone would result in an x contraction $-\nu\sigma_y/E$ and a y strain σ_y/E. The simultaneous action of σ_x and σ_y, applying the principle of superposition, leads to the following strains:

$$\varepsilon_x = \frac{\sigma_x}{E} - \nu\frac{\sigma_y}{E}, \qquad \varepsilon_y = \frac{\sigma_y}{E} - \nu\frac{\sigma_x}{E}$$

For pure shear (Fig. 2.7b), it is found in experiments that in the linearly elastic range, stress and strain are related by

$$\gamma_{xy} = \frac{\tau_{xy}}{G}$$

where G is the *shear modulus of elasticity*.

Similar analysis enables one to express the components $\varepsilon_z, \gamma_{yz}, \gamma_{xz}$ of strain in terms of stress and material properties. In the case of a three-dimensional state of stress, the above procedure leads to the *generalized Hooke's law*, valid for an isotropic homogeneous material:

$$\varepsilon_x = \frac{1}{E}\left[\sigma_x - \nu(\sigma_y + \sigma_z)\right], \qquad \gamma_{xy} = \frac{\tau_{xy}}{G}$$

$$\varepsilon_y = \frac{1}{E}\left[\sigma_y - \nu(\sigma_x + \sigma_z)\right], \qquad \gamma_{yz} = \frac{\tau_{yz}}{G} \qquad (2.15)$$

$$\varepsilon_z = \frac{1}{E}\left[\sigma_z - \nu(\sigma_x + \sigma_y)\right], \qquad \gamma_{xz} = \frac{\tau_{xz}}{G}$$

It is demonstrated below that the elastic constants E, ν, G are related, serving to reduce the number of independent constants in Eq. (2.15) to two. For this purpose, refer again the element subjected to pure shear (Fig. 2.7b). In accordance with Sec. 1.5, a pure shearing stress τ_{xy} can be expressed in terms of the principal stresses acting on planes (in the x' and

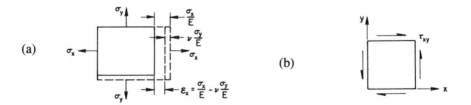

(a)

(b)

Figure 2.7

y' directions) making an angle of 45° with the shear planes: $\sigma_{x'} = \tau_{xy}$ and $\sigma_{y'} = -\tau_{xy}$. Then, applying Hooke's law, we find that

$$\varepsilon_{x'} = \frac{\sigma_{x'}}{E} - \nu \frac{\sigma_{y'}}{E} = \frac{\tau_{xy}}{E}(1+\nu) \tag{b}$$

On the other hand, as $\varepsilon_x = \varepsilon_y = 0$ for pure shear, Eq. (2.9) yields, for $\theta = 45°$, $\varepsilon_{x'} = \gamma_{xy}/2$, or

$$\varepsilon_{x'} = \frac{\tau_{xy}}{2G} \tag{c}$$

Equating the alternative relations for $\varepsilon_{x'}$ in Eqs. (b) and (c), we find that

$$G = \frac{E}{2(1+\nu)} \tag{2.16}$$

It can be shown that for any angle θ, the same result is obtained, and it is seen that when any two of the constants ν, E, G are determined experimentally, the third may be found from Eq. (2.16). From Eq. (2.15) together with Eq. (2.16), we obtain the following stress–strain relationships:

$$\begin{aligned} \sigma_x &= 2G\varepsilon_x + \lambda e, & \tau_{xy} &= G\gamma_{xy} \\ \sigma_y &= 2G\varepsilon_y + \lambda e, & \tau_{yz} &= G\gamma_{yz} \\ \sigma_z &= 2G\varepsilon_z + \lambda e, & \tau_{xz} &= G\gamma_{xz} \end{aligned} \tag{2.17}$$

Here

$$e = \varepsilon_x + \varepsilon_y + \varepsilon_z \tag{2.18}$$

and

$$\lambda = \frac{\nu E}{(1+\nu)(1-2\nu)} \tag{2.19}$$

The shear modulus G and the quantity λ, are referred to as the Lamé constants.

The physical significance of e, defined by Eq. (2.18), becomes clear upon examination of the volumetric change ΔV of an element of initial volume $V_0 = dx\, dy\, dz$, subsequent to straining: $V_f = (1+\varepsilon_x)\, dx \cdot (1+\varepsilon_y)\, dy \cdot (1+\varepsilon_z)\, dz$. Here V_f is the final volume of the element, $V_0 + \Delta V$. Expanding the right-hand side of the above expression, neglecting higher order terms, and substituting e from Eq. (2.18), we obtain

$$V_f = V_0 + V_0 e \qquad \text{or} \qquad e = \frac{\Delta V}{V_0} \tag{d}$$

It is observed that e, the *dilatation*, represents the change in volume per unit volume. The dilatation may also be expressed in terms of components of stress by combining Eq. (2.18) with Hooke's law:

$$e = \frac{1-2\nu}{E}(\sigma_x + \sigma_y + \sigma_z) \tag{2.20}$$

Consider, for example, the case of a cubic element subjected to hydrostatic

Table 2.1. Average Mechanical Properties of Metals

Material	Specific weight (kN/m³)	Modulus of elasticity (GPa) tension	shear	Yield stress (MPa) tension	shear	Ultimate stress (MPa) tension	shear	Coef. of thermal expansion (10⁻⁶ per °C)
Aluminum alloy								
6061-T6	26.6	70	25.9	241	138	290	186	23.6
2024-T4	27.2	73	27.6	290	172	441	276	23.2
Brass	82.5	103	41	103		276	193	18.9
Bronze	87	103	45	138		345	241	18
Copper	80.6	117	41	245		345	345	16.7
Cast iron	72.3	103	41			138	207	10.8
Magnesium alloy	17.6	45	16.5	138		262	131	25.2
Steel								
mild	77	200	79	248	165	410–550	331	11.7
high strength	77	200	79	345	172	483		11.7

pressure p. As the stress field is described by $\sigma_x = \sigma_y = \sigma_z = -p$, and $\tau_{xy} = \tau_{xz} = \tau_{yz} = 0$, Eq. (2.20) reduces to $e = -3(1-2\nu)p/E$ or

$$K = -\frac{p}{e} = \frac{E}{3(1-2\nu)} \qquad (2.21)$$

Here K is the modulus of volumetric expansion or *bulk modulus* of elasticity. It is seen that the unit volume contraction is proportional to the pressure and inversely proportional to K. Equation (2.21) also indicates that for incompressible materials, for which $e = 0$, Poisson's ratio is $\frac{1}{2}$. For all common materials however, $\nu < \frac{1}{2}$, since they demonstrate some change in volume, $e \neq 0$.

Table 2.1 lists average mechanical properties of a number of common metals. Exact values may *vary widely* with composition, cold working, and heat treatment. The relationships between the elastic constants introduced in this section are given by Eqs. (P2.17).

2.7   Measurement of Strain. Bonded Strain Gages

A wide variety of mechanical, electrical, and optical systems have been developed for measuring the average strain at a point on a surface.* The method in widest use employs the bonded electric wire or foil resistance strain gages. The *bonded wire gage* consists of a grid of fine wire filament cemented between two sheets of treated paper or plastic backing (Fig. 2.8). The backing serves to insulate the grid from the metal surface on which it

---

*See, for example, M. Hetényi, *Handbook of Experimental Stress Analysis*, New York: Wiley, 1957.

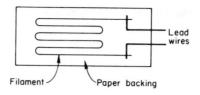

Figure 2.8

is to be bonded, and functions also as a carrier so that the filament may be conveniently handled. Generally, 0.025-mm diameter wire is used. The grid in the case of *bonded foil gages* is constructed of very thin metal foil (approximately 0.0025 mm) rather than wire. Because the filament cross section of a foil gage is rectangular, the ratio of surface area to cross-sectional area is higher than that of a round wire. This results in increased heat dissipation and improved adhesion between the grid and the backing material. Foil gages are readily manufactured in a variety of configurations. In general, the selection of a particular bonded gage will depend upon the specific service application.

The ratio of the unit change in the resistance of the gage to the unit change in length (strain) of the gage is called the *gage factor*. The metal of which the filament element is made is the principal factor determining the magnitude of this factor. *Constantan*, an alloy composed of 60 percent copper and 40 percent nickel, produces wire or foil gages with a gage factor of approximately 2.

The operation of the bonded strain gage is based upon the change in electrical resistance of the filament that accompanies a change in the strain. Deformation of the surface on which the gage is bonded results in a deformation of the backing and the grid as well. Thus, with straining, a variation in the resistance of the grid will manifest itself as a change in the voltage across the grid. An electrical bridge circuit, attached to the gage by means of lead wires, is then used to translate electrical changes into strains. The *Wheatstone bridge*, one of the most accurate and convenient systems of this type employed, is capable of measuring strains as small as 10^{-6}.

Special combination gages are available for the measurement of the state of strain at a point on a surface simultaneously in three or more directions. Generally, these consist of three gages whose axes are either 45° or 60° apart. Using the reading of the three gages together with Mohr's circle for strain, the principal strains and their orientations may readily be obtained as is illustrated in the following example.

Example 2.2 Strain rosette readings are made at a critical point in a loaded structural steel member. The 60° rosette contains three wire gages positioned at 0°, 60°, and 120°. The readings are

$$\varepsilon_0 = 190 \times 10^{-6}, \qquad \varepsilon_{60} = 200 \times 10^{-6}, \qquad \varepsilon_{120} = -300 \times 10^{-6} \qquad \text{(a)}$$

Determine the principal stresses and their directions. The material properties are $E=200$ GPa and $v=0.3$.

SOLUTION. For the situation described, Eq. (2.9) provides three simultaneous expressions:

$$\varepsilon_0 = \varepsilon_x$$

$$\varepsilon_{60} = \frac{1}{2}(\varepsilon_x + \varepsilon_y) - \frac{1}{4}(\varepsilon_x - \varepsilon_y) + \frac{\sqrt{3}}{4}\gamma_{xy}$$

$$\varepsilon_{120} = \frac{1}{2}(\varepsilon_x + \varepsilon_y) - \frac{1}{4}(\varepsilon_x - \varepsilon_y) - \frac{\sqrt{3}}{4}\gamma_{xy}$$

From the above,

$$\varepsilon_x = \varepsilon_0$$

$$\varepsilon_y = \frac{1}{3}[2(\varepsilon_{60} + \varepsilon_{120}) - \varepsilon_0] \tag{b}$$

$$\gamma_{xy} = \frac{2}{\sqrt{3}}(\varepsilon_{60} - \varepsilon_{120})$$

Upon substituting numerical values we obtain $\varepsilon_x = 190 \times 10^{-6}$, $\varepsilon_y = -130 \times 10^{-6}$, and $\gamma_{xy} = 577 \times 10^{-6}$. Then, from Eq. (2.11), the principal strains are

$$\varepsilon_{1,2} = \frac{190 - 130}{2}(10^{-6}) \pm (10^{-6})\left[\left(\frac{190 + 130}{2}\right)^2 + \left(\frac{577}{2}\right)^2\right]^{1/2}$$

$$= 30 \times 10^{-6} \pm 330 \times 10^{-6}$$

or

$$\varepsilon_1 = 360 \times 10^{-6}, \qquad \varepsilon_2 = -300 \times 10^{-6} \tag{c}$$

The maximum shear strain is found from

$$\gamma_{max} = 2 \times 10^{-6}\left[\left(\frac{190 + 130}{2}\right)^2 + \left(\frac{577}{2}\right)^2\right]^{1/2} = \pm 660 \times 10^{-6}$$

The orientations of the principal axes are given by Eq. (2.10):

$$2\theta_p = \tan^{-1}\frac{577}{320} = 61° \qquad \text{or} \qquad \theta_p' = 30.5°, \quad \theta_p'' = 120.5° \tag{d}$$

When θ_p' is substituted into Eq. (2.9) together with Eq. (b), we obtain 360×10^{-6}. Therefore 30.5° and 120.5° are the respective directions of ε_1 and ε_2, measured from the horizontal axis in a counterclockwise direction. The principal stresses may now be found from the generalized Hooke's law. Thus, the first two equations of (2.15), for plane stress, letting $\sigma_z = 0$, $\sigma_x = \sigma_1$, and $\sigma_y = \sigma_2$, together with Eqs. (c), yield

$$\sigma_1 = \frac{200 \times 10^9}{1 - 0.09}[360 + 0.3(-300)](10^{-6}) = 59.341 \text{ MPa}$$

$$\sigma_2 = \frac{200 \times 10^3}{0.91}[-300 + 0.3(360)] = -42.198 \text{ MPa}$$

The directions of σ_1 and σ_2 are given by Eq. (d). From Eq. (2.17) the maximum shear stress is

$$\tau_{max} = \frac{200 \times 10^9}{2(1 + 0.3)}660 \times 10^{-6} = 50.769 \text{ MPa}$$

Note as a check that $(\sigma_1 - \sigma_2)/2$ yields the same result.

2.8 Strain Energy

The work done by external forces in causing deformation is stored within the body in the form of *strain energy*. In an ideal elastic process, no dissipation of energy takes place, and all of the stored energy is recoverable upon unloading.

We begin our analysis by considering a rectangular prism of dimensions dx, dy, dz subjected to uniaxial tension. The front view of the prism is represented in Fig. 2.9a. If the stress is applied very slowly, as is generally the case in this text, it is reasonable to assume that equilibrium is maintained at all times. In evaluating the work done by stresses σ_x on either side of the element, it is noted that each stress acts through a different displacement. Clearly, the work done by oppositely directed forces ($\sigma_x \, dy \, dz$) through positive displacement (u) cancel one another. The *net* work done on the element by force ($\sigma_x \, dy \, dz$) is therefore

$$dW = dU = \int_0^{\varepsilon_x} \sigma_x d\left(\frac{\partial u}{\partial x} dx \right) dy \, dz = \int_0^{\varepsilon_x} \sigma_x \, d\varepsilon_x (dx \, dy \, dz)$$

where $\partial u / \partial x = \varepsilon_x$. Note that dW is the work done on $dx \, dy \, dz$, and dU is the corresponding increase in strain energy. Designating the *strain energy per unit volume* (strain energy density) as U_0, for a linearly elastic material we have

$$U_0 = \int_0^{\varepsilon_x} \sigma_x \, d\varepsilon_x = \int_0^{\varepsilon_x} E\varepsilon_x \, d\varepsilon_x \tag{a}$$

After integration, Eq. (a) yields

$$U_0 = \tfrac{1}{2} E\varepsilon_x^2 = \tfrac{1}{2}\sigma_x \varepsilon_x \tag{2.22}$$

This quantity represents the shaded area in Fig. 2.9b. The area above the stress–strain curve, termed the complementary energy density, may be determined from

$$U_0^* = \int_0^{\sigma_x} \varepsilon_x \, d\sigma_x \tag{2.23}$$

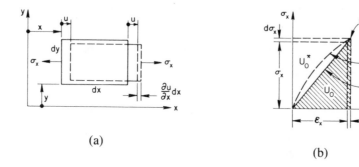

(a)

(b)

Figure 2.9

For a linearly elastic material, $U_0 = U_0^*$, but for a nonlinearly elastic material U_0 and U_0^* will differ as seen in the figure.

When σ_x, σ_y, and σ_z act simultaneously, the total work done by these normal stresses is simply the sum of expressions similar to Eq. (2.22) for each direction. This is because an x-directed stress does no work in the y or z directions, etc. The total strain energy per unit volume is thus

$$U_0 = \tfrac{1}{2}(\sigma_x \varepsilon_x + \sigma_y \varepsilon_y + \sigma_z \varepsilon_z) \tag{b}$$

The elastic strain energy associated with shear deformation is now analyzed by considering an element of thickness dz subject only to shearing stresses τ_{xy} (Fig. 2.10). From the figure, we note that shearing force $\tau_{xy}\,dx\,dz$ causes a displacement of $\gamma_{xy}\,dy$. The strain energy due to shear is $\tfrac{1}{2}(\tau_{xy}\,dx\,dz)(\gamma_{xy}\,dy)$, where the factor $\tfrac{1}{2}$ arises because the stress varies linearly with strain from zero to its final value as before. The strain energy density is therefore

$$U_0 = \frac{1}{2}\tau_{xy}\gamma_{xy} = \frac{1}{2G}\tau_{xy}^2 = \frac{1}{2}G\gamma_{xy}^2 \tag{2.24}$$

Because the work done by τ_{xy} accompanying perpendicular strains γ_{yz} and γ_{xz} is zero, the total strain energy density attributable to shear alone is found by superposition of three terms identical in form with Eq. (2.24):

$$U_0 = \tfrac{1}{2}(\tau_{xy}\gamma_{xy} + \tau_{yz}\gamma_{yz} + \tau_{xz}\gamma_{xz}) \tag{c}$$

Given a general state of stress, the strain energy density is found by adding Eqs. (b) and (c):

$$U_0 = \tfrac{1}{2}(\sigma_x \varepsilon_x + \sigma_y \varepsilon_y + \sigma_z \varepsilon_z + \tau_{xy}\gamma_{xy} + \tau_{yz}\gamma_{yz} + \tau_{xz}\gamma_{xz}) \tag{2.25}$$

Introducing Hooke's law into the above expression leads to the following form involving only stresses and elastic constants:

$$U_0 = \frac{1}{2E}(\sigma_x^2 + \sigma_y^2 + \sigma_z^2) - \frac{\nu}{E}(\sigma_x \sigma_y + \sigma_y \sigma_z + \sigma_x \sigma_z)$$
$$+ \frac{1}{2G}(\tau_{xy}^2 + \tau_{yz}^2 + \tau_{xz}^2) \tag{2.26}$$

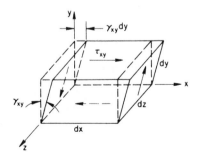

Figure 2.10

An alternate form of Eq. (2.25), written in terms of strains, is

$$U_0 = \tfrac{1}{2}\left[\lambda e^2 + 2G\left(\varepsilon_x^2 + \varepsilon_y^2 + \varepsilon_z^2\right) + G\left(\gamma_{xy}^2 + \gamma_{yz}^2 + \gamma_{xz}^2\right)\right] \tag{2.27}$$

The quantities λ and e are defined by Eqs. (2.18) and (2.19). It is interesting to observe that

$$\frac{\partial U_0(\tau)}{\partial \tau_{ij}} = \varepsilon_{ij}, \qquad \frac{\partial U_0(\varepsilon)}{\partial \varepsilon_{ij}} = \tau_{ij} \qquad (i, j = x, y, z) \tag{2.28}$$

where $U_0(\tau)$ and $U_0(\varepsilon)$ designate the strain energy densities expressed in terms of stress and strain, respectively [Eqs. (2.26) and (2.27)]. Derivatives of this type will be discussed again in connection with energy methods in Chapter 10.

In order to ascertain the energy stored within an entire body, the elastic energy density is integrated over the original or undeformed volume V:

$$U = \int_V U_0 \, dV = \int \int \int U_0 \, dx \, dy \, dz \tag{2.29}$$

This expression permits the strain energy to be readily evaluated for a number of commonly encountered geometries and loadings. Note especially that the strain energy is a nonlinear (quadratic) function of load or deformation. The principle of superposition is thus *not* valid for the strain energy.

Example 2.3. Derive an expression for the strain energy stored in a circular uniform bar subjected to oppositely directed twisting moments M_t at both ends.

SOLUTION. Let the axial direction coincide with x axis. Since the state of stress is pure shear, the strain energy density is expressed by Eq. (2.24), as $U_0 = \tau^2/2G$. Here, according to the torsion formula (Sec. 6.1), $\tau = M_t r/J$. The total strain energy U in the bar is evaluated by integrating U_0 over the volume of the bar:

$$U = \int \frac{M_t^2}{2GJ^2}\left[\int \int r^2 \, dz \, dy\right] dx$$

By definition, the term in the brackets is the polar moment of inertia, J, of the cross-sectional area. We thus have

$$U = \int \frac{M_t^2 \, dx}{2GJ} \tag{2.30}$$

where the integration proceeds over the length of the bar.

2.9   Components of Strain Energy

A new perspective on the strain energy may be gained by viewing the general state of stress (Fig. 2.11a) in terms of the superposition shown in Fig. 2.11. The state of stress in Fig. 2.11b, represented by

$$\begin{bmatrix} \sigma_m & 0 & 0 \\ 0 & \sigma_m & 0 \\ 0 & 0 & \sigma_m \end{bmatrix} \tag{a}$$

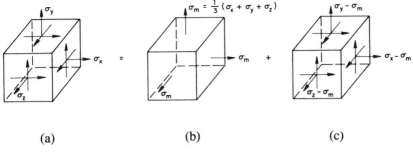

(a)   (b)   (c)

Figure 2.11

results in volume change without distortion, and is termed the *dilatational* stress tensor. Here $\sigma_m = \frac{1}{3}(\sigma_x + \sigma_y + \sigma_z)$ is the mean stress defined by Eq. (1.26). Associated with σ_m is the *mean strain*, $\varepsilon_m = \frac{1}{3}(\varepsilon_x + \varepsilon_y + \varepsilon_z)$. The sum of the normal strains accompanying the application of the dilatational stress tensor is the dilatation $e = \varepsilon_x + \varepsilon_y + \varepsilon_z$, representing a change in volume only. Thus, the dilatational strain energy absorbed per unit volume is given by

$$U_{0v} = \tfrac{1}{2}\sigma_m \varepsilon_m = \frac{\sigma_m^2}{2K} = \frac{1}{18K}(\sigma_x + \sigma_y + \sigma_z)^2 \qquad (2.31)$$

where K is defined by Eq. (2.21).

The state of stress in Fig. 2.11c, represented by

$$\begin{bmatrix} \sigma_x - \sigma_m & \tau_{xy} & \tau_{xz} \\ \tau_{xy} & \sigma_y - \sigma_m & \tau_{yz} \\ \tau_{xz} & \tau_{yz} & \sigma_z - \sigma_m \end{bmatrix} \qquad \text{(b)}$$

is called the *deviator* or distortional stress tensor. This produces deviator strains or distortion without change in volume. This is because the sum of the normal strains is $(\varepsilon_x - \varepsilon_m) + (\varepsilon_y - \varepsilon_m) + (\varepsilon_z - \varepsilon_m) = 0$. The distortional energy per unit volume, U_{0d}, associated with the deviator stress tensor is attributable to the change of shape of the unit volume, while the volume remains constant. Since U_{0v} and U_{0d} are the only components of the strain energy, we have $U_0 = U_{0v} + U_{0d}$. By subtracting Eq. (2.31) from Eq. (2.26) the distortional energy is readily found to be

$$U_{0d} = \frac{3}{4G}\tau_{oct}^2 \qquad (2.32)$$

This is the elastic strain energy absorbed by the unit volume as a result of its change in shape (distortion). In the above, the octahedral shearing stress τ_{oct}, is given by

$$\tau_{oct} = \tfrac{1}{3}\left[(\sigma_x - \sigma_y)^2 + (\sigma_y - \sigma_z)^2 + (\sigma_z - \sigma_x)^2 + 6(\tau_{xy}^2 + \tau_{xz}^2 + \tau_{yz}^2)\right]^{1/2} \quad (2.33)$$

The planes where the τ_{oct} acts are shown in Fig. 1.11b of Sec. 1.10. The strain energy of distortion plays an important role in the theory of failure of a ductile metal under any condition of stress. This is discussed further in Chapter 4. The stresses and strains associated with both components of the strain energy are also very useful in describing the plastic deformation (Chapter 12).

Example 2.4. A mild steel bar of uniform cross section A is subjected to an axial tensile load N. Derive an expression for the strain energy density, its components, and the total strain energy stored in the bar. Let $\nu = 0.25$.

SOLUTION. The state of stress at any point in the bar is axial tension, $\tau_{xy} = \tau_{xz} = \tau_{yz} = \sigma_y = \sigma_z = 0$, $\sigma_x = \sigma = N/A$ (Fig. 2.11a). We therefore have the stresses associated with volume change $\sigma_m = \sigma/3$ and shape change $\sigma_x - \sigma_m = 2\sigma/3$, $\sigma_y - \sigma_m = \sigma_z - \sigma_m = -\sigma/3$ (Fig. 2.11b, c). The strain energy densities for the state of stress in cases a, b, and c are found, respectively, as follows:

$$U_0 = \frac{\sigma^2}{2E}$$

$$U_{0v} = \frac{(1-2\nu)\sigma^2}{6E} = \frac{\sigma^2}{12E}$$

$$U_{0d} = \frac{(1+\nu)\sigma^2}{3E} = \frac{5\sigma^2}{12E}$$

It is observed from the above expressions that $U_0 = U_{0v} + U_{0d}$ and that $5U_{0v} = U_{0d}$. Thus, we observe that in changing the shape of a unit volume element under uniaxial stressing, five times more energy is absorbed than in changing the volume. The total strain energy stored in the bar is

$$U = \int \frac{N^2\,dx}{2AE} \tag{2.34}$$

where the integration is carried out over the beam length.

2.10   Effect of Local Stress and Strain. St. Venant's Principle

The reader will recall from a study of Newtonian mechanics that for purposes of analyzing the statics or dynamics of a body, one force system may be replaced by an equivalent force system whose force and moment resultants are identical. It is often added in discussing this point that the force resultants, while equivalent, need not cause an identical distribution of strain, owing to difference in the *arrangement* of the forces. *St. Venant's principle* permits the use of an equivalent loading for the calculation of stress and strain. This principle states that if an actual distribution of forces is replaced by a statically equivalent system, the distribution of stress and strain throughout the body is altered only near the regions of load application.* The contribution of St. Venant's principle to the solution of engineering problems is quite important, for it often frees the

---

*See, for example, E. Sternberg, On St. Venant's principle, *Quart. Appl. Math.* 11:393 (1954); E. Sternberg and W. T. Koiter, The wedge under a concentrated couple, *J. Appl. Mech.* 25:575–581 (1958).

analyst of the burden of prescribing the boundary conditions very precisely when it is very difficult to do so. Furthermore, where a certain solution is predicated upon a particular boundary loading, the solution can serve equally for another type of boundary loading, not quite the same as the first. That is, when an analytical solution calls for a certain distribution of stress on a boundary (such as σ_x in Sec. 5.5), we need not discard the solution merely because the boundary distribution is not quite the same as that required by the solution. The value of existing solutions is thus greatly extended. Consider, for example, the substitution of a uniform distribution of stress at the ends of a tensile test specimen for the actual irregular distribution which results from end clamping. If we require the stress in a region away from the ends, the stress variation at the ends need not be of concern, since it does not lead to significant variation in the region of interest. As a further example, according to St. Venant's principle the complex distribution of force supplied by the wall to a cantilever beam may be replaced by vertical and horizontal forces and a couple moment for purposes of determining the stresses some distance from the wall.

Chapter 2—Problems

Secs. 2.1 to 2.5

2.1. Determine whether the following strain fields are possible in a continuous material:

$$\text{(a)} \begin{bmatrix} c(x^2+y^2) & cxy \\ cxy & y^2 \end{bmatrix}, \quad \text{(b)} \begin{bmatrix} cz(x^2+y^2) & cxyz \\ cxyz & y^2z \end{bmatrix}$$

Here c is a small constant, and it is assumed that $\varepsilon_z = \gamma_{xz} = \gamma_{yz} = 0$.

2.2. A 1000×1500 mm rectangular plate $OABC$ is deformed into a shape $O'A'B'C'$ shown in Fig. P2.2. Find (a) the strain components ε_x, ε_y, γ_{xy} and (b) the principal strains and the direction of the principal axes.

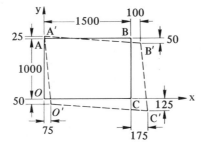

Figure P2.2

2.3. A displacement field is given by

$$u=(x^2+10)\times 10^{-2}$$
$$v=(2yz)\times 10^{-2}$$
$$w=(z^2-xy)\times 10^{-2}$$

Determine the state of strain of an element positioned at $(0,2,1)$.

2.4. The principal strains at a point are $\varepsilon_1=400\times 10^{-5}$ and $\varepsilon_2=200\times 10^{-5}$. Determine (a) the maximum shear strain and the direction along which it occurs and (b) the strains in directions at $\theta=30°$ from the principal axes. Solve the problem by using the formulas developed and check the results by employing Mohr's circle.

2.5. If the strains at a point are $\varepsilon_x=-900\times 10^{-6}$, $\varepsilon_y=-300\times 10^{-6}$, and $\gamma_{xy}=900\times 10^{-6}$, what are the principal strains, and in what direction do they occur? Use Mohr's circle of strain.

2.6. Solve the preceding problem for $\varepsilon_x=300\times 10^{-6}$, $\varepsilon_y=900\times 10^{-6}$, and $\gamma_{xy}=-900\times 10^{-6}$.

2.7. Show that for plane strain, $\varepsilon_x+\varepsilon_y=\varepsilon_{x'}+\varepsilon_{y'}$.

2.8. At a point in a stressed body the strains, related to the coordinate set xyz, are given by

$$\begin{bmatrix} 2 & 3 & 2 \\ 3 & -1 & 5 \\ 2 & 5 & -4 \end{bmatrix}\times 10^{-6}$$

Determine, referring to Secs. 2.4 and 1.9, (a) the strain invariants, (b) the normal strain in the x' direction, which is directed at an angle $\theta=30°$ from the x axis, (c) the principal strains ε_1, ε_2, and ε_3, and (d) the maximum shear strain.

2.9. Solve the preceding problem for a state of strain given by

$$\begin{bmatrix} 4 & 1 & 0 \\ 1 & 0 & -2 \\ 0 & -2 & 6 \end{bmatrix}\times 10^{-6}$$

Secs. 2.6 to 2.10

2.10. A 12-mm-diameter specimen is subjected to tensile loading. The increase in length resulting from a load of 9 kN is 0.025 mm for an original length L of 75 mm. What are the true and conventional strains and stresses? Calculate the modulus of elasticity.

2.11. A 50-mm square plate is subjected to the stresses (in MPa) shown in Fig. P2.11. What deformation is experienced by diagonal BD? Express the solution in terms of E, using two approaches: (a) determine the components of strain along the x and y directions and then employ the equations governing the transformation of strain; (b) determine the stress on planes perpendicular and parallel to BD and then employ the generalized Hooke's law.

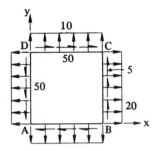

Figure P2.11

2.12. At a critical point P in a loaded beam, a $45°$ rosette measures normal strains. The readings are $\varepsilon_0 = -100 \times 10^{-6}$, $\varepsilon_{45} = 50 \times 10^{-6}$, $\varepsilon_{90} = 100 \times 10^{-6}$. Calculate the principal strains and the principal directions at P.

2.13. For a given steel, $E = 200$ GPa and $G = 80$ GPa. If the state of strain at a point within this material is given by

$$\begin{bmatrix} 0.002 & 0.001 & 0 \\ 0.001 & 0.003 & 0.004 \\ 0 & 0.004 & 0 \end{bmatrix}$$

ascertain the corresponding components of the stress tensor.

2.14. For a material with $G = 80$ GPa psi and $E = 200$ GPa, determine the strain tensor for a state of stress given by

$$\begin{bmatrix} 20 & -4 & 5 \\ -4 & 0 & 10 \\ 5 & 10 & 15 \end{bmatrix} \text{MPa}$$

2.15. At a point in an elastic body the principal strains $\varepsilon_3, \varepsilon_2, \varepsilon_3$ are in the ratio $3:4:5$; the largest principal stress is $\sigma_1 = 140$ MPa. Determine the ratio $\sigma_3 : \sigma_2 : \sigma_1$ and the values of σ_2, σ_3. Take $\nu = 0.3$ and $E = 200$ GPa.

2.16. The stress field in an elastic body is given by

$$\begin{bmatrix} ky^2 & 0 \\ 0 & -kx^2 \end{bmatrix}$$

where k is a constant. Derive expressions for the displacement components $u(x, y)$ and $v(x, y)$ in the body.

2.17. Derive the following relations involving the elastic constants:

$$K = \lambda + \frac{2}{3}G = \frac{2G(1+\nu)}{3(1-2\nu)} = \frac{E}{3(1-2\nu)}$$

$$G = \frac{\lambda(1-2\nu)}{2\nu} = \frac{3K(1-2\nu)}{2(1+\nu)} = \frac{3K}{9K-E}$$

$$E = \frac{G(3\lambda+2G)}{\lambda+G} = 2G(1+\nu) = 3K(1-2\nu) = \frac{9KG}{3K+G} \qquad \text{(P2.17)}$$

$$\nu = \frac{\lambda}{2(\lambda+G)} = \frac{E}{2G} - 1 = \frac{3K-2G}{2(3K+G)} = \frac{3K-E}{6K}$$

2.18. As shown in Fig. P2.18, a thin prismatical bar of specific weight γ and constant cross section hangs in the vertical plane. Under the effect of its own weight, the displacement field is described by

$$u = \frac{\gamma}{2E}(2xa - x^2 - \nu y^2), \qquad v = -\frac{\nu\gamma}{E}(a-x)y$$

The z displacement and stresses may be neglected. Find the strain and stress components in the bar. Check to see whether the boundary conditions [Eq. (1.28)] are satisfied by the stresses found.

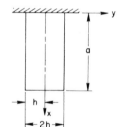

Figure P2.18

2.19. The state of stress at a point is

$$\begin{bmatrix} 200 & 20 & 10 \\ 20 & -50 & 0 \\ 10 & 0 & 40 \end{bmatrix} \text{MPa}$$

Decompose this array into a set of dilatational stresses and a set of deviator stresses. Determine the values of *principal deviator stress*.

2.20. A circular prismatic cantilever is subjected to a twisting moment M_t and an axial force N at its free end. The length of the bar is L, the radius is r, and the modulus of elasticity of the material is E. Determine the total strain energy stored in the bar, and its components. Assume $\nu = \frac{1}{4}$ for the material.

Chapter 3

Two-Dimensional Problems in Elasticity

3.1 Introduction

The approaches in widespread use for determining the influence of applied loads upon elastic bodies are the *mechanics of materials* or *elementary theory* (also known as *technical theory*) and the *theory of elasticity*. Both must, of course, rely upon the fundamental laws of equilibrium. The difference between these methods lies primarily in the extent to which the strain is described and the nature of the simplifications. The theory of elasticity is preferred when critical design constraints such as minimum weight, minimum cost, or high reliability dictate more exact treatment, or when prior experience is limited and intuition does not serve adequately to supply the needed simplifications with a degree of assurance. If properly applied, the theory of elasticity should yield solutions more closely approximating the actual distributions of strain, stress, and displacement. We emphasize, however, that both techniques cited are approximations to nature. The influences of material anisotropy, the extent to which boundary conditions depart from reality, and a host of other factors all contribute to error.

To ascertain the distribution of stress, strain, and displacement within an elastic body subject to a prescribed system of forces requires consideration of a number of fundamental conditions relating to certain physical laws, material properties, geometry, and surface forces:

1. The equations of equilibrium must be satisfied throughout the body.
2. The linear elastic stress–strain relations (Hooke's law) must apply to the material.
3. The components of strain, related to the derivatives of displacement, must be compatible with one another, i.e., the distribution of strain must be consistent with the preservation of body continuity. (The matter of compatibility is not always broached in mechanics of materials analysis.)
4. The stress, strain, and displacement fields must be such as to conform to the conditions of loading imposed at the boundaries.

The conditions described above, stated mathematically in the previous chapters, are used to derive the equations of elasticity. In the case of a *three-dimensional* problem in elasticity, it is required that the following 15 quantities be ascertained: six stress components, six strain components, and three displacement components. These components must satisfy 15 governing equations throughout the body in addition to the boundary conditions: three equations of equilibrium, six stress–strain relations, and six strain-displacement relations. It is to be noted that the equations of compatibility are derived from the strain-displacement relations, which are already included in the above description. Thus, if the 15 expressions are satisfied, the equations of compatibility will also be satisfied. Three-dimensional problems, often quite complex, are not treated in this text.

In many engineering applications, ample justification may be found for simplifying assumptions with respect to the distribution of strain and stress. Of special importance, because of the resulting decrease in complexity, are those reducing a three-dimensional problem to one involving only two dimensions. In this regard, we shall discuss plane strain and plane stress problems.

3.2   Plane Strain Problems

Consider a long prismatic member subject to lateral loading (e.g., a cylinder under pressure), held between *fixed*, smooth, rigid planes (Fig. 3.1). Assume the external forces to be functions of the x and y coordinates only. As a consequence, we expect all cross sections to experience identical deformation, including those sections near the ends. The frictionless nature of the end constraint permits x, y deformation, but precludes z displacement, i.e., $w = 0$ at $z = \pm L/2$. Considerations of symmetry dictate that w must also be zero at midspan. Symmetry arguments can again be used to infer that $w = 0$ at $\pm L/4$, and so on, until every cross section is taken into

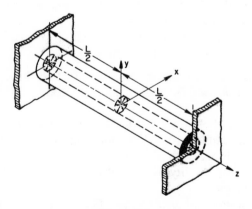

Figure 3.1

account. For the case described, the strain depends upon x and y only:

$$\varepsilon_x = \frac{\partial u}{\partial x}, \qquad \varepsilon_y = \frac{\partial v}{\partial y}, \qquad \gamma_{xy} = \frac{\partial u}{\partial y} + \frac{\partial v}{\partial x} \tag{3.1}$$

$$\varepsilon_z = \frac{\partial w}{\partial z} = 0, \qquad \gamma_{xz} = \frac{\partial w}{\partial x} + \frac{\partial u}{\partial z} = 0, \qquad \gamma_{yz} = \frac{\partial w}{\partial y} + \frac{\partial v}{\partial z} = 0 \tag{3.2}$$

The latter expressions depend upon $\partial u/\partial z$ and $\partial v/\partial z$ vanishing, since w and its derivatives are zero. A state of plane strain has thus been described wherein each point remains within its transverse plane, following application of the load. We next proceed to develop the equations governing the behavior of bodies under plane strain.

Substitution of $\varepsilon_z = \gamma_{yz} = \gamma_{xz} = 0$ into Eq. (2.17) provides the following *stress–strain relationships*:

$$\sigma_x = 2G\varepsilon_x + \lambda(\varepsilon_x + \varepsilon_y)$$

$$\sigma_y = 2G\varepsilon_y + \lambda(\varepsilon_x + \varepsilon_y) \tag{3.3a}$$

$$\tau_{xy} = G\gamma_{xy}$$

and

$$\tau_{xz} = \tau_{yz} = 0, \qquad \sigma_z = \lambda(\varepsilon_x + \varepsilon_y) = \nu(\sigma_x + \sigma_y) \tag{a}$$

Because σ_z is not contained in the other governing expressions for plane strain, it is determined independently by applying Eq. (a). The strain–stress relations, Eqs. (2.15), for this case become

$$\varepsilon_x = \frac{1 - \nu^2}{E}\left(\sigma_x - \frac{\nu}{1 - \nu}\sigma_y\right)$$

$$\varepsilon_y = \frac{1 - \nu^2}{E}\left(\sigma_y - \frac{\nu}{1 - \nu}\sigma_x\right) \tag{3.3b}$$

$$\gamma_{xy} = \frac{\tau_{xy}}{G}$$

Inasmuch as the above stress components are functions of x and y only, the first two equations of (1.5) yield the following *equations of equilibrium of plane strain*:

$$\frac{\partial \sigma_x}{\partial x} + \frac{\partial \tau_{xy}}{\partial y} + F_x = 0$$

$$\frac{\partial \sigma_y}{\partial y} + \frac{\partial \tau_{xy}}{\partial x} + F_y = 0 \tag{3.4}$$

The third equation of (1.5) is satisfied if $F_z = 0$. In the case of plane strain, therefore, no body force in the axial direction can exist.

A similar restriction is imposed on the surface forces. That is, plane strain will result in a prismatic body if the surface forces T_x and T_y are

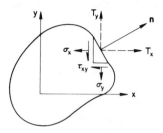

Figure 3.2

each functions of x and y and $T_z = 0$. On the lateral surface, $n = 0$ (Fig. 3.2). The *boundary conditions*, from the first two equations of (1.28), are thus given by

$$T_x = \sigma_x l + \tau_{xy} m$$
$$T_y = \tau_{xy} l + \sigma_y m$$

(3.5)

Clearly, the last equation of (1.28) is also satisfied.

In the case of a plane strain problem, therefore, eight quantities, σ_x, σ_y, τ_{xy}, ε_x, ε_y, γ_{xy}, u, *and* v, must be determined so as to satisfy Eqs. (3.1), (3.3), and (3.4) and the boundary conditions (3.5). How eight governing equations, (3.1), (3.3a), and (3.4), may be reduced to three is now discussed.

Three expressions for two-dimensional strain at a point [Eq. (3.1)] are functions of only *two* displacements, u and v, and therefore a compatibility relationship exists among the strains [Eq. (2.7)]:

$$\frac{\partial^2 \varepsilon_x}{\partial y^2} + \frac{\partial^2 \varepsilon_y}{\partial x^2} = \frac{\partial^2 \gamma_{xy}}{\partial x \, \partial y}$$

(3.6)

This equation must be satisfied in order that the strain components will be related to the displacements as in Eqs. (3.1). The condition as expressed by Eq. (3.6) may be transformed into one involving components of stress by substituting the strain–stress relations and employing the equations of equilibrium. Performing the operations indicated, using Eqs. (3.3a) and (3.6), we have

$$\frac{\partial^2}{\partial y^2} \left[(1-\nu)\sigma_x - \nu\sigma_y \right] + \frac{\partial^2}{\partial x^2} \left[(1-\nu)\sigma_y - \nu\sigma_x \right] = 2 \frac{\partial^2 \tau_{xy}}{\partial x \, \partial y}$$

(b)

Next, the first and second equations of (3.4) are differentiated with respect to x and y, respectively, and added to yield

$$2 \frac{\partial^2 \tau_{xy}}{\partial x \, \partial y} = - \left(\frac{\partial^2 \sigma_x}{\partial x^2} + \frac{\partial^2 \sigma_y}{\partial y^2} \right) - \left(\frac{\partial F_x}{\partial x} + \frac{\partial F_y}{\partial y} \right)$$

Finally, substitution of the above into Eq. (b) results in

$$\left(\frac{\partial^2}{\partial x^2} + \frac{\partial^2}{\partial y^2}\right)(\sigma_x + \sigma_y) = -\frac{1}{1-\nu}\left(\frac{\partial F_x}{\partial x} + \frac{\partial F_y}{\partial y}\right) \qquad (3.7)$$

This is the *equation of compatibility* in terms of stress. We now have three expressions, Eqs. (3.4) and (3.7), in terms of three unknown quantities: σ_x, σ_y, and τ_{xy}. This set of equations, together with the boundary conditions (3.5), is used in the solution of plane strain problems. For a given situation, after determining the stress, Eqs. (3.3b) and (3.1) yield the strain and displacement, respectively. In Sec. 3.4, Eqs. (3.4) and (3.7) will further be reduced to one equation, containing a single variable.

3.3 Plane Stress Problems

There are many problems of practical importance in which the stress condition is one of *plane stress*. The basic definition of this state of stress has already been given in Sec. 1.4. In this section we shall present the governing equations for the solution of plane stress problems.

To exemplify the case of plane stress, consider a thin plate, as in Fig. 3.3, wherein the loading is uniformly distributed over the thickness, parallel to the plane of the plate. This geometry contrasts with that of the long prism previously discussed, which is in a state of plane strain. To arrive at some tentative conclusions with regard to the stress within the plate, consider the fact that σ_z, τ_{xz}, and τ_{yz} are zero on both faces of the plate. Because the plate is thin, the stress distribution may be very closely approximated by assuming that the foregoing is likewise true throughout the plate. We shall, as a condition of the problem, take the body force $F_z = 0$, and F_x and F_y each to be functions of x and y only. As a consequence of the foregoing, the stress is specified by

$$\sigma_x, \qquad \sigma_y, \qquad \tau_{xy}$$
$$\sigma_z = \tau_{xz} = \tau_{yz} = 0 \qquad (3.8)$$

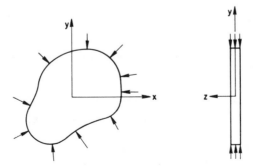

Figure 3.3

The nonzero stress components remain constant over the thickness of the plate, and are functions of x and y only. The above situation describes a state of plane stress. Equations (1.5) and (1.28), together with this combination of stress, again reduce to the forms found in the previous article. Thus, Eqs. (3.4) and (3.5) describe the equations of equilibrium and the boundary conditions in this case, as in the case of plane strain.

Substitution of Eq. (3.8) into Eq. (2.15) yields the following *stress–strain relations* for plane stress:

$$\varepsilon_x = \frac{1}{E}(\sigma_x - \nu\sigma_y)$$

$$\varepsilon_y = \frac{1}{E}(\sigma_y - \nu\sigma_x) \tag{3.9}$$

$$\gamma_{xy} = \frac{\tau_{xy}}{G}$$

and

$$\gamma_{xz} = \gamma_{yz} = 0, \qquad \varepsilon_z = -\frac{\nu}{E}(\sigma_x + \sigma_y) \tag{a}$$

As ε_z is not contained in the other governing expressions for plane stress, it can be obtained independently from Eqs. (a); then $\varepsilon_z = \partial w/\partial z$ may be applied to yield w. That is, only u and v are considered as independent variables in the governing equations. In the case of plane stress, therefore, the basic strain-displacement relations are again given by Eqs. (3.1). Exclusion from Eq. (2.3) of $\varepsilon_z = \partial w/\partial z$ makes the plane stress equations approximate, as is demonstrated in the section that follows.

The governing equations of plane stress will now be reduced, as in the case of plane strain, to three equations, involving stress components only. Since Eqs. (3.1) apply to plane strain and plane stress, the compatibility condition represented by Eq. (3.6) applies in both cases. The latter expression may be written as follows, substituting strains from Eqs. (3.9) and employing Eqs. (3.4):

$$\left(\frac{\partial^2}{\partial x^2} + \frac{\partial^2}{\partial y^2}\right)(\sigma_x + \sigma_y) = -(1+\nu)\left(\frac{\partial F_x}{\partial x} + \frac{\partial F_y}{\partial y}\right) \tag{3.10}$$

This *equation of compatibility*, together with the equations of equilibrium, represents a useful form of the governing equations for problems of plane stress.

In summary of the two-dimensional situations discussed, the equations of equilibrium [Eqs. (3.4)], together with those of compatibility [Eq. (3.7)

Table 3.1

Solution	To convert to	E is replaced by	ν is replaced by
Plane stress	Plane strain	$\dfrac{E}{1-\nu^2}$	$\dfrac{\nu}{1-\nu}$
Plane strain	Plane stress	$\dfrac{1+2\nu}{(1+\nu)^2}E$	$\dfrac{\nu}{1+\nu}$

for plane strain and Eq. (3.10) for plane stress] and the boundary conditions [Eqs. (3.5)], provide a system of equations sufficient for determination of the complete stress distribution. It can be shown that a solution satisfying all of these equations is, for a given problem, unique.* That is, it is the *only* solution to the problem.

In the absence of body forces or in the case of constant body forces, the compatibility equations for plane strain and plane stress are the same. In these cases, the equations governing the distribution of stress do not contain the elastic constants. Given identical geometry and loading, a bar of steel and one of Lucite should thus display identical stress distributions. This characteristic is important in that any convenient isotropic material may be used to substitute for the actual material, as, for example, in *photoelastic* studies.

It is of interest to note that by comparing Eqs. (3.3b) with Eqs. (3.9) one can form Table 3.1, which facilitates the conversion of a plane stress solution into a plane strain solution and vice versa. For instance, conditions of plane stress and plane strain prevail in a *narrow* beam and a *very wide* beam, respectively. Hence, in a result pertaining to a thin beam, EI would become $EI(1-\nu^2)$ for the case of a wide beam. The stiffness in the latter case is greater owing to the prevention of sidewise displacement (Secs. 5.2 and 13.2).

3.4 The Stress Function

The preceding sections have demonstrated that the solution of two-dimensional problems in elasticity requires integration of the differential equations of equilibrium [Eqs. (3.4)], together with the compatibility equation [Eq. (3.7) or (3.10)] and the boundary conditions [Eqs. (3.5)]. In the

---

*See I. S. Sokolnikoff, *Mathematical Theory of Elasticity*, New York: McGraw-Hill, 1956, Sec. 27.

event that the body forces F_x and F_y are negligible, these equations reduce to

$$\frac{\partial \sigma_x}{\partial x} + \frac{\partial \tau_{xy}}{\partial y} = 0, \qquad \frac{\partial \sigma_y}{\partial y} + \frac{\partial \tau_{xy}}{\partial x} = 0 \qquad \text{(a)}$$

$$\left(\frac{\partial^2}{\partial x^2} + \frac{\partial^2}{\partial y^2} \right)(\sigma_x + \sigma_y) = 0 \qquad \text{(b)}$$

together with the boundary conditions (3.5). The equations of equilibrium are identically satisfied by the *stress function*, $\phi(x, y)$, introduced by G. B. Airy, related to the stresses as follows:

$$\sigma_x = \frac{\partial^2 \phi}{\partial y^2}, \qquad \sigma_y = \frac{\partial^2 \phi}{\partial x^2}, \qquad \tau_{xy} = -\frac{\partial^2 \phi}{\partial x \, \partial y} \qquad (3.11)$$

Substitution of the above expressions into the compatibility equation, Eq. (b), yields

$$\frac{\partial^4 \phi}{\partial x^4} + 2\frac{\partial^4 \phi}{\partial x^2 \partial y^2} + \frac{\partial^4 \phi}{\partial y^4} = \nabla^4 \phi = 0 \qquad (3.12)$$

What has been accomplished is the formulation of a two-dimensional problem in which body forces are absent, in such a way as to require the solution of a single *biharmonic equation*, which must of course satisfy the boundary conditions.

It should be noted that in the case of plane stress, we have $\sigma_z = \tau_{xz} = \tau_{yz} = 0$ and $\sigma_x, \sigma_y, \tau_{xy}$ independent of z. As a consequence, $\gamma_{xz} = \gamma_{yz} = 0$, and $\varepsilon_x, \varepsilon_y, \varepsilon_z, \gamma_{xy}$ are independent of z. In accordance with the foregoing, from Eq. (2.8), it is seen that in addition to Eq. (3.12), the following compatibility equations also hold:

$$\frac{\partial^2 \varepsilon_z}{\partial x^2} = 0, \qquad \frac{\partial^2 \varepsilon_z}{\partial y^2} = 0, \qquad \frac{\partial^2 \varepsilon_z}{\partial x \, \partial y} = 0$$

Clearly, these additional conditions will not be satisfied in a case of plane stress by a solution of Eq. (3.12) alone. Therefore, such a solution of a plane stress problem has an approximate character. However, it can be shown that for thin plates, the error introduced is negligibly small.

3.5   Methods for Solution of Two-Dimensional Problems

Unfortunately, solving directly the equations of elasticity derived may be a formidable task, and it is often advisable to attempt a solution by the *inverse* or *semi-inverse method*. The inverse method requires examination of the assumed solutions with a view toward finding one which will satisfy the governing equations and boundary conditions. The semi-inverse method requires the assumption of a partial solution, formed by expressing stress, strain, displacement, or stress function in terms of known or undetermined coefficients. The governing equations are thus rendered more manageable. Here it is important to note that these assumptions, based upon the mechanics of a particular problem, are subject to later verification. This is in contrast with the mechanics of materials approach, in which analytical verification does not occur. The applications of inverse, semi-inverse, and *direct* methods are found in examples to follow and in Chapters 5, 6, and 8.

A number of problems may be solved by using a linear combination of *polynomials* in x and y and undetermined coefficients of the stress function ϕ. Clearly, an assumed polynomial form must satisfy the biharmonic equation and must be of second degree or higher in order to yield a nonzero stress solution of Eq. (3.11) as described in the following paragraphs. In general, finding the desirable polynomial form is laborious and requires a systematic approach.* The *Fourier series*, indispensible in the analytical treatment of many problems in the field of applied mechanics, is also often employed (Secs. 10.9 and 13.5). Another way to overcome the difficulty involved in the solution of Eq. (3.12) is to use the method of finite differences. Here the governing equation is replaced by series of finite difference equations (Sec. 7.4), which relate the stress function at stations that are removed from one another by finite distances. These equations, although not exact, frequently lead to solutions that are close to the exact solution. The results obtained are, however, applicable only to specific numerical problems.

Polynomial Solutions

An elementary approach to obtaining solutions of the biharmonic equation uses polynomial functions of various degree with their coefficients adjusted so that $\nabla^4\phi = 0$ is satisfied. A brief discussion of this procedure follows.

A polynomial of the *second* degree

$$\phi_2 = \frac{a_2}{2}x^2 + b_2 xy + \frac{c_2}{2}y^2 \tag{3.13}$$

satisfies Eq. (3.12). The associated stresses are

$$\sigma_x = c_2, \qquad \sigma_y = a_2, \qquad \tau_{xy} = -b_2$$

For a rectangular plate, it is apparent that the foregoing may be adapted to represent *simple tension* ($c_2 \neq 0$), *double tension* ($c_2 \neq 0$, $a_2 \neq 0$), or *pure shear* ($b_2 \neq 0$).

---

*See S. Timoshenko and J. Goodier, *Theory of Elasticity*, New York: McGraw-Hill, 1970, Chapter 3; C. Y. Neous, Direct method of determining Airy polynomial stress functions, *J. Appl. Math.* 24 3:387 (September 1957).

A polynomial of the *third* degree

$$\phi_3 = \frac{a_3}{6}x^3 + \frac{b_3}{2}x^2y + \frac{c_3}{2}xy^2 + \frac{d_3}{6}y^3 \tag{3.14}$$

fulfills Eq. (3.12). It leads to stresses

$$\sigma_x = c_3x + d_3y, \qquad \sigma_y = a_3x + b_3y, \qquad \tau_{xy} = -b_3x - c_3y$$

For $a_3 = b_3 = c_3 = 0$, these expressions reduce to

$$\sigma_x = d_3y, \qquad \sigma_y = \tau_{xy} = 0$$

representing the case of *pure bending* of the rectangular plate.

A polynomial of the *fourth* degree

$$\phi_4 = \frac{a_4}{12}x^4 + \frac{b_4}{6}x^3y + \frac{c_4}{2}x^2y^2 + \frac{d_4}{6}xy^3 + \frac{e_4}{12}y^4 \tag{3.15}$$

satisfies Eq. (3.12), if $e_4 = -(2c_4 + a_4)$. The corresponding stresses are

$$\sigma_x = c_4x^2 + d_4xy - (2c_4 + a_4)y^2$$

$$\sigma_y = a_4x^2 + b_4xy + c_4y^2$$

$$\tau_{xy} = -\frac{b_4}{2}x^2 - 2c_4xy - \frac{d_4}{2}y^2$$

A polynomial of the *fifth* degree

$$\phi_5 = \frac{a_5}{20}x^5 + \frac{b_5}{12}x^4y + \frac{c_5}{6}x^3y^2 + \frac{d_5}{6}x^2y^3 + \frac{e_5}{12}xy^4 + \frac{f_5}{20}y^5 \tag{3.16}$$

fulfills Eq. (3.12) provided that

$$(3a_5 + 2c_5 + e_5)x + (b_5 + 2d_5 + 3f_5)y = 0$$

It follows that

$$e_5 = -3a_5 - 2c_5, \qquad b_5 = -2d_5 - 3f_5$$

The components of stress are then

$$\sigma_x = \frac{c_5}{3}x^3 + d_5x^2y - (3a_5 + 2c_5)xy^2 + f_5y^3$$

$$\sigma_y = a_5x^3 - (3f_5 + 2d_5)x^2y + c_5xy^2 + \frac{d_5}{3}y^3$$

$$\tau_{xy} = \tfrac{1}{3}(3f_5 + 2d_5)x^3 - c_5x^2y - d_5xy^2 + \tfrac{1}{3}(3d_5 + 2c_5)y^3$$

Problems of practical importance may be solved by combining functions (3.13) through (3.16), as required. With experience, the analyst begins to understand the types of stress distributions arising from a variety of polynomials.

Example 3.1. A narrow cantilever of rectangular cross section is loaded by a concentrated force at its free end of such magnitude that the beam weight may be neglected (Fig. 3.4a). Determine the stress distribution in the beam.

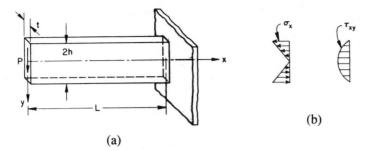

Figure 3.4

SOLUTION. The situation described may be regarded as a case of plane stress provided that the beam thickness t is small relative to the beam depth $2h$.

The following boundary conditions are consistent with the coordinate system in Fig. 3.4a:

$$(\tau_{xy})_{y=\pm h}=0, \qquad (\sigma_y)_{y=\pm h}=0 \tag{a}$$

These conditions simply express the fact that the top and bottom edges of the beam are not loaded. In addition to Eq. (a) it is necessary, on the basis of zero external loading in the x direction at $x=0$, that $\sigma_x=0$ along the vertical surface at $x=0$. Finally, the applied load P must be equal to the resultant of the shearing forces distributed across the free end:

$$P=-\int_{-h}^{+h}\tau_{xy}t\,dy \tag{b}$$

The negative sign agrees with the convention for stress discussed in Sec. 1.3.

For purposes of illustration, three approaches will be employed to determine the distribution of stress within the beam.

Method 1: Inasmuch as the bending moment varies linearly with x, and σ_x at any section depends upon y, it is reasonable to assume a general expression of the form

$$\sigma_x=\frac{\partial^2\phi}{\partial y^2}=c_1xy \tag{c}$$

in which c_1 represents a constant. Integrating twice with respect to y,

$$\phi=\tfrac{1}{6}c_1xy^3+yf_1(x)+f_2(x) \tag{d}$$

where $f_1(x)$ and $f_2(x)$ are functions of x to be determined. Introducing the ϕ thus obtained into Eq. (3.12), we have

$$y\frac{d^4f_1}{dx^4}+\frac{d^4f_2}{dx^4}=0$$

Since the second term is independent of y, a solution exists for all x and y provided that $d^4f_1/dx^4=0$ and $d^4f_2/dx^4=0$, which upon integrating, leads to

$$f_1(x)=c_2x^3+c_3x^2+c_4x+c_5$$

$$f_2(x)=c_6x^3+c_7x^2+c_8x+c_9$$

where c_2, c_3, etc. are constants of integration. Substitution of $f_1(x)$ and $f_2(x)$ into Eq. (d) gives

$$\phi=\tfrac{1}{6}c_1xy^3+\left(c_2x^3+c_3x^2+c_4x+c_5\right)y$$

$$+c_6x^3+c_7x^2+c_8x+c_9$$

Expressions for σ_y and τ_{xy} follow from Eq. (3.11):

$$\sigma_y=\frac{\partial^2\phi}{\partial x^2}=6(c_2y+c_6)x+2(c_3y+c_7) \tag{e}$$

$$\tau_{xy}=-\frac{\partial^2\phi}{\partial x\,\partial y}=-\tfrac{1}{2}c_1y^2-3c_2x^2-2c_3x-c_4$$

At this point, we are prepared to apply the boundary conditions. Substituting Eqs. (a) into (e), we obtain $c_2=c_3=c_6=c_7=0$ and $c_4=-\tfrac{1}{2}c_1h^2$. The final condition, Eq. (b), may now be written

$$-\int_{-h}^{h}\tau_{xy}t\,dy=\int_{-h}^{h}\tfrac{1}{2}c_1t(y^2-h^2)\,dy=P$$

from which

$$c_1=-\frac{3P}{2th^3}=-\frac{P}{I}$$

where $I=\tfrac{2}{3}th^3$ is the moment of inertia of the cross section about the neutral axis. From Eqs. (c) and (e), together with the values of the constants, the stresses are found to be

$$\sigma_x=-\frac{Pxy}{I},\qquad \sigma_y=0,\qquad \tau_{xy}=-\frac{P}{2I}(h^2-y^2) \tag{3.17}$$

The distribution of these stresses at sections away from the ends is shown in Fig. 3.4b.

Method 2: Beginning with bending moments $M_z=Px$, we may assume a stress field similar to that for the case of pure bending:

$$\sigma_x=-\left(\frac{Px}{I}\right)y,\qquad \tau_{xy}=\tau_{xy}(x,y),\qquad \sigma_y=\sigma_z=\tau_{xz}=\tau_{yz}=0 \tag{f}$$

The equations of compatibility are satisfied by these equations. On the basis of Eqs. (f), the equations of equilibrium lead to

$$\frac{\partial\sigma_x}{\partial x}+\frac{\partial\tau_{xy}}{\partial y}=0,\qquad \frac{\partial\tau_{xy}}{\partial x}=0 \tag{g}$$

From the second expression above, τ_{xy} can depend only upon y. The first equation of (g) together with Eqs. (f) yields

$$\frac{d\tau_{xy}}{dy}=\frac{Py}{I}$$

from which

$$\tau_{xy} = \frac{Py^2}{2I} + c$$

Here c is determined on the basis of $(\tau_{xy})_{y=\pm h} = 0$: $c = -Ph^2/2I$. The resulting expression for τ_{xy} satisfies Eq. (b) and is identical with the result previously obtained.

Method 3: The problem may be treated by superimposing the polynomials ϕ_2 and ϕ_4,

$$a_2 = c_2 = a_4 = b_4 = c_4 = e_4 = 0$$

Thus,

$$\phi = \phi_2 + \phi_4 = b_2 xy + \frac{d_4}{6} xy^3$$

The corresponding stress components are

$$\sigma_x = d_4 xy, \qquad \sigma_y = 0, \qquad \tau_{xy} = -b_2 - \frac{d_4}{2} y^2$$

It is observed that the foregoing satisfies the second condition of Eqs. (a). The first of Eqs. (a) leads to $d_4 = -2b_2/h^2$. We then obtain

$$\tau_{xy} = -b_2 \left(1 - \frac{y^2}{h^2} \right)$$

which when substituted into condition (b) results in $b_2 = -3P/4ht = Ph^2/2I$. As before, τ_{xy} is as given in Eqs. (3.17).

It is observed that the stress distribution obtained is the same as that found by employing the elementary theory. If the boundary forces result in a stress distribution as indicated in Fig. 3.4b, the solution is exact. Otherwise, the solution is not exact. In any case, recall, however, that St. Venant's principle permits us to regard the result as quite accurate for sections away from the ends.

In Sec. 5.4 is illustrated the determination of the displacement field after derivation of the curvature-moment relation.

3.6 Basic Relations in Polar Coordinates

Geometrical considerations related either to the loading or to the boundary of a loaded system often make it preferable to employ polar coordinates, rather than the Cartesian system used exclusively thus far. In general, polar coordinates are used advantageously where a degree of axial symmetry exists. Examples include a cylinder, a disk, a curved beam, and a large thin plate containing a circular hole.

The polar coordinate system (r, θ) and the Cartesian system (x, y) are related by the following expressions:

$$x = r\cos\theta, \qquad r^2 = x^2 + y^2$$
$$y = r\sin\theta, \qquad \theta = \tan^{-1}\frac{y}{x} \tag{a}$$

Equations of Equilibrium

Consider the state of stress on an infinitesimal element *abcd* of unit thickness, described by polar coordinates (Fig. 3.5). The r and θ directed body forces are denoted by F_r and F_θ. Equilibrium of radial forces requires that

$$\left(\sigma_r + \frac{\partial \sigma_r}{\partial r} dr\right)(r+dr)\,d\theta - \sigma_r r\,d\theta - \left(\sigma_\theta + \frac{\partial \sigma_\theta}{\partial \theta} d\theta\right) dr \sin\frac{d\theta}{2}$$

$$- \sigma_\theta\, dr \sin\frac{d\theta}{2} + \left(\tau_{r\theta} + \frac{\partial \tau_{r\theta}}{\partial \theta} d\theta\right) dr \cos\frac{d\theta}{2} - \tau_{r\theta}\, dr \cos\frac{d\theta}{2}$$

$$+ F_r r\,dr\,d\theta = 0$$

Inasmuch as $d\theta$ is small, $\sin(d\theta/2)$ may be replaced by $d\theta/2$ and $\cos(d\theta/2)$ by 1. Additional simplification is achieved by dropping terms containing higher order infinitesimals. A similar analysis may be performed for the tangential direction. When both equilibrium equations are divided by $r\,dr\,d\theta$, the results are

$$\frac{\partial \sigma_r}{\partial r} + \frac{1}{r}\frac{\partial \tau_{r\theta}}{\partial \theta} + \frac{\sigma_r - \sigma_\theta}{r} + F_r = 0$$

$$\frac{1}{r}\frac{\partial \sigma_\theta}{\partial \theta} + \frac{\partial \tau_{r\theta}}{\partial r} + \frac{2\tau_{r\theta}}{r} + F_\theta = 0$$

(3.18)

In the absence of body forces, Eqs. (3.18) are satisfied by a stress function $\phi(r, \theta)$ for which

$$\sigma_r = \frac{1}{r}\frac{\partial \phi}{\partial r} + \frac{1}{r^2}\frac{\partial^2 \phi}{\partial \theta^2}$$

$$\sigma_\theta = \frac{\partial^2 \phi}{\partial r^2}$$

(3.19)

$$\tau_{r\theta} = \frac{1}{r^2}\frac{\partial \phi}{\partial \theta} - \frac{1}{r}\frac{\partial^2 \phi}{\partial r\,\partial \theta} = -\frac{\partial}{\partial r}\left(\frac{1}{r}\frac{\partial \phi}{\partial \theta}\right)$$

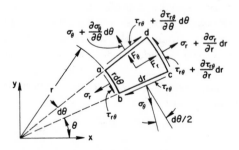

Figure 3.5

Strain-Displacement Relations

Consider now the deformation of the infinitesimal element *abcd*, denoting the *r* and *θ* displacements by *u* and *v*, respectively. The general deformation experienced by an element may be regarded as composed of (1) a change in length of the sides, as in Figs. 3.6a and b, and (2) rotation of the sides, as in Figs. 3.6c and d.

In the analysis which follows, the small angle approximation $\sin\theta \approx \theta$ is employed, and arcs *ab* and *cd* are regarded as straight lines. Referring to Fig. 3.6a, it is observed that a *u* displacement of side *ab* results in both radial and tangential strain. The radial strain ε_r, the deformation per unit length of side *ad*, is associated only with the *u* displacement:

$$\varepsilon_r = \frac{\partial u}{\partial r} \tag{3.20a}$$

The tangential strain owing to *u*, the deformation per unit length of *ab*, is

$$(\varepsilon_\theta)_u = \frac{(r+u)\,d\theta - r\,d\theta}{r\,d\theta} = \frac{u}{r} \tag{b}$$

Clearly, a *v* displacement of element *abcd* (Fig. 3.6b) also produces a tangential strain

$$(\varepsilon_\theta)_v = \frac{(\partial v/\partial\theta)\,d\theta}{r\,d\theta} = \frac{1}{r}\frac{\partial v}{\partial\theta} \tag{c}$$

since the increase in length of *ab* is $(\partial v/\partial\theta)\,d\theta$. The resultant tangential strain, combining Eqs. (b) and (c), is

$$\varepsilon_\theta = \frac{1}{r}\frac{\partial v}{\partial\theta} + \frac{u}{r} \tag{3.20b}$$

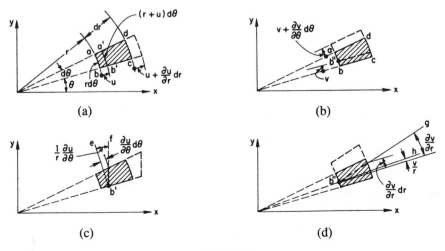

(a)   (b)

(c)   (d)

Figure 3.6

In Fig. 3.6c is seen the angle of rotation $eb'f$ of side $a'b'$ due to a u displacement. The associated strain is

$$(\gamma_{r\theta})_u = \frac{(\partial u/\partial\theta)\,d\theta}{r\,d\theta} = \frac{1}{r}\frac{\partial u}{\partial\theta} \tag{d}$$

The rotation of side bc associated with a v displacement alone is shown in Fig. 3.6d. Since an initial rotation of b'' through an angle v/r has occurred, the relative rotation $gb''h$ of side bc is

$$(\gamma_{r\theta})_v = \frac{\partial v}{\partial r} - \frac{v}{r} \tag{e}$$

The sum of Eqs. (d) and (e) provides the total shearing strain

$$\gamma_{r\theta} = \frac{\partial v}{\partial r} + \frac{1}{r}\frac{\partial u}{\partial\theta} - \frac{v}{r} \tag{3.20c}$$

The strain-displacement relationships in polar coordinates are thus given by Eqs. (3.20).

Hooke's Law

To write Hooke's law in polar coordinates, one need only replace subscripts x by r and y by θ in the appropriate Cartesian equations. In the case of plane stress, from Eqs. (3.9), we have

$$\varepsilon_r = \frac{1}{E}(\sigma_r - \nu\sigma_\theta)$$

$$\varepsilon_\theta = \frac{1}{E}(\sigma_\theta - \nu\sigma_r) \tag{3.21}$$

$$\gamma_{r\theta} = \frac{1}{G}\tau_{r\theta}$$

For plane strain, Eqs. (3.3) lead to

$$\varepsilon_r = \frac{1+\nu}{E}\left[(1-\nu)\sigma_r - \nu\sigma_\theta\right]$$

$$\varepsilon_\theta = \frac{1+\nu}{E}\left[(1-\nu)\sigma_\theta - \nu\sigma_r\right] \tag{3.22}$$

$$\gamma_{r\theta} = \frac{1}{G}\tau_{r\theta}$$

Compatibility Equation

It can be shown that Eqs. (3.20) result in the following form of the equation of compatibility:

$$\frac{\partial^2\varepsilon_\theta}{\partial r^2} + \frac{1}{r^2}\frac{\partial^2\varepsilon_r}{\partial\theta^2} + \frac{2}{r}\frac{\partial\varepsilon_\theta}{\partial r} - \frac{1}{r}\frac{\partial\varepsilon_r}{\partial r} = \frac{1}{r}\frac{\partial^2\gamma_{r\theta}}{\partial r\,\partial\theta} + \frac{1}{r^2}\frac{\partial\gamma_{r\theta}}{\partial\theta} \tag{3.23}$$

To arrive at a compatibility equation expressed in terms of the stress function ϕ, it is necessary to evaluate the partial derivatives $\partial^2\phi/\partial x^2$ and $\partial^2\phi/\partial y^2$ in terms of r and θ by means of the chain rule together with Eqs. (a). These derivatives lead to the Laplacian operator:

$$\nabla^2\phi = \frac{\partial^2\phi}{\partial x^2} + \frac{\partial^2\phi}{\partial y^2} = \frac{\partial^2\phi}{\partial r^2} + \frac{1}{r}\frac{\partial\phi}{\partial r} + \frac{1}{r^2}\frac{\partial^2\phi}{\partial\theta^2} \tag{f}$$

The equation of compatibility in alternative form is thus

$$\nabla^4\phi = \left(\frac{\partial^2}{\partial r^2} + \frac{1}{r}\frac{\partial}{\partial r} + \frac{1}{r^2}\frac{\partial^2}{\partial\theta^2}\right)(\nabla^2\phi) = 0 \tag{3.24}$$

For the axisymmetrical, zero body force case, the compatibility equation is, from Eq. (3.7) [referring to (f)],

$$\nabla^2(\sigma_r + \sigma_\theta) = \frac{d^2(\sigma_r + \sigma_\theta)}{dr^2} + \frac{1}{r}\frac{d(\sigma_r + \sigma_\theta)}{dr} = 0 \tag{3.25}$$

The remaining relationships appropriate to two-dimensional elasticity are found in a manner similar to that outlined in the foregoing discussion.

Example 3.2 A large thin plate is subjected to uniform tensile stress σ_0 at its ends, as shown in Fig. 3.7. Determine the field of stress existing within the plate.

SOLUTION.   For purposes of this analysis, it will prove convenient to locate the origin of coordinate axes at the center of the plate as shown. The state of stress in the plate is expressed by

$$\sigma_x = \sigma_0, \qquad \sigma_y = \tau_{xy} = 0$$

The stress function, $\phi = \sigma_0 y^2/2$, satisfies the biharmonic equation, Eq. (3.12). The geometry suggests polar form. The stress function ϕ may be transformed by substituting $y = r\sin\theta$, with the following result:

$$\phi = \tfrac{1}{4}\sigma_0 r^2(1 - \cos 2\theta) \tag{g}$$

The stresses in the plate now follow from Eqs. (g) and (3.19):

$$\sigma_r = \tfrac{1}{2}\sigma_0(1 + \cos 2\theta)$$

$$\sigma_\theta = \tfrac{1}{2}\sigma_0(1 - \cos 2\theta) \tag{3.26}$$

$$\tau_{r\theta} = -\tfrac{1}{2}\sigma_0 \sin 2\theta$$

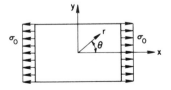

Figure 3.7

Clearly, replacement of the subscripts x' and y' by r and θ could have led directly to the foregoing result, using the transformation expressions of stress, Eqs. (1.7).

Example 3.3. Shown in Figs. 3.8a and b are a *knife edge* or *pivot* and a *wedge-cantilever*, respectively. Both members are of *unit thickness* and are subjected to *line loads* per unit thickness as shown. Determine the distribution of stress in each.

SOLUTION.

Pivot (Fig. 3.8a). Assume the stress function

$$\phi = cPr\theta \sin\theta \qquad\qquad (h)$$

where c is a constant. It can be verified that Eq. (h) satisfies Eq. (3.24) and compatibility is ensured. For equilibrium, the stresses from Eqs. (3.19) are

$$\sigma_r = 2cP\frac{\cos\theta}{r}, \qquad \sigma_\theta = 0, \qquad \tau_{r\theta} = 0 \qquad\qquad (i)$$

Referring to the figure, the boundary conditions are expressed by

$$\sigma_\theta = \sigma_{r\theta} = 0 \qquad (\theta = \pm\alpha) \qquad\qquad (j)$$

$$2\int_0^\alpha \sigma_r r\cos\theta\, d\theta = -P \qquad\qquad (k)$$

Conditions (j) are fulfilled by the last two of Eqs. (i). Substituting the first of Eqs. (i) into condition (k) results in

$$4cP\int_0^\alpha \cos^2\theta\, d\theta = -P$$

Integrating and solving for c: $c = 1/(2\sigma + \sin 2\alpha)$. The stress distribution in the knife edge is therefore

$$\sigma_r = -\frac{2P}{2\alpha + \sin 2\alpha}\frac{\cos\theta}{r}, \qquad \sigma_\theta = 0, \qquad \tau_{r\theta} = 0 \qquad (3.27a)$$

By letting $\alpha = \pi/2$ in the above, the result

$$\sigma_r = -\frac{2P}{\pi}\frac{\cos\theta}{r}, \qquad \sigma_\theta = 0, \qquad \tau_{r\theta} = 0 \qquad (3.27b)$$

is an expression for stress in a very large or so-called *semi-infinite plate* under a compressive load (Fig. P3.3).

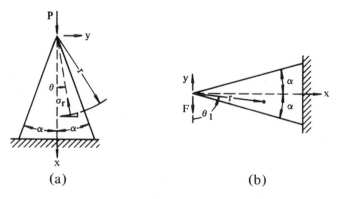

(a)   (b)

Figure 3.8

Wedge-Cantilever (Fig. 3.8b). Now we employ $\phi = cFr\theta_1 \sin \theta_1$. The equilibrium condition at 0 is

$$\int_{\pi/2-\alpha}^{\pi/2+\alpha} \sigma_r r \cos \theta_1 \, d\theta_1 = 2cF \int_{\pi/2-\alpha}^{\pi/2+\alpha} \cos^2 \theta_1 \, d\theta_1 = F$$

from which, after integration, $c = 1/(2\alpha - \sin 2\alpha)$. Thus,

$$\sigma_r = \frac{2F}{2\alpha - \sin 2\alpha} \frac{\cos \theta_1}{r}, \qquad \sigma_\theta = 0, \quad \tau_{r\theta} = 0 \qquad (3.27c)$$

In the case of loading in both compression and bending, superposition of the effects of P and F results in the expression

$$\sigma_r = -\frac{2P}{2\alpha + \sin 2\alpha} \frac{\cos \theta}{r} + \frac{2F}{2\alpha - \sin 2\alpha} \frac{\cos \theta_1}{r}, \qquad \sigma_\theta = 0, \quad \tau_{r\theta} = 0 \quad (3.27d)$$

for combined stress in a pivot or in a wedge-cantilever.

It is noted that Eqs. (3.27) represent exact results only if at the supported ends the members are held by radially directed forces distributed as defined by the solutions.

3.7 Stress Concentration

For situations in which the cross section of a load carrying member varies gradually, reasonably accurate results can be expected if one applies equations derived on the basis of constant section. On the other hand, where abrupt changes in the cross section exist, the elementary or mechanics of materials approach cannot predict the high values of stress which actually exist. The condition referred to occurs in such frequently encountered configurations as holes, notches, and fillets, which are likely starting points of material failure. While the stresses in these regions can in some cases (e.g., Table 3.2) be analyzed by applying the theory of elasticity, it is more usual to rely upon experimental techniques, and in particular, photoelastic methods. The finite element method (Chapter 7) is also very efficient for this purpose. Large values of stress are found not only when cross sections manifest pronounced changes, but also when loads are applied over very small areas. Such loading, even when it occurs on a member of uniform section, results in a severely stressed region subject to what are termed *contact stresses*, discussed in the next section.

It is usual to specify the high local stresses owing to geometrical irregularities in terms of a *stress concentration factor*: the ratio of the disturbed stress (the high, localized stress) to the uniform stress in the remainder of the body (or the average or nominal stress). The technical literature contains an abundance of specialized information on stress concentration factors in the form of graphs, tables, and formulas.*

---

*See, for example, R. E. Peterson, *Stress Concentration Design Factors*, New York: Wiley, 1974; H. P. Neuber, *Kerbspannungslehre*, 2nd edition, New York: Springer, 1958; R. J. Roark and W. C. Young, *Formulas for Stress and Strain*, New York: McGraw-Hill, 1975.

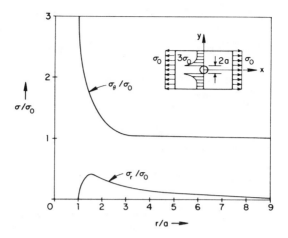

Figure 3.9

Example 3.4 A large, thin plate containing a small circular hole of radius a is subjected to simple tension (Fig. 3.9). Determine the field of stress and compare with those of Example 3.2.

SOLUTION.   The boundary conditions appropriate to the circumference of the hole are

$$\sigma_r = \tau_{r\theta} = 0 \qquad (r = a) \tag{a}$$

For large distances away from the origin we set σ_r, σ_θ, $\tau_{r\theta}$ equal to the values found for a solid plate in Example 3.2. Thus, from Eq. (3.26), for $r = \infty$,

$$\sigma_r = \tfrac{1}{2}\sigma_0(1 + \cos 2\theta)$$

$$\sigma_\theta = \tfrac{1}{2}\sigma_0(1 - \cos 2\theta), \qquad \tau_{r\theta} = -\tfrac{1}{2}\sigma_0 \sin 2\theta \tag{b}$$

For this case, we assume a stress function analogous to Eq. (g) of Example 3.2,

$$\phi = f_1(r) + f_2(r) \cos 2\theta \tag{c}$$

in which f_1 and f_2 are yet to be determined. Substituting Eq. (c) into the biharmonic equation (3.24) and noting the validity of the resulting expression for all θ, we have

$$\left(\frac{d^2}{dr^2} + \frac{1}{r}\frac{d}{dr} \right)\left(\frac{d^2 f_1}{dr^2} + \frac{1}{r}\frac{df_1}{dr} \right) = 0 \tag{d}$$

$$\left(\frac{d^2}{dr^2} + \frac{1}{r}\frac{d}{dr} - \frac{4}{r^2} \right)\left(\frac{d^2 f_2}{dr^2} + \frac{1}{r}\frac{df_2}{dr} - \frac{4f_2}{r^2} \right) = 0 \tag{e}$$

The solutions of Eqs. (d) and (e) are (Problem 3.15)

$$f_1 = c_1 r^2 \ln r + c_2 r^2 + c_3 \ln r + c_4 \tag{f}$$

$$f_2 = c_5 r^2 + c_6 r^4 + \frac{c_7}{r^2} + c_8 \tag{g}$$

where the c's are the constants of integration. The stress function is then obtained by introducing Eqs. (f) and (g) into (c). By substituting ϕ into Eq. (3.19), the stresses are found to be

$$\sigma_r = c_1(1+2\ln r) + 2c_2 + \frac{c_3}{r^2} - \left(2c_5 + \frac{6c_7}{r^4} + \frac{4c_8}{r^2}\right)\cos 2\theta$$

$$\sigma_\theta = c_1(3+2\ln r) + 2c_2 - \frac{c_3}{r^2} + \left(2c_5 + 12c_6 r^2 + \frac{6c_7}{r^4} + \frac{4c_8}{r^2}\right)\cos 2\theta \qquad \text{(h)}$$

$$\tau_{r\theta} = \left(2c_5 + 6c_6 r^2 - \frac{6c_7}{r^4} - \frac{2c_8}{r^2}\right)\sin 2\theta$$

The absence of c_4 indicates that it has no influence upon the solution.

According to the boundary conditions (b), $c_1 = c_6 = 0$ in Eq. (h), because as $r \to \infty$, the stresses must assume finite values. Then, according to the conditions (a), the equations (h) yield

$$2c_2 + \frac{c_3}{a^2} = 0, \qquad 2c_5 + \frac{6c_7}{a^4} + \frac{4c_8}{a^2} = 0, \qquad 2c_5 - \frac{6c_7}{a^4} - \frac{2c_8}{a^2} = 0$$

Also from Eqs. (b) and (h) we have

$$\sigma_0 = -4c_5, \qquad \sigma_0 = 4c_2$$

Solving the above five expressions, we obtain $c_2 = \sigma_0/4$, $c_3 = -a^2\sigma_0/2$, $c_7 = -a^4\sigma_0/4$, and $c_8 = a^2\sigma_0/2$. The determination of the stress distribution in an infinite plate containing a circular hole is completed by substituting these constants into Eq. (h):

$$\sigma_r = \tfrac{1}{2}\sigma_0\left[\left(1 - \frac{a^2}{r^2}\right) + \left(1 + \frac{3a^4}{r^4} - \frac{4a^2}{r^2}\right)\cos 2\theta\right]$$

$$\sigma_\theta = \tfrac{1}{2}\sigma_0\left[\left(1 + \frac{a^2}{r^2}\right) - \left(1 + \frac{3a^4}{r^4}\right)\cos 2\theta\right] \qquad (3.28)$$

$$\tau_{r\theta} = -\tfrac{1}{2}\sigma_0\left(1 - \frac{3a^4}{r^4} + \frac{2a^2}{r^2}\right)\sin 2\theta$$

We observe that σ_θ is a maximum for $\theta = \pm\pi/2$ and that $(\sigma_\theta)_{max} = 3\sigma_0$ for $r = a$. On the other hand, Eqs. (3.26) indicate that for $\theta = \pm\pi/2, (\sigma_\theta)_{max} = \sigma_0$. The stress concentration factor, defined as the ratio of the maximum stress at the hole to the nominal stress σ_0, is therefore $k = 3\sigma_0/\sigma_0 = 3$. To depict the variation of $\sigma_r(r, \pi/2)$ and $\sigma_\theta(r, \pi/2)$ over the distance from the origin, dimensionless stresses are plotted against dimensionless radius in Fig. 3.9. The shear stress $\tau_{r\theta}(r, \pi/2) = 0$. At a distance $r = 9a$, we have $\sigma_\theta \approx 1.006\sigma_0$ and $\sigma_r = 0.018\sigma_0$, as is observed in the figure. Thus, simple tension prevails at a distance of approximately nine radii; the hole has a *local effect* upon the distribution of stress.

The results expressed by Eq. (3.28) are applied, together with the method of superposition, to the case of biaxial loading. Distributions of maximum stress $\sigma_\theta(r, \pi/2)$, obtained in this way (Problem 3.16), are given in Fig. 3.10.

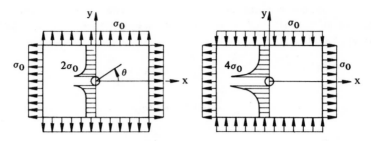

Figure 3.10

Neuber's Diagram

Several geometries of practical importance, given in Table 3.2, were the subject of stress concentration determination by Neuber on the basis of mathematical analysis, as in the above example. *Neuber's diagram* (a nomograph), which is used with the table for determining the stress concentration factor k for the configurations shown, is plotted in Fig. 3.11. In applying Neuber's diagram, the first step is the calculation of the values of $\sqrt{h/a}$ and $\sqrt{b/a}$. Given a value of $\sqrt{b/a}$, one proceeds vertically upward to cut the appropriate curve designated by the number found in column 5 of the table, then horizontally to the left to the ordinate axis. This point is then connected by a straight line to a point on the left-hand abscissa representing $\sqrt{h/a}$, according to either scale e or f as indicated in column 4 of the table. The value of k is read off on the circular scale at a point located on a normal from the origin. [The values of (theoretical) stress concentration factors obtained from Neuber's nomograph agree satisfactorily with those found by the photoelastic method.]

Consider, for example, the case of a member with a single notch (Fig. B in the table), and assume that it is subjected to axial tension P only. For given $a = 7.925$ mm, $h = 44.450$ mm, and $b = 266.700$ mm, $\sqrt{h/a} = 2.37$ and $\sqrt{b/a} = 5.80$. Table 3.2 indicates that scale f and curve 3 are applicable. Then, as described above, the stress concentration factor is found to be $k = 3.25$. The path followed is denoted by the broken lines in the diagram. The nominal stress P/bt, multiplied by k, yields maximum theoretical stress, found on the edge of the notch.

3.8 Contact Stresses

As has been pointed out in the previous section, the application of a load over a small area of contact results in unusually high stresses. Situations of this nature are found on a microscopic scale whenever force is transmitted through bodies in contact. There are important practical cases when the

Table 3.2

Type of change of section	Type of loading	Nominal stress	Scale for $\sqrt{h/a}$	Curve for k
A (diagram)	Tension	$\dfrac{P}{2bt}$	f	1
	Bending	$\dfrac{3M}{2b^2t}$	f	2
B (diagram)	Tension	$\dfrac{P}{bt}$	f	3
	Bending	$\dfrac{6M}{b^2t}$	f	4
C (diagram)	Tension	$\dfrac{P}{2bt}$	f	5
	Bending	$\dfrac{3Mh}{2t(c^3-h^3)}$	e	5
D (diagram)	Tension	$\dfrac{P}{\pi b^2}$	f	6
	Bending	$\dfrac{4M}{\pi b^3}$	f	7
	Direct Shear	$\dfrac{1.23V}{\pi b^2}$	e	8
	Torsional Shear	$\dfrac{2M_t}{\pi b^3}$	e	9

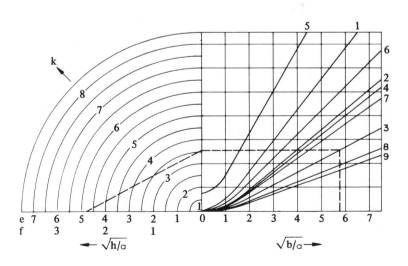

Figure 3.11

geometry of the contacting bodies results in large stresses, disregarding the stresses associated with the asperities found on any nominally smooth surface. The so-called *Hertz problem* relates to the stresses owing to the contact of a sphere on a plane, a sphere on a sphere, a cylinder on a cylinder, etc. The practical implications with respect to ball and roller bearings, locomotive wheels, and a host of machine components, are apparent.

Consider, in this regard, the contact without deformation of two bodies having spherical surfaces of radii r_1 and r_2, in the vicinity of contact. If now a collinear pair of forces P acts to press the bodies together, as in Fig. 3.12a, deformation will occur, and the point of contact O will be replaced by a small area of contact. A common tangent plane and common normal axis are denoted Ox and Oy, respectively. The first steps taken toward the solution of this problem are the determination of the size and shape of the contact area as well as the distribution of normal pressure acting on the area. The stresses and deformations resulting from the interfacial pressure are then evaluated.

The following assumptions are generally made in the solution of the contact problem:

1. The contacting bodies are isotropic and elastic.
2. The contact areas are essentially flat and small relative to the radii of curvature of the undeformed bodies in the vicinity of the interface.
3. The contacting bodies are perfectly smooth, and therefore only normal pressures need be taken into account.

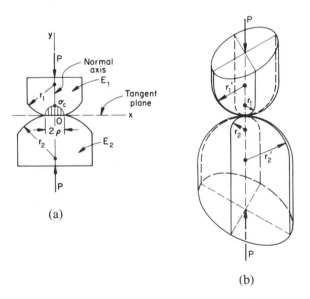

(a)

(b)

Figure 3.12

The foregoing set of assumptions enables an elastic analysis to be conducted. Without going into the derivations, we shall, in the following paragraphs, introduce some of the results.*

For the *two spherical surfaces* in contact because of forces P (Fig. 3.12a), the contact pressure is distributed over a small circle of radius ρ given by

$$\rho = 0.88 \left[\frac{P(E_1 + E_2) r_1 r_2}{E_1 E_2 (r_1 + r_2)} \right]^{1/3} \tag{3.29}$$

where E_1 and E_2 (r_1 and r_2) are the respective moduli of elasticity (radii) of the spheres. The force P—causing the contact pressure—acts in the direction of the normal axis, perpendicular to tangent plane passing through the contact area. The *maximum contact pressure* is found to be

$$\sigma_c = 1.5 \frac{P}{\pi \rho^2} \tag{3.30}$$

This is the maximum principal stress owing to the fact that at the center of the contact area, material is compressed not only in the normal direction but also in the lateral directions. The relationship between the force of contact, P, and the relative displacement of the centers of the two elastic spheres, owing to local deformation, is

$$e = 0.77 \left[P^2 \left(\frac{1}{E_1} + \frac{1}{E_2} \right)^2 \left(\frac{1}{r_1} + \frac{1}{r_2} \right) \right]^{1/3} \tag{3.31}$$

In the special case of a *sphere* of radius r_1 contacting a body of the same material but having a *flat surface*, substitution of $r_2 = \infty$ and $E_1 = E_2 = E$ in Eqs. (3.29) through (3.31) leads to

$$\rho = 0.88 \left(\frac{2 P r_1}{E} \right)^{1/3}, \quad \sigma_c = 0.62 \left(\frac{P E^2}{4 r_1^2} \right)^{1/3}, \quad e = 1.54 \left(\frac{P^2}{2 E^2 r_1} \right)^{1/3} \tag{3.32}$$

For the case of a *sphere* in a *spherical seat* of the same material, substituting $r_2 = -r_2$ and $E_1 = E_2 = E$ in Eqs. (3.29) through (3.31), we obtain

$$\rho = 0.88 \left[\frac{2 P r_1 r_2}{E(r_2 - r_1)} \right]^{1/3}, \quad \sigma_c = 0.62 \left[P E^2 \left(\frac{r_2 - r_1}{2 r_1 r_2} \right)^2 \right]^{1/3},$$

$$e = 1.54 \left[\frac{P^2 (r_2 - r_1)}{2 E^2 r_1 r_2} \right]^{1/3} \tag{3.33}$$

---

*A summary and complete list of references dealing with contact stress problems is given by W. Flügge, Ed., *Handbook of Engineering Mechanics*, New York: McGraw-Hill, 1962, Chapter 42. See also S. P. Timoshenko, *Strength of Materials—Part* II, Princeton, NJ: Van Nostrand, 1965, pp. 300–344.

In all instances, the maximum contact pressure σ_c is located at the center of the area of contact.

Consider now two rigid bodies of equal elastic moduli E, compressed by force P (Fig. 3.12b). The load lies along the axis passing through the centers of the bodies and through the point of contact, and is perpendicular to the plane tangent to both bodies at the point of contact. The minimum and maximum radii of curvature of the surface of the upper body are r_1 and r_1'; those of the lower body are r_2 and r_2' at the point of contact. Thus $1/r_1$, $1/r_1'$, $1/r_2$, and $1/r_2'$ are the principal curvatures. The *sign convention* of the *curvature* is such that it is *positive* if the corresponding center of curvature is *inside* the body. If the center of the curvature is *outside* the body the curvature is *negative*. (For example, in Fig. 3.13a, r_1, r_1' are positive, while r_2, r_2' are negative.) Let θ be the angle between the normal planes in which radii r_1 and r_2 lie. Subsequent to loading the area of contact will be an *ellipse* with semiaxes a and b. The maximum pressure occurs at the center of the contact area and is

$$\sigma_c = 1.5 \frac{P}{\pi ab} \tag{3.34}$$

In this expression the semiaxes are given by

$$a = c_a \sqrt[3]{\frac{Pm}{n}}, \qquad b = c_b \sqrt[3]{\frac{Pm}{n}} \tag{3.35}$$

in which

$$m = \frac{4}{\dfrac{1}{r_1} + \dfrac{1}{r_1'} + \dfrac{1}{r_2} + \dfrac{1}{r_2'}}, \qquad n = \frac{4E}{3(1-\nu^2)} \tag{a}$$

The constants c_a and c_b are read in Table 3.3. In the first column of the table are listed values of α, calculated from

$$\cos \alpha = \frac{B}{A} \tag{b}$$

where

$$A = \frac{2}{m}, \qquad B = \frac{1}{2}\left[\left(\frac{1}{r_1} - \frac{1}{r_1'}\right)^2 + \left(\frac{1}{r_2} - \frac{1}{r_2'}\right)^2\right.$$

$$\left. + 2\left(\frac{1}{r_1} - \frac{1}{r_1'}\right)\left(\frac{1}{r_2} - \frac{1}{r_2'}\right)\cos 2\theta\right]^{1/2} \tag{c}$$

By applying Eq. (3.34) many problems of practical importance may be treated, e.g., contact stresses in ball bearings (Fig. 3.13a), contact stresses

Table 3.3

α (degrees)	c_a	c_b	α (degrees)	c_a	c_b
20	3.778	0.408	60	1.486	0.717
30	2.731	0.493	65	1.378	0.759
35	2.397	0.530	70	1.284	0.802
40	2.136	0.567	75	1.202	0.846
45	1.926	0.604	80	1.128	0.893
50	1.754	0.641	85	1.061	0.944
55	1.611	0.678	90	1.000	1.000

between a cylindrical wheel and a rail (Fig. 3.13b), and contact stresses in cam and push rod mechanisms.

Example 3.5 A railway car wheel rolls on a rail. Both rail and wheel are made of steel for which $E=210$ GPa and $\nu=0.3$. The wheel has a radius of $r_1=0.4$ m, and the cross radius of the rail top surface is $r_2=0.3$ m (Fig. 3.13b). Determine the size of the contact area and the maximum contact pressure, given a compression load of $P=90$ kN.

SOLUTION. For the situation described, $1/r_1'=1/r_2'=0$, and as the axes of the members are mutually perpendicular, $\theta=\pi/2$. The first equation of (a) and the two equations (c) reduce to

$$m=\frac{4}{1/r_1+1/r_2}, \qquad A=\frac{1}{2}\left(\frac{1}{r_1}+\frac{1}{r_2}\right), \qquad B=\pm\frac{1}{2}\left(\frac{1}{r_1}-\frac{1}{r_2}\right) \qquad (d)$$

The proper sign in B must be chosen in order that its values be positive. Now Eq. (b) has the form

$$\cos\alpha=\pm\frac{1/r_1-1/r_2}{1/r_1+1/r_2} \qquad (e)$$

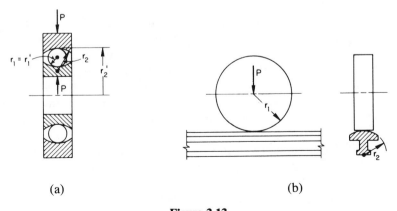

(a)   (b)

Figure 3.13

Substituting the given numerical values into Eqs. (d), (e), and the second of (a), we obtain

$$m = \frac{4}{1/0.4 + 1/0.3} = 0.6857 \qquad n = \frac{4(210 \times 10^9)}{3(0.91)} = 3.07692 \times 10^{11}$$

$$\cos \alpha = \pm \frac{1/0.4 - 1/0.3}{1/0.4 + 1/0.3} = 0.1428 \quad \text{or} \quad \alpha = 81.79°$$

Corresponding to this value of α, interpolating in Table 3.2, we have

$$c_a = 1.1040, \qquad c_b = 0.9112$$

The semiaxes of the elliptical contact area are found by applying Eqs. (3.35):

$$a = 1.1040 \left[\frac{90,000 \times 0.6857}{3.07692 \times 10^{11}} \right]^{1/3} = 0.00646 \text{ m}$$

$$b = 0.9112 \left[\frac{90,000 \times 0.6857}{3.07692 \times 10^{11}} \right]^{1/3} = 0.00533 \text{ m}$$

The maximum contact pressure, or maximum principal stress, is thus

$$\sigma_c = 1.5 \frac{90,000}{\pi(0.00646 \times 0.00533)} = 1248 \text{ MPa}$$

A hardened steel material is capable of resisting this or somewhat higher stress levels for the body geometries and loading conditions described in this section.

3.9 Thermal Stresses

Consider the consequences of increasing or decreasing the *uniform* temperature of an entirely unconstrained elastic body. The resultant expansion or contraction occurs in such a way as to cause a cubic element of the solid to remain cubic, while experiencing changes of length on each of its sides. Normal strains occur in each direction, unaccompanied by normal stresses. In addition, there are neither shear strains nor shear stresses. If the body is heated in such a way as to produce a nonuniform temperature field, or if the thermal expansions are prohibited from taking place freely because of restrictions placed on the boundary even if the temperature is uniform, or if the material exhibits an isotropy in a uniform temperature field, thermal stresses will occur. The effects of such stresses can be severe, especially since the most adverse thermal environments are often associated with design requirements involving unusually stringent constraints as to weight and volume. This is especially true in aerospace applications, but is of considerable importance too in many everyday machine design applications.

Solution of thermal stress problems requires reformulation of the stress–strain relationships, accomplished by superposition of the strain attributable to stress and that due to temperature. For a change in temperature $T(x, y)$, the change of length, δL, of a small linear element of length L in an unconstrained body is $\delta L = \alpha L T$. Here α, a positive number, is termed the coefficient of linear thermal expansion. The thermal strain ε_t associated with the free expansion at a point is then

$$\varepsilon_t = \alpha T \tag{3.36}$$

The total x and y strains, ε_x and ε_y, are obtained by adding to the thermal strains of the type described above, strains due to stress resulting from external forces:

$$\varepsilon_x = \frac{1}{E}(\sigma_x - \nu\sigma_y) + \alpha T$$

$$\varepsilon_y = \frac{1}{E}(\sigma_y - \nu\sigma_x) + \alpha T \tag{3.37}$$

$$\gamma_{xy} = \frac{\tau_{xy}}{G}$$

Because free thermal expansion results in no angular distortion in an isotropic material, the shearing strain is unaffected, as indicated above. Equations (3.37) represent modified strain–stress relations for plane stress. Similar expressions may be written for the case of plane strain. The differential equations of equilibrium (3.4) are based upon purely mechanical considerations and are unchanged for *thermoelasticity*. The same is true of the strain-displacement relations (2.3) and the compatibility equation (3.6) which are geometrical in character. Thus, for given boundary conditions (expressed either as surface forces or displacements) and temperature distribution, thermoelasticity and ordinary elasticity differ only to the extent of the strain–stress relationship.

By substituting the strains given by Eq. (3.37) into the equation of compatibility (3.6), employing Eq. (3.4) as well, and neglecting body forces, a compatibility equation is derived in terms of stress:

$$\left(\frac{\partial^2}{\partial x^2} + \frac{\partial^2}{\partial y^2}\right)(\sigma_x + \sigma_y + \alpha ET) = 0 \tag{3.38}$$

Introducing Eq. (3.11), we now have

$$\nabla^4 \phi + \alpha E \nabla^2 T = 0 \tag{3.39}$$

The above expression is valid for plane strain or plane stress, provided that the body forces are negligible.

It has been implicit in treating the matter of thermoelasticity as a superposition problem that the distribution of stress or strain plays a negligible role in influencing the temperature field.* This lack of coupling enables the temperature field to be determined independently of any

---

*See B. A. Boley and J. H. Weiner, *Theory of Thermal Stresses*, New York: Wiley, 1960, Chapter 2.

consideration of stress or strain. If the effect of the temperature distribution upon material properties cannot be disregarded, the equations become coupled and analytical solutions significantly more complex, occupying an area of considerable interest and importance. Numerical solutions can, however, be obtained in a relatively simple manner through the use of finite difference methods.

Example 3.6. Consider a thin rectangular beam of small thickness t, depth $2h$, and Length L. Locate the origin of coordinates at the center of the beam, x being the axial direction. Find the distribution of stress and strain associated with an arbitrary variation in temperature throughout the depth, $T = T(y)$. Let the beam be entirely free of surface forces, and assume body forces to be negligible.

SOLUTION.   The beam geometry indicates a problem of plane stress. We begin with the assumptions

$$\sigma_x = \sigma_x(y), \qquad \sigma_y = \tau_{xy} = 0 \tag{a}$$

Direct substitution of Eqs. (a) into Eqs. (3.4) indicates that the equations of equilibrium are satisfied. Equations (a) reduce the compatibility equation (3.38) to the form

$$\frac{d^2}{dy^2}(\sigma_x + \alpha ET) = 0 \tag{b}$$

from which

$$\sigma_x = -\alpha ET + c_1 y + c_2 \tag{c}$$

where c_1 and c_2 are constants of integration. The requirement that faces $y = \pm h$ be free of surface forces is obviously fulfilled by Eq. (b). Finally, the boundary conditions at the end faces are satisfied by determining the constants which assure zero resultant force and moment at $x = \pm L/2$:

$$\int_{-h}^{h} \sigma_x t \, dy = 0, \qquad \int_{-h}^{h} \sigma_x y t \, dy = 0 \tag{d}$$

Substituting Eq. (c) into Eqs. (d), it is found that $c_1 = (3/2h^3) \int_{-h}^{h} \alpha ETy \, dy$ and $c_2 = (1/2h) \int_{-h}^{h} \alpha ET \, dy$. The normal stress, upon substituting the values of the constants obtained, together with the moment of inertia $I = 2h^3 t/3$ and area $A = 2ht$, into Eq. (c), is thus

$$\sigma_x = E\alpha \left[-T + \frac{t}{A} \int_{-h}^{h} T \, dy + \frac{yt}{I} \int_{-h}^{h} Ty \, dy \right] \tag{3.40}$$

The corresponding strains are

$$\varepsilon_x = \frac{\sigma_x}{E} + \alpha T, \qquad \varepsilon_y = -\frac{\nu \sigma_x}{E} + \alpha T, \qquad \gamma_{xy} = 0 \tag{e}$$

The displacements can readily be determined from Eqs. (3.1).

From Eq. (3.40) it is observed that the temperature distribution for $T = $ constant results in zero stress, as expected. Of course, the strains (e) and the displacements will, in this case, not be zero. It is also noted that when the temperature is symmetrical about the midsurface ($y = 0$), i.e., $T(y) = T(-y)$, the final integral in Eq. (3.40) vanishes. For an antisymmetrical temperature distribution about the midsurface, $T(y) = -T(-y)$, and the first integral in Eq. (3.40) is zero.

Chapter 3—Problems

Secs. 3.1 to 3.5

3.1. A stress distribution is given by

$$\sigma_x = c_1 y x^3 - 2c_2 xy + c_3 y$$

$$\sigma_y = c_1 xy^3 - 2c_1 x^3 y$$

$$\tau_{xy} = -\tfrac{3}{2}c_1 x^2 y^2 + c_2 y^2 + \tfrac{1}{2}c_1 x^4 + c_4$$

where the c's are constants.

(a) Verify that this field represents a solution for a thin plate (Fig. P3.1); (b) obtain the corresponding stress function; (c) find the surface forces along the edges $y=0$ and $y=b$ of the plate.

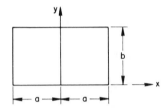

Figure P3.1

3.2. In Fig. P3.2 is shown a long, thin, steel plate of thickness t, width $2h$, and length $2a$. The plate is subjected to loads which produce the uniform stresses σ_0 at the ends. The edges at $y = \pm h$ are fixed between the two rigid walls. Show that, by using an inverse method, the displacements are expressed by

$$u = -\frac{1-\nu^2}{E}\sigma_0 x, \qquad v = 0, \qquad w = \frac{\nu(1+\nu)}{E}\sigma_0 z$$

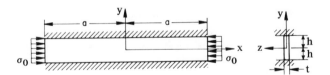

Figure P3.2

3.3. A vertical force P per unit thickness is applied on the horizontal boundary of a semi-infinite solid plate of unit thickness (Fig. P3.3). Show that the stress function $\phi = -(P/\pi)y\tan^{-1}(y/x)$ results in the following stress field within the plate:

$$\sigma_x = -\frac{2P}{\pi}\frac{x^3}{(x^2+y^2)^2}, \qquad \sigma_y = -\frac{2P}{\pi}\frac{xy^2}{(x^2+y^2)^2}, \qquad \tau_{xy} = -\frac{2P}{\pi}\frac{yx^2}{(x^2+y^2)^2}$$

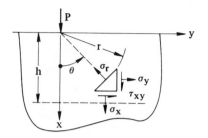

Figure P3.3

Also plot the resulting stress distribution for σ_x and τ_{xy} at a constant depth h below the boundary.

3.4. The thin cantilever shown in Fig. P3.4 is subjected to uniform shearing stress τ_0 along its upper surface ($y = +h$) while surfaces $y = -h$ and $x = L$ are free of stress. Determine whether the Airy stress function

$$\phi = \frac{1}{4}\tau_0\left(xy - \frac{xy^2}{h} - \frac{xy^3}{h^2} + \frac{Ly^2}{h} + \frac{Ly^3}{h^2}\right)$$

satisfies the required conditions for this problem.

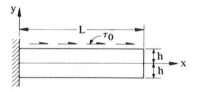

Figure P3.4

3.5. Consider a thin square plate with sides a. For a stress function $\phi = (p/a^2)(\frac{1}{2}x^2y^2 - \frac{1}{6}y^4)$, determine the stress field and sketch it along the boundaries of the plate. Here p represents a uniformly distributed loading per unit length. Note that the origin of the x, y coordinate system is located at the lower left corner of the plate.

3.6. Show that for the case of plane stress, in the absence of body forces, the equations of equilibrium may be expressed in terms of displacements u and v as follows:

$$\frac{\partial^2 u}{\partial x^2} + \frac{\partial^2 u}{\partial y^2} + \frac{1+\nu}{1-\nu}\frac{\partial}{\partial x}\left(\frac{\partial u}{\partial x} + \frac{\partial v}{\partial y}\right) = 0$$

$$\frac{\partial^2 v}{\partial y^2} + \frac{\partial^2 v}{\partial x^2} + \frac{1+\nu}{1-\nu}\frac{\partial}{\partial y}\left(\frac{\partial v}{\partial y} + \frac{\partial v}{\partial x}\right) = 0$$

[Hint: Substitute Eqs. (3.9) together with (2.3) into (3.4).]

Secs. 3.6 to 3.9

3.7. Demonstrate that the biharmonic equation $\nabla^4\phi=0$ in polar coordinates can be written

$$\left(\frac{\partial^2}{\partial r^2}+\frac{1}{r}\frac{\partial}{\partial r}+\frac{1}{r^2}\frac{\partial^2}{\partial\theta^2}\right)\left(\frac{\partial^2\phi}{\partial r^2}+\frac{1}{r}\frac{\partial\phi}{\partial r}+\frac{1}{r^2}\frac{\partial^2\phi}{\partial\theta^2}\right)=0$$

[Hint: Employ the chain rule for $\partial\phi/\partial x$ and $\partial\phi/\partial y$ together with the expressions $\partial r/\partial x=x/r=\cos\theta$, $\partial r/\partial y=y/r=\sin\theta$, $\partial\theta/\partial x=-y/r^2=-(\sin\theta)/r$, $\partial\theta/\partial y=x/r^2=(\cos\theta)/r$, found from Eqs. (a) of Sec. 3.6.]

3.8. Assume that moment M acts in the plane and at the vertex of the wedge-cantilever shown in Fig. 3.8a and that $P=0$. For a stress function

$$\phi=-\frac{M(\sin 2\theta-2\theta\cos 2\alpha)}{2(\sin 2\alpha-2\alpha\cos 2\alpha)}$$

verify that (a) ϕ satisfies Eq. (3.24) and that (b) the expressions

$$\sigma_r=\frac{2M\sin 2\theta}{\pi r^2},\qquad \sigma_\theta=0,\qquad \tau_{r\theta}=\frac{2M\cos^2 2\theta}{\pi r^2}\qquad\text{(P3.8)}$$

represent the stress field in a semi-infinite plate (i.e., for $\alpha=\pi/2$).

3.9. Show that the integral of the radial stress σ_r, found in Problem 3.3 along any semicircle about the origin (Fig. P3.3) is equal to P.

3.10. Consider the pivot of unit thickness subject to force P at its vertex, Fig. 3.8a. Determine $\sigma_{x,\max}$ and $\tau_{xy,\max}$ on a plane a distance h (measured as in Fig. P3.3) from the apex, through the use of σ_r given by Eq. (3.27a). Also obtain a solution using the elementary theory: (a) take $\alpha=15°$; (b) take $\alpha=60°$; (c) compare the results given by the two approaches.

3.11. Show that the solution of Problem 3.3 reduces to Eq. (3.27b).

3.12. Redo Problem 3.10 in its entirety for the wedge-cantilever shown in Fig. 3.8b.

3.13. The shaft shown in Fig. D of Table 3.2 has the following dimensions: $a=6$ mm, $h=12$ mm, and $b=200$ mm. If the shaft is subjected simultaneously to M_t, V, M, and P, as indicated, determine the maximum stress concentration factor.

3.14. Determine the maximum principal stress (pressure) at the contact point between the outer race and a ball in the single-row ball bearing assembly shown in Fig. 3.13a. The ball diameter is 50 mm; the radius of the grooves, 30 mm; the diameter of the outer race, 250 mm; and the highest compressive force on the ball, $P=1.8$ kN. Take $E=200$ GPa and $\nu=0.3$.

3.15. Verify the result given by Eqs. (f) and (g) of Sec. 3.7 (a) by rewriting Eqs. (d) and (e) in the following forms, respectively:

$$\frac{1}{r}\frac{d}{dr}\left\{r\frac{d}{dr}\left[\frac{1}{r}\frac{d}{dr}\left(r\frac{df_1}{dr}\right)\right]\right\}=0$$

$$r\frac{d}{dr}\left(\frac{1}{r^3}\frac{d}{dr}\left\{r^3\frac{d}{dr}\left[\frac{1}{r^3}\frac{d}{dr}(r^2f_2)\right]\right\}\right)=0$$

and by integrating the above; (b) by expanding Eqs. (d) and (e), setting $t=\ln r$, and thereby transforming the resulting expressions into two ordinary differential equations with constant coefficients.

3.16. Verify the results given in Fig. 3.10 by employing Eq. (3.28) together with the method of superposition.

3.17. A prismatic bar is restrained in the x (axial) and y directions, but free to expand in z direction. Determine the stress and strains in the bar for a temperature rise of T_1 degrees.

3.18. Show that the compatibility equation in polar coordinates, for the axisymmetrical problem of thermal elasticity, is given by

$$\frac{1}{r}\frac{d}{dr}\left(r\frac{d\phi}{dr}\right)+E\alpha T=0$$

3.19. Under free thermal expansion, the strain components within a given elastic solid are $\varepsilon_x=\varepsilon_y=\varepsilon_z=\alpha T$, $\gamma_{xy}=\gamma_{yz}=\gamma_{xz}=0$. Show that the temperature field associated with this condition is of the form

$$\alpha T=c_1x+c_2y+c_3z+c_4$$

in which the c's are constants.

Chapter 4

Mechanical Behavior of Materials

4.1 Introduction

The efficiency of design relies in great measure upon an ability to predict the circumstances under which failure is likely to occur. The important variables connected with structural failure include the nature of the material; the load configuration; the rate of loading; the shape, surface peculiarities, and temperature of the member; and the characteristics of the medium surrounding the member (environmental conditions). Exact quantitative formulation of the problem of failure and accurate means for predicting failure represent areas of current research. In Chapter 2 the stress–strain properties and basic characteristics of engineering materials were presented. We now discuss the mechanical behavior of materials associated with failure, theories of failure for *static* loading, and the response of materials to *dynamic* loading and *temperature* change.

In the most general terms, failure refers to any action leading to an inability on the part of the structure or machine to function in the manner intended. It follows that permanent deformation, fracture, or even excessive linear elastic deflection may be regarded as modes of failure, the last being the most easily predicted. In this chapter, the failure of homogeneous materials by yielding or permanent deformation and by fracture are given particular emphasis.*

Among the variables cited above, one of the most important factors in regard to influencing the threshold of failure is the rate at which the load is applied. Loading at high rate—i.e., dynamic loading—may lead to a variety of adverse phenomena associated with impact, acceleration, and vibration, with the concomitant high levels of stress and strain as well as

---

*For details see: A. Nadai, *Theory of Flow and Fracture of Solids*, New York: McGraw-Hill, 1950, pp. 175–228; also texts on materials science, for example, N. H. Polakowsky and E. J. Ripling, *Strength and Structure of Engineering Materials*, Englewood Cliffs, NJ: Prentice-Hall, 1966; J. Marin, *Mechanical Behavior of Engineering Materials*, Englewood Cliffs, NJ: Prentice -Hall, 1962.

rapid reversal of stress. In a conventional tension test, the rate referred to may relate either to the application of load or to changes in strain. Ordinarily, strain rates on the order of 10^{-4} sec^{-1} are regarded as "static" loading.

Our primary concern in this chapter and in this text is with metals, which are composed of crystals or grains built up of atoms. It is reasonable to expect that very small volumes of a given metal will not exhibit isotropy in such properties as elastic modulus. Nevertheless, we adhere to the basic assumptions of isotropy and homogeneity because we deal primarily with an entire body or a large enough segment of the body to contain many randomly distributed crystals, behaving as an isotropic material would.

The brittle or ductile character of a metal has relevance to the mechanism of failure. If a metal is capable of undergoing an appreciable amount of yielding or permanent deformation, it is regarded as *ductile*. If prior to fracture, the material can suffer only small yielding (less than 5%), the material is classified as *brittle*. The distinction between ductile and brittle materials is not as simple as might be inferred from the above discussion. The nature of the stress, the temperature, and the material itself all play a role, as discussed in Sec. 4.14, in defining the boundary between ductility and brittleness.

4.2   Failure by Yielding and Fracture

Yielding

Whether because of material inhomogeneity or nonuniformity of loading, regions of high stress may be present in which *localized* yielding occurs. As the load increases, the inelastic action becomes more widespread, resulting eventually in a state of *general* yielding. The rapidity with which the transition from localized to general yielding occurs is dependent upon the service conditions as well as the distribution of stress and the properties of the materials. Among the various service conditions, temperature represents a particularly significant factor.

The relative motion or *slip* between two planes of atoms (and the relative displacement of two sections of a crystal which results) represents the most common mechanism of yielding. Slip occurs most readily along certain crystallographic planes termed slip or shear planes. The planes along which slip takes place easily are generally those containing the largest number of atoms per unit area. Inasmuch as the gross yielding of material represents the total effect of slip occurring along many randomly oriented planes, the yield strength is clearly a statistical quantity, as are other material properties such as the modulus of elasticity. If a metal fails by yielding, one can, on the basis of the above considerations, expect the shearing stress to play an important role.

It is characteristic of most ductile materials that after yielding has taken place, the load must be increased to produce further deformation. In other words, the material exhibits a strengthening termed *strain hardening* or *cold working*. The slip occurring on intersecting planes of randomly oriented crystals and their resulting interaction is believed to be a factor of prime importance in strain hardening.

The deformation of a material under short time loading (as occurs in a simple tension test) is simultaneous with the increase in load. Under certain circumstances, deformation may continue with time while the load remains constant. This deformation, beyond that experienced as the material is initially loaded, is termed *creep*. In materials such as lead, rubber, and certain plastics, creep may occur at ordinary temperatures. Most metals, on the other hand, begin to evidence a loss of strain hardening and manifest appreciable creep only when the absolute temperature is roughly 35 to 50% of the melting temperature. The rate at which creep proceeds in a given material is dependent upon the stress, temperature, and history of loading. A deformation time curve (creep curve), as in Fig. 4.1, typically displays a segment of decelerating *creep rate* (stage 0–1), a segment of essentially constant deformation or minimum creep rate (stage 1–2), and finally a segment of accelerating creep rate (stage 2–3). In the figure, curve A might correspond to a condition of either higher stress or higher temperature than curve B. Both curves terminate in fracture at point 3. The *creep strength* refers to the maximum employable strength of the material at a prescribed elevated temperature. This value of stress corresponds to a given rate of creep in the second stage (2–3), e.g., 1% creep in 10,000 hours. Inasmuch as the creep stress and creep strain are not linearly related, calculations involving such material behavior are generally not routine.

Fracture

Separation of a material under stress into two or more parts (thereby creating new surface area) is referred to as *fracture*. The determination of the conditions of combined stress which lead to either elastic or inelastic

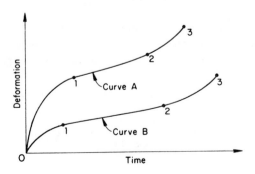

Figure 4.1

termination of deformation, that is, predicting the failure strength of a material, is quite difficult. Griffith* was the first to equate the strain energy associated with material failure to that required for the formation of new surfaces. He also concluded that with respect to its capacity to cause failure, tensile stress represents a more important influence than does compressive stress. The Griffith theory assumes the presence in brittle materials of minute cracks, which as a result of applied stress are caused to grow to macroscopic size, leading eventually to failure. While Griffith's experiments dealt primarily with glass, his results have been widely applied to other materials. Application to metals requires modification of the theory, however, because failure does not occur in an entirely brittle manner.

Brittle materials most commonly fracture through the grains, in what is termed a transcrystalline failure. Here the tensile stress is usually regarded as playing the most significant role. Examination of the failed material reveals very little deformation prior to fracture.

Machine and structural members subjected to repeated, fluctuating, or alternating stresses, which are below the ultimate tensile strength or even the yield strength, may nevertheless manifest diminished strength and ductility. Since the phenomenon described, termed *fatigue*, is difficult to predict and is often influenced by factors eluding recognition, increased uncertainty in strength and in service life must be dealt with. As is true for brittle behavior in general, fatigue is importantly influenced by minor structural discontinuities, the quality of surface finish, and the chemical nature of the environment. A fatigue crack is generally observed to have, as its origin, a point of high stress concentration, e.g., the corner of a keyway or a groove. This failure, through the involvement of slip planes and spreading cracks, is progressive in nature. For this reason, *progressive fracture* is probably a more appropriate term than fatigue failure. Tensile stress, and to a lesser degree shearing stress, lead to fatigue crack propagation, while compressive stress probably does not. The *fatigue life* or *endurance* of a material is defined as the number of stress repetitions or *cycles* prior to fracture. The fatigue life of a given material is dependent upon the magnitudes (and the algebraic signs) of the stresses at the extremes of a stress cycle (Sec. 4.14).

Experimental determination is made of the number of cycles (N) required to break a specimen at a particular stress level (S) under a fluctuating load. From such tests, called *fatigue tests*, curves termed *S–N*

---

*A. A. Griffith, The phenomena of rupture and flow in solids, *Phil. Trans. Roy. Soc., Lond.* A221:163 (1920). For a discussion of fracture mechanics see, for example, P. S. Nam and A. P. L. Turner, *Elements of the Mechanical Behavior of Solids*, New York: McGraw-Hill, 1975, Chapter 8.

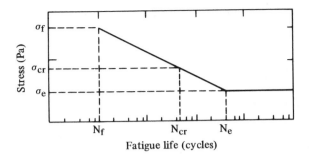

Figure 4.2

diagrams can be constructed. Various types of simple fatigue stress testing machines have been developed. Detailed information on this kind of equipment may be found in publications such as cited in the footnote of Sec. 4.1. The simplest is a rotating bar fatigue-testing machine on which a specimen (usually of circular cross section) is held so that it rotates while under condition of alternating *pure bending*. A *complete reversal* (tension to compression) of stress thus results.

It is usual practice to plot stress versus number of cycles with semilogarithmic scales (i.e., σ against $\log N$). For most steels, the *S–N* diagram obtained in a simple fatigue test performed on a number of nominally identical specimens loaded at different stress levels, has the appearance shown in Fig. 4.2. The stress at which the curve levels off is called the *endurance limit* strength σ_e. Beyond the point (σ_e, N_e) failure *will not* take place no matter how great the number of cycles. For a lower number of cycles $N < N_f$, the loading is regarded as *static*. At $N = N_f$ cycles, failure occurs at static tensile *fracture stress* σ_f. The fatigue strength for *complete stress reversal* at a specified number of cycles N_{cr}, is designated σ_{cr} on the diagram. The *S–N* curve relationships are utilized in Secs. 4.13 and 4.14, in which combined stress fatigue properties are discussed.

While yielding and fracture may well depend upon the rate of load application or the rate at which the small permanent strains form, we shall, with the exception of Secs. 4.15, 4.16, and 12.10, assume that yielding and fracture in solids are functions solely of the states of stress or strain.

4.3 Yielding Theories of Failure

As the tensile loading of a ductile member is increased, a point is eventually reached at which changes in geometry are no longer entirely reversible. The beginning of inelastic behavior is thus marked. The extent of the inelastic deformation preceding fracture is very much dependent upon the material involved.

Table 4.1

Quantity	Tension test	Torsion test
Maximum principal stress	σ_{yp}	$\sigma_{yp} = \tau_{yp}$
Maximum shear stress	$\tau_{yp} = \sigma_{yp}/2$	τ_{yp}
Maximum principal strain	$\varepsilon_{yp} = \sigma_{yp}/E$	$\varepsilon_{yp} = \tau_{yp}(1+\nu)/E$
Maximum energy of distortion	$U_{0d} = [(1+\nu)/3E]\sigma_{yp}^2$	$U_{0d} = \tau_{yp}^2(1+\nu)/E$
Maximum octahedral shear stress	$\tau_{oct} = (\sqrt{2}/3)\sigma_{yp}$	$\tau_{oct} = (\sqrt{2}/\sqrt{3})\tau_{yp}$

Consider an element subjected to a general state of stress, where $\sigma_1 > \sigma_2 > \sigma_3$. Recall that subscripts 1, 2, 3 refer to the principal directions. The state of stress in uniaxial loading is described by σ_1, equal to the normal force divided by the cross-sectional area, and $\sigma_2 = \sigma_3 = 0$. Corresponding to the start of the yielding event in this *simple tension test* are the quantities pertinent to stress, strain, and strain energy shown in the second column of Table 4.1. Note that the items listed in this column, expressed in terms of the uniaxial yield point stress σ_{yp}, have special significance in predicting failure involving multiaxial states of stress. In the case of a material in *simple torsion*, the state of stress is given by $\tau = \sigma_1 = -\sigma_3$ and $\sigma_2 = 0$. Here τ is calculated using the standard torsion formula. Corresponding to this case of pure shear, at the onset of yielding, are the quantities shown in the third column of the table, expressed in terms of yield point stress in torsion, τ_{yp}.

The behavior of materials subjected to uniaxial normal stresses or pure shearing stresses is readily presented on stress–strain diagrams. The onset of yielding in these cases is considerably more apparent than in situations involving combined stress. From the viewpoint of mechanical design, it is imperative that some practical guides be available to predict yielding under the conditions of stress as they are likely to exist in service. To meet this need and to understand the basis of material failure, a number of failure theories have been developed. Unfortunately, no theory can claim to be the final answer. In the development which follows, the yield stress obtained in a simple tension test is denoted by σ_{yp}', and in a simple compression test by σ_{yp}''.

4.4 The Maximum Principal Stress Theory

According to the maximum principal stress theory, credited to W. J. M. Rankine (1802–1872), a material fails by yielding when the maximum principal stress exceeds the tensile yield strength, or when the minimum principal stress exceeds the compressive yield strength. That is, at the onset of yielding,

$$|\sigma_1| = \sigma_{yp}' \qquad \text{or} \qquad |\sigma_3| = \sigma_{yp}'' \tag{4.1}$$

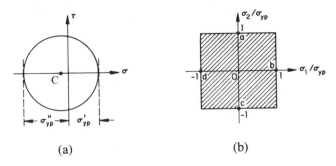

Figure 4.3

The region of failure is shown graphically by referring to the Mohr's circle representation of Fig. 4.3a. Failure occurs when the circle extends beyond either of the dashed vertical lines.

The theory has at least two limitations. First, there is ample experimental evidence that while a material may be weak in simple compression, it may nevertheless sustain very high hydrostatic pressure without yielding or fracturing. This is inconsistent with the maximum stress theory. Furthermore, failure in ductile materials is fundamentally a shearing phenomenon, and one would therefore assume the failure criteria for such materials to rely upon shearing rather than tensile stresses. The theory is nevertheless presented because of its widespread application in the design of members in a uniaxial state of stress.

For materials possessing the same yield stress in tension and compression ($\sigma'_{yp} = \sigma''_{yp} = \sigma_{yp}$), in the case of plane stress ($\sigma_3 = 0$), Eq. (4.1) becomes

$$|\sigma_1| = \sigma_{yp} \qquad \text{or} \qquad |\sigma_2| = \sigma_{yp} \qquad (4.2a)$$

This may be rewritten

$$\frac{\sigma_1}{\sigma_{yp}} = \pm 1 \qquad \text{or} \qquad \frac{\sigma_2}{\sigma_{yp}} = \pm 1 \qquad (4.2b)$$

The foregoing is depicted in Fig. 4.3b, points a, b and c, d indicating the tensile and compressive principal stresses, respectively. For this case, the boundaries represent the onset of failure due to yielding. The area within the boundary of the figure is thus a region of no yielding.

4.5   The Maximum Shear Stress Theory

The maximum shear stress theory is an outgrowth of the experimental observation that a ductile material yields as a result of slip or shear along crystalline planes. Proposed by C. A. Coulomb (1736–1806) it is also referred to as the Tresca or Guest theory in recognition of the contribution of these men to its application. This theory predicts that yielding will start when the maximum shear stress in the material equals the maximum shear stress at yielding in a simple tension test. Thus, by applying Eq. (1.27) and Table 4.1, we obtain

$$\tfrac{1}{2}|\sigma_1 - \sigma_3| = \tau_{yp} = \tfrac{1}{2}\sigma_{yp}$$

or

$$|\sigma_1 - \sigma_3| = \sigma_{yp} \tag{4.3}$$

In the case of plane stress, $\sigma_3 = 0$, there are two combinations of stresses to be considered. When σ_1 and σ_2 are of *opposite sign*, i.e., one tensile, the other compressive, the maximum shearing stress is $(\sigma_1 - \sigma_2)/2$. Thus, the yield condition is given by

$$|\sigma_1 - \sigma_2| = \sigma_{yp} \tag{4.4a}$$

which may be restated

$$\frac{\sigma_1}{\sigma_{yp}} - \frac{\sigma_2}{\sigma_{yp}} = \pm 1 \tag{4.4b}$$

When σ_1 and σ_2 carry the *same sign*, the maximum shearing stress equals $(\sigma_1 - \sigma_3)/2 = \sigma_1/2$. Then, for $|\sigma_1| > |\sigma_2|$ and $|\sigma_2| > |\sigma_1|$, we have the following yield conditions, respectively:

$$|\sigma_1| = \sigma_{yp} \qquad \text{and} \qquad |\sigma_2| = \sigma_{yp} \tag{4.5}$$

Fig. 4.4 is a plot of Eqs. (4.4) and (4.5). Note that Eq. (4.4) applies to the second and fourth quadrants, while Eq. (4.5) applies to the first and third quadrants. The boundary of the hexagon thus marks the onset of yielding, with points outside the shaded region representing a yielded state. The foregoing describes the Tresca yield condition. Good agreement with experiment has been realized for ductile materials. The theory offers an additional advantage in its ease of application.

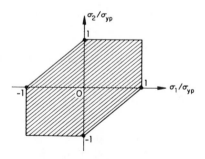

Figure 4.4

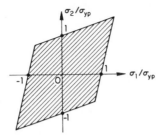

Figure 4.5

4.6 The Maximum Principal Strain Theory

According to the maximum principal strain theory, often referred to as the St. Venant theory in honor of its originator (1797–1866), a material fails by yielding when the maximum principal strain exceeds the tensile yield strain (ε'_{yp}) or when the minimum principal strain exceeds the compressive yield strain (ε''_{yp}). Applying the generalized Hooke's law, Eq. (2.15), we have after canceling E,

$$|\sigma_1 - \nu(\sigma_2 + \sigma_3)| = \sigma'_{yp}$$
$$|\sigma_3 - \nu(\sigma_1 + \sigma_2)| = \sigma''_{yp}$$

$$(4.6)$$

For plane stress, $\sigma_3 = 0$, and the yield conditions are described by

$$|\sigma_1 - \nu\sigma_2| = \sigma'_{yp}$$
$$|\sigma_2 - \nu\sigma_1| = \sigma''_{yp}$$

These may be restated for $\sigma'_{yp} = \sigma''_{yp} = \sigma_{yp}$:

$$\frac{\sigma_1}{\sigma_{yp}} - \nu\frac{\sigma_2}{\sigma_{yp}} = \pm 1$$

$$(4.7)$$

$$\frac{\sigma_2}{\sigma_{yp}} - \nu\frac{\sigma_1}{\sigma_{yp}} = \pm 1$$

The above expression, for $\nu = 0.3$, is plotted in Fig. 4.5, where again the region outside the boundary represents the states for which, according to this theory, yielding may be expected. The maximum strain theory of failure has been used with some success in the design of thick-walled cylinders.

4.7 The Maximum Distortion Energy Theory

The maximum distortion energy theory was proposed by M. T. Huber in 1904 and further developed by R. von Mises (1913) and H. Hencky (1925). In this theory, failure by yielding occurs when, at any point in the body, the distortion energy per unit volume in a state of combined stress becomes equal to that associated with yielding in a simple tension test. Equation (2.32) and Table 4.1 thus lead to

$$(\sigma_x - \sigma_y)^2 + (\sigma_y - \sigma_z)^2 + (\sigma_z - \sigma_x)^2 + 6(\tau_{xy}^2 + \tau_{yz}^2 + \tau_{xz}^2) = 2\sigma_{yp}^2 \qquad (4.8a)$$

or, in terms of principal stresses,

$$(\sigma_1 - \sigma_2)^2 + (\sigma_2 - \sigma_3)^2 + (\sigma_3 - \sigma_1)^2 = 2\sigma_{yp}^2 \qquad (4.8b)$$

For plane stress $\sigma_3 = 0$, and the criterion for yielding becomes

$$\sigma_1^2 - \sigma_1\sigma_2 + \sigma_2^2 = \sigma_{yp}^2 \qquad (4.9a)$$

or, alternatively,

$$\left(\frac{\sigma_1}{\sigma_{yp}}\right)^2 - \left(\frac{\sigma_1}{\sigma_{yp}}\right)\left(\frac{\sigma_2}{\sigma_{yp}}\right) + \left(\frac{\sigma_2}{\sigma_{yp}}\right)^2 = 1 \qquad (4.9b)$$

The above expression defines the ellipse shown in Fig. 4.6a

Returning to Eq. (4.8b), it is observed that only the *differences* of the principal stresses are involved. Consequently, addition of an equal amount to each stress does not affect the conclusion with respect to whether or not yielding will occur. In other words, yielding does not depend upon hydrostatic tensile or compressive stresses. Now consider Fig. 4.6b, in which a state of stress is defined by the position $P(\sigma_1, \sigma_2, \sigma_3)$ in a principal stress coordinate system as shown. It is clear that a hydrostatic alteration of the stress at point P requires shifting of this point along a direction parallel to direction n, making equal angles with coordinate axes. This is because changes in hydrostatic stress involve changes of the normal stresses by equal amounts. On the basis of the foregoing, it is concluded that the yield criterion is properly described by the cylinder shown in Fig. 4.6c, and that the surface of the cylinder is the yield surface. Points within the surface represent states of nonyielding. The ellipse of Fig. 4.6a is defined by the intersection of the cylinder with the σ_1, σ_2 plane. Note that the yield surface appropriate to the maximum shearing stress criterion (shown by the dashed

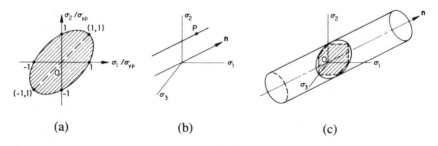

(a)   (b)   (c)

Figure 4.6

lines for plane stress) is described by a hexagonal surface placed within the cylinder.

The maximum distortion energy theory of failure finds considerable experimental support in situations involving ductile materials and plane stress. For this reason, it is in common use in design.

4.8 The Octahedral Shearing Stress Theory

The octahedral shearing stress theory (also referred to as the Mises–Hencky or simply the von Mises criterion) predicts failure by yielding when the octahedral shearing stress at a point achieves a particular value. This value is determined by the relationship of τ_{oct} to σ_{yp} in a simple tension test. Referring to Table 4.1, we obtain

$$\tau_{oct} = 0.47\sigma_{yp} \tag{4.10}$$

where τ_{oct} for a general state of stress is given by Eq. (2.33).

The Mises–Hencky criterion may also be viewed in terms of distortion energy [Eq. (2.32)]:

$$U_{0d} = \frac{3}{2} \frac{1+\nu}{E} \tau_{oct}^2 \tag{a}$$

If it is now asserted that yielding will, in a general state of stress, occur when U_{0d} as defined by Eq. (a) is equal to the value given in Table 4.1, then Eq. (4.10) will again be obtained. We conclude, therefore, that the octahedral shearing stress theory enables us to apply the distortion energy theory while dealing with stress rather than energy.

Example 4.1. A circular bar of tensile yield strength $\sigma_{yp} = 350$ MPa is subjected to a combined state of loading defined by bending moment $M = 8$ kN·m and torque $M_t = 24$ kN·m. Calculate the diameter d which the bar must have in order to achieve a factor of safety $N = 2$. Apply the following theories: (a) maximum principal stress, (b) maximum shearing stress, (c) maximum principal strain, (d) maximum energy of distortion.

SOLUTION. For the situation described, the principal stresses are

$$\sigma_{1,2} = \frac{\sigma}{2} \pm \frac{1}{2}\sqrt{\sigma^2 + 4\tau^2}, \quad \sigma_3 = 0 \tag{b}$$

where (see Secs. 5.2 and 6.1),

$$\sigma = \frac{My}{I} = \frac{32M}{\pi d^3}, \quad \tau = \frac{M_t r}{J} = \frac{16M_t}{\pi d^3}$$

Thus,

$$\sigma_{1,2} = \frac{16}{\pi d^3}\left(M \pm \sqrt{M^2 + M_t^2}\right) \tag{c}$$

(a) *Maximum principal stress theory*: On the basis of Eqs. (c) and (4.2a),

$$\frac{16}{\pi d^3}\left(M + \sqrt{M^2 + M_t^2}\right) = \frac{\sigma_{yp}}{N}$$

Substituting the data given, we have

$$\frac{16}{\pi d^3}\left(8+\sqrt{(8)^2+(24)^2}\right)=175\times10^6$$

from which $d=0.0989$ m $=98.9$ mm.

(b) *Maximum shearing stress theory*: For the state of stress under consideration it may be observed from a Mohr's circle construction that σ_1 is tensile and σ_2 is compressive. Thus, through the use of Eqs. (b) and (4.4a),

$$\left(\frac{32}{\pi d^3}\right)\sqrt{M^2+M_t^2}=\frac{\sigma_{yp}}{N}$$

After substitution of the numerical values, the above gives $d=113.7$ mm.

(c) *Maximum principal strain theory*: On the basis of Hooke's law and Table 4.1,

$$\varepsilon_1=\frac{1}{E}[\sigma_1-\nu(\sigma_2+\sigma_3)]=\frac{\varepsilon_{yp}}{N}=\frac{\sigma_{yp}}{EN}$$

or

$$\sigma_1-\nu(\sigma_2+\sigma_3)=\frac{\sigma_{yp}}{N}$$

Introducing Eq. (c) and $\nu=0.3$, inserting the data, and solving for d, we obtain $d=103.8$ mm.

(d) *Maximum energy of distortion theory*: From Eqs. (4.9a) and (c),

$$\left(\frac{16}{\pi d^3}\right)\sqrt{4M^2+3M_t^2}=\frac{\sigma_{yp}}{N}$$

This result may also be obtained from the octahedral shearing stress theory by applying Eqs. (4.10) and (c). Substitution of the data into the above equation, yields $d=109$ mm.

Example 4.2. A steel conical tank, supported at its edges, is filled with a liquid of density γ (Fig. P13.8). The yield point stress (σ_{yp}) of the material is known. The cone angle is 2α. Determine the required wall thickness t of the tank, based upon a factor of safety N. Apply (a) the maximum shear stress theory and (b) the maximum energy of distortion theory.

SOLUTION. The variations of the circumferential and longitudinal stresses in the tank are, respectively (Problem 13.8)

$$\sigma_1=\gamma(a-y)y\frac{\tan\alpha}{t\cos\alpha}, \qquad \sigma_2=\gamma\left(a-\frac{2}{3}y\right)y\frac{\tan\alpha}{2t\cos\alpha} \tag{d}$$

The principal stresses have the largest magnitude

$$\sigma_{1,\max}=\frac{\gamma a^2}{4t}\frac{\tan\alpha}{\cos\alpha} \qquad \text{at} \quad y=\frac{a}{2}$$

$$\sigma_{2,\max}=\frac{3\gamma a^2}{16t}\frac{\tan\alpha}{\cos\alpha} \qquad \text{at} \quad y=\frac{3a}{4} \tag{e}$$

(a) *Maximum shear stress theory*: As σ_1 and σ_2 are of the same sign and $|\sigma_1| > |\sigma_2|$, we have, from the first equation of (4.5) together with (e),

$$\frac{\sigma_{yp}}{N} = \frac{\gamma a^2}{4t} \frac{\tan \alpha}{\cos \alpha}$$

The thickness of the tank is found from the above to be

$$t = 0.250 \frac{\gamma a^2 N}{\sigma_{yp}} \frac{\tan \alpha}{\cos \alpha}$$

(b) *Maximum distortion energy theory*: It is observed in Eq. (e) that the largest values of principal stress are found at different locations. We shall therefore first locate the section at which the combined principal stresses are at a critical value. For this purpose, we insert Eq. (d) into Eq. (4.9a):

$$\sigma_{yp}^2 = \left[\gamma(a-y)y \frac{\tan \alpha}{t \cos \alpha} \right]^2 + \left[\left(a - \frac{2}{3}y \right)y \frac{\tan \alpha}{2t \cos \alpha} \right]^2$$
$$- \left[\gamma(a-y)y \frac{\tan \alpha}{t \cos \alpha} \right] \left[\left(a - \frac{2}{3}y \right)y \frac{\tan \alpha}{2t \cos \alpha} \right] \tag{f}$$

Upon differentiating Eq. (f) with respect to the variable y, and equating the result to zero, we obtain

$$y = 0.52a$$

Upon substitution of this value of y into Eq. (f), the thickness of the tank is determined:

$$t = 0.225 \frac{\gamma a^2 N}{\sigma_{yp}} \frac{\tan \alpha}{\cos \alpha}$$

The thickness based upon the maximum shear stress theory is thus 10% larger than that based upon the maximum energy of distortion theory.

4.9 Mohr's Theory

The Mohr theory of failure makes use of the well-known Mohr circles of stress. As discussed in Sec. 1.11, in a Mohr's circle representation, the shear and normal components of stress acting on a particular plane are specified by the coordinates of a point within the shaded area of Fig. 4.7a. Note that τ depends upon σ, i.e., $|\tau| = f(\sigma)$.

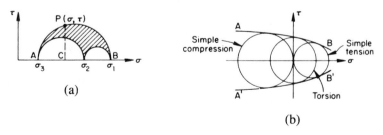

Figure 4.7

The figure indicates that a vertical line such as PC represents the states of stress on planes with the same σ but with differing τ. It follows that the weakest of all these planes is the one on which the maximum shearing stress acts, designated P. The same conclusion can be drawn regardless of the position of the vertical line between A and B; the points on the outer circle correspond to the weakest planes. On these planes, the *maximum* and *minimum* principal stresses alone are sufficient to decide whether or not failure will occur, because these stresses determine the outer circle shown in Fig. 4.7a. Using these extreme values of principal stress thus enables us to apply the Mohr approach to either two- or three-dimensional situations.

The foregoing serves to provide background for the Mohr theory of failure, which relies upon stress plots in σ, τ coordinates. The particulars of the Mohr approach are presented below.

Experiments are performed on a given material to determine the states of stress which result in failure. Each such stress state defines a Mohr's circle. If the data describing states of limiting stress are derived from only simple tension, simple compression, and pure shear tests, the three resulting circles are adequate to construct the *envelope*, denoted by lines AB and $A'B'$ in Fig. 4.7b. The Mohr envelope thus represents the locus of all possible failure states. Many solids, particularly those which are brittle, exhibit greater resistance to compression than to tension. As a consequence, higher limiting shear stresses will, for these materials, be found to the left of the origin, as shown in the figure. For the case of equal yield stresses in tension and compression, $\sigma'_{yp} = \sigma''_{yp}$, the Mohr envelope is represented by a pair of horizontal τ lines. The failure theory now reduces to the maximum shear theory.

4.10   The Coulomb–Mohr Theory

The *Coulomb–Mohr* or *internal friction* theory assumes that the critical shearing stress is related to internal friction. If the frictional force is regarded as a function of the normal stress acting on a shear plane, the critical shearing stress and normal stress can be connected by an equation of the following form (Fig. 4.8a):

$$\tau = a\sigma + b \tag{a}$$

The constants a and b represent properties of the particular material. The above expression may also be viewed as a straight line version of the Mohr envelope.

For the case of plane stress, $\sigma_3 = 0$ when σ_1 is tensile and σ_2 is compressive. The maximum shearing stress τ and the normal stress σ acting on the shear plane are, from Eqs. (1.10) and (1.11), given by

$$\tau = \frac{\sigma_1 - \sigma_2}{2}, \qquad \sigma = \frac{\sigma_1 + \sigma_2}{2} \tag{b}$$

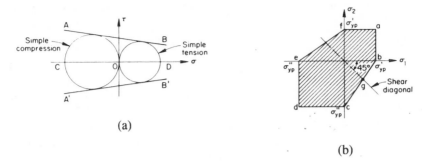

Figure 4.8

Introducing these expressions into Eq. (a), we obtain

$$\sigma_1(1-a) - \sigma_2(1+a) = 2b \tag{c}$$

In order to evaluate the material constants, the following conditions are applied:

$$\begin{aligned}
\sigma_1 &= \sigma'_{yp} && \text{when } \sigma_2 = 0 \\
\sigma_2 &= \sigma''_{yp} && \text{when } \sigma_1 = 0
\end{aligned} \tag{d}$$

If now Eqs. (d) are inserted into Eq. (c), the results are

$$\sigma'_{yp}(1-a) = 2b \qquad \text{and} \qquad \sigma''_{yp}(1+a) = 2b$$

from which

$$a = \frac{\sigma'_{yp} - \sigma''_{yp}}{\sigma'_{yp} + \sigma''_{yp}}, \qquad b = \frac{\sigma'_{yp}\sigma''_{yp}}{\sigma'_{yp} + \sigma''_{yp}} \tag{e}$$

The above constants are now introduced into Eq. (c) to complete the equation of the envelope of failure by yielding. When this is done, the following expression is obtained, applicable for $\sigma_1 > 0, \sigma_2 < 0$:

$$\frac{\sigma_1}{\sigma'_{yp}} - \frac{\sigma_2}{\sigma''_{yp}} = 1 \tag{4.11a}$$

For any given ratio σ_1/σ_2, the individual stresses at yielding, σ_1 and σ_2, can be calculated by applying the above expression (Problem 4.10).

Relationships for the case where the principal stresses have the same sign ($\sigma_1 > 0, \sigma_2 > 0$ or $\sigma_1 < 0, \sigma_2 < 0$) may be deduced from Fig. 4.8a without resort to the above procedure. In the case of *biaxial tension* (now $\sigma_{min} = \sigma_3 = 0$, σ_1 and σ_2 are tensile), the corresponding Mohr's circle is represented by diameter OD. Therefore, yielding occurs if either of the two tensile stresses achieves the value σ'_{yp}. That is,

$$\sigma_1 = \sigma'_{yp}, \qquad \sigma_2 = \sigma'_{yp} \tag{4.11b}$$

For biaxial compression (now $\sigma_{max} = \sigma_3 = 0$, σ_1 and σ_2 are compressive), a Mohr's circle of diameter OC is obtained. Failure by yielding occurs if either of the compressive stresses attains the value σ''_{yp}:

$$\sigma_2 = -\sigma''_{yp}, \qquad \sigma_1 = -\sigma''_{yp} \qquad (4.11c)$$

Figure 4.8b is a graphical representation of the Coulomb–Mohr theory plotted in the σ_1, σ_2 plane. Lines ab and af represent Eq. (4.11b), and lines dc and de, Eq. (4.11c). The boundary bc is obtained through the application of Eq. (4.11a). Line ef completes the hexagon in a manner analogous to Fig. 4.4. Points lying within the shaded area should not represent yielding, according to the theory. In the case of pure shear, the corresponding limiting point is g. The magnitude of the limiting shear stress may be graphically determined from the figure or calculated from Eq. (4.11a) by letting $\sigma_1 = -\sigma_2$:

$$\sigma_1 = \tau_{yp} = \frac{\sigma'_{yp}\sigma''_{yp}}{\sigma'_{yp} + \sigma''_{yp}} \qquad (4.12)$$

4.11 Comparison of the Yielding Theories

Two approaches may be employed for the purpose of comparing the theories of yielding heretofore discussed, for a material with $\sigma'_{yp} = \sigma''_{yp}$. The first comparison is made by equating for each theory the critical values corresponding to uniaxial loading and torsion for Poisson's ratio $\nu = 0.3$. Referring to Table 4.1, we have:

Maximum principal stress theory: $\tau_{yp} = \sigma_{yp}$

Maximum shearing stress theory: $\tau_{yp} = 0.50\sigma_{yp}$

Maximum principal strain theory: $\tau_{yp} = 0.77\sigma_{yp}$

Maximum energy of distortion theory: $\tau_{yp} = 0.577\sigma_{yp}$

Maximum octahedral shearing stress theory: $\tau_{yp} = 0.577\sigma_{yp}$

It is observed that the differences in strength predicted by the various theories are substantial. A second comparison may be made by means of a superposition of Figs. 4.3 to 4.6. This is left as an exercise for the reader.

Experiment shows that for ductile materials, the yield stress obtained in a torsion test is 0.5 to 0.6 times that determined from a simple tension test. We conclude therefore that the energy of distortion theory or its equivalent, the octahedral shearing stress theory, is most suitable for ductile materials. The shearing stress theory, which gives $\tau_{yp} = 0.50\sigma_{yp}$, is in widespread use, however, because it is simple to apply and offers a conservative result in design.

4.12   Theories of Fracture

The foregoing theories represent attempts to predict failure due to yielding. Certain brittle materials, including a variety of metals and plastics, display little yielding, and a yield stress does not, in practical terms, exist. Failure in such brittle materials is predicated upon the ultimate strength, by assuming the ultimate and fracture stresses to be equal. Those theories of failure by yielding which do *not* involve Hooke's law can readily be coverted so as to apply to fracture by simply replacing the stresses σ'_{yp}, σ''_{yp}, and τ_{yp} with the uniaxial ultimate stresses σ'_u, σ''_u, and τ_u, respectively. The maximum principal stress, maximum shearing stress, octahedral shearing stress, Mohr, and Coulomb–Mohr theories may be so treated.

It is to be noted that, in addition to the theories of failures, failure is sometimes predicted conveniently through the use of the interaction curves discussed in Sec. 12.5.

Example 4.3. A thin-walled tube is fabricated of a brittle metal having ultimate tensile and compressive strengths $\sigma'_u = 300$ MPa and $\sigma''_u = 700$ MPa. The outer and inner radii are $b = 105$ mm and $a = 100$ mm. Calculate the limiting torque that can be applied without causing failure by fracture. Apply three criteria: (a) the maximum principal stress theory, (b) the maximum shearing stress theory, (c) the Coulomb–Mohr theory.

SOLUTION. The torque and maximum shearing stress are related by the torsion formula (see Sec. 6.1):

$$M_t = \frac{J}{r}\tau = \frac{\pi(b^4 - a^4)}{2b}\tau = \frac{\pi[(0.105)^4 - (0.1)^4]}{2(0.105)}\tau = 3.224 \times 10^{-4}\tau \qquad (a)$$

The state of stress is described by $\sigma_1 = -\sigma_2 = \tau$, $\sigma_3 = 0$.

(a) *Maximum principal stress theory*: Equations (4.1) are applied, with σ_3 replaced by σ_2 because the latter is negative: $\sigma_1 = \sigma'_u$ or $\sigma_2 = \sigma''_u$. As we have $\sigma_1 = \sigma'_u = 300 \times 10^6 = \tau$, from Eq. (a),

$$M_t = 3.224 \times 10^{-4}(300 \times 10^6) = 96{,}720 \ N \cdot m$$

(b) *Maximum shearing stress theory*: This theory requires that $\sigma'_u = \sigma''_u = \sigma_u$. Applying Eq. (4.4a),

$$|\sigma_1 - \sigma_2| = \sigma'_u; \qquad 2\sigma_1 = \sigma'_u = 2\tau \quad \text{or} \quad \tau = \sigma'_u/2$$

Substituting the above into Eq. (a), we have

$$M_t = 1.612 \times 10^{-4}\sigma'_u = 48{,}360 \ N \cdot m$$

(c) *Coulomb–Mohr theory*: Applying Eq. (4.11a),

$$\frac{\tau}{300 \times 10^6} - \frac{-\tau}{700 \times 10^6} = 1$$

from which $\tau = 210$ MPa. Equation (a) gives $M_t = 3.224 \times 10^{-4}(210 \times 10^6) = 67{,}704$ $N \cdot m$, which is intermediate between the values found by the maximum principal stress theory and the maximum shearing stress theory.

4.13   Failure Criteria for Metal Fatigue

A very common type of fatigue loading consists of an alternating sinusoidal stress superimposed upon a uniform stress, Fig. 4.9. Such variation of stress with time occurs, for example, if a forced vibration of constant amplitude is applied to a structural member already transmitting a constant load. Referring to the figure, we define the *average* or *mean* stress and the *alternating, fluctuating*, or *fatigue* stress as follows:

$$\sigma_m = \tfrac{1}{2}(\sigma_{max} + \sigma_{min})$$

$$\sigma_a = \tfrac{1}{2}(\sigma_{max} - \sigma_{min})$$

(4.13)

In the case of complete stress reversal, it is clear that the average stress equals zero. The alternating stress component is the most important factor in determining the number of cycles of load the material can withstand before fracture; the mean stress level is less important, particularly, if σ_m is negative (compressive).

As mentioned in Sec. 4.3, the local character of fatigue phenomena makes it necessary to analyze carefully the stress field within an element. A fatigue crack can start in one small region of high alternating stress and propagate, producing complete failure regardless of how adequately proportioned the remainder of the member may be. To predict whether the state of stress at a critical point will result in failure, a criterion is employed which is based upon the mean and fluctuating stresses and which utilizes the simple *S–N* curve relationships.

Uniaxial State of Stress. Many approaches have been suggested for interpreting fatigue data. Table 4.2 lists commonly employed *criteria*, also referred to as *mean stress–alternating stress relations*. In each case, fatigue strength for complete stress reversal at a specified number of cycles, may have a value between the fracture stress and endurance stress (i.e., $\sigma_f \leqslant \sigma_{cr} \leqslant \sigma_e$), Fig. 4.2.

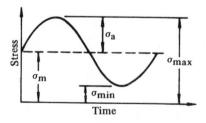

Figure 4.9

Table 4.2^a

Fatigue criterion	Modified Goodman	Soderberg	Gerber	SAE
Equation	$\dfrac{\sigma_a}{\sigma_{cr}} + \dfrac{\sigma_m}{\sigma_u} = 1$	$\dfrac{\sigma_a}{\sigma_{cr}} + \dfrac{\sigma_m}{\sigma_{yp}} = 1$	$\dfrac{\sigma_a}{\sigma_{cr}} + \left(\dfrac{\sigma_m}{\sigma_u}\right)^2 = 1$	$\dfrac{\sigma_a}{\sigma_{cr}} + \dfrac{\sigma_m}{\sigma_f} = 1$

^{a}Subscript notations: a = alternating, yp = static tensile yield, m = mean, u = static tensile ultimate, cr = completely reversed, f = fracture.

Experience has shown that for steel, the *Soderberg* or *Modified Goodman* relations are the most reliable for predicting fatigue failure. The *Gerber* criterion leads to more liberal results and hence less safe to use. For hard steels, the SAE and modified Goodman relations result in identical solutions, since for brittle materials $\sigma_u = \sigma_f$.

Relationships presented in Table 4.2 together with specified material properties form the basis for practical fatigue calculations for members under uniaxial states of stress.

Example 4.4. A square prismatic bar of sides 0.05 m is subjected to an axial thrust (tension) $F_m = 90$ kN. The fatigue strength for completely reversed stress at 10^6 cycles is 210 MPa and the static tensile yield strength is 280 MPa. Apply the Soderberg criterion to determine the limiting value of completely reversed axial load F_a that can be superimposed to F_m at the midpoint of a side of the cross section without causing fatigue failure at 10^6 cycles.

SOLUTION. The alternating and mean stresses are given by

$$\sigma_a = \frac{M_a c}{I} = \frac{0.025 F_a (0.025)}{(0.05)^4/12} = 1200 F_a$$

$$\sigma_m = \frac{F_m}{A} = \frac{90,000}{0.05 \times 0.05} = 144 \text{ MPa}$$

Applying the Soderberg criterion,

$$\frac{1200 F_a}{210 \times 10^6} + \frac{144 \times 10^6}{280 \times 10^6} = 1$$

we obtain $F_a = 85$ kN.

Combined State of Stress. Quite often structural and machine elements are subjected to *combined* fluctuating bending, torsion, and axial loading. Examples include crankshafts, propeller shafts, and aircraft wings. Under conditions of a general state of stress that are cyclic, it is common practice to modify the static failure theories for purposes of analysis, substituting the subscripts a, m, and e in the expressions developed in the preceeding sections. In so doing, the maximum distortion energy theory, for example, is expressed

$$\left(\sigma_{xa}-\sigma_{ya}\right)^2+\left(\sigma_{ya}-\sigma_{za}\right)^2+\left(\sigma_{za}-\sigma_{xa}\right)^2+6\left(\tau_{xya}^2+\tau_{xza}^2+\tau_{yza}^2\right)=2\sigma_{ea}^2$$

$$\left(\sigma_{xm}-\sigma_{ym}\right)^2+\left(\sigma_{ym}-\sigma_{zm}\right)^2+\left(\sigma_{zm}-\sigma_{xm}\right)^2+6\left(\tau_{xym}^2+\tau_{xzm}^2+\tau_{yzm}^2\right)=2\sigma_{em}^2$$

$$\text{(4.14a)}$$

or

$$\left(\sigma_{1a}-\sigma_{2a}\right)^2+\left(\sigma_{2a}-\sigma_{3a}\right)^2+\left(\sigma_{3a}-\sigma_{1a}\right)^2=2\sigma_{ea}^2$$

$$\left(\sigma_{1m}-\sigma_{2m}\right)^2+\left(\sigma_{2m}-\sigma_{3m}\right)^2+\left(\sigma_{3m}-\sigma_{1m}\right)^2=2\sigma_{em}^2$$

$$\text{(4.14b)}$$

Here σ_{ea} and σ_{em}, the *equivalent alternating* stress and *equivalent mean* stress, respectively, replace the quantity σ_{yp} (or σ_u) used thus far. Relations for other fatigue failure theories can be written in a like manner.

The equivalent mean stress-equivalent alternating stress fatigue failure relations are represented in Table 4.2, replacing σ_a and σ_m by σ_{ea} and σ_{em}. These criteria together with modified static failure theories are used to compute fatigue strength under combined loading.

Example 4.5. Consider a thin-walled cylinder of radius $r=0.04$ m and thickness $t=5$ mm, subject to an internal pressure varying from a value of $-p/4$ to p. Employ the octahedral shear theory together with the Soderberg criterion to compute the value of p producing failure after 10^8 cycles. The material tensile yield strength is 300 MPa and the fatigue strength is $\sigma_e=250$ MPa at 10^8 cycles.

SOLUTION. The maximum and minimum values of the tangential and axial principal stresses are given by

$$\sigma_{\theta,\max}=\frac{pr}{t}=8p, \qquad \sigma_{\theta,\min}=\frac{(-p/4)r}{t}=-2p$$

$$\sigma_{z,\max}=\frac{pr}{2t}=4p, \qquad \sigma_{z,\min}=\frac{(-p/4)r}{2t}=-p$$

The alternating and mean stresses are therefore

$$\sigma_{\theta a}=\tfrac{1}{2}(\sigma_{\theta,\max}-\sigma_{\theta,\min})=5p, \qquad \sigma_{\theta m}=\tfrac{1}{2}(\sigma_{\theta,\max}+\sigma_{\theta,\min})=3p$$

$$\sigma_{za}=\tfrac{1}{2}(\sigma_{z,\max}-\sigma_{z,\min})=2.5p, \qquad \sigma_{zm}=\tfrac{1}{2}(\sigma_{z,\max}+\sigma_{z,\min})=1.5p$$

The octahedral shear theory, Eq. (4.10), for cyclic combined stress is expressed

$$\left(\sigma_{\theta a}^2-\sigma_{\theta a}\sigma_{za}+\sigma_{za}^2\right)^{1/2}=\sigma_{ea}$$

$$\left(\sigma_{\theta m}^2-\sigma_{\theta m}\sigma_{zm}+\sigma_{zm}^2\right)^{1/2}=\sigma_{em}$$

$$\text{(4.15)}$$

In terms of computed alternating and mean stresses, Eqs. (4.15) appear as

$$(25p^2-12.5p^2+6.25p^2)^{1/2}=\sigma_{ea}$$

$$(9p^2-4.5p^2+2.25p^2)^{1/2}=\sigma_{em}$$

from which $\sigma_{ea}=4.33p$ and $\sigma_{em}=2.60p$.

The Soderberg relation then leads to

$$\frac{4.33p}{250\times 10^6} + \frac{2.60p}{300\times 10^6} = 1$$

Solving the above, $p = 38.43$ MPa.

4.14 Fatigue Life Under Combined Loading

Combined stress conditions can lower fatigue life appreciably. The approach described here predicts the durability of a structural or machine element loaded in fatigue. The procedure applies to any uniaxial, biaxial, or general state of stress. According to the method, *fatigue life* N_{cr} (Fig. 4.2) is defined by the formula*

$$N_{cr} = N_f \left(\frac{\sigma_{cr}}{\sigma_f} \right)^{1/b} \tag{4.16}$$

in which

$$b = \frac{\ln(\sigma_f / \sigma_e)}{\ln(N_f / N_e)} \tag{4.17}$$

Here the values of σ_e and σ_f are specified in terms of material static tensile strengths, while N_e and N_f are given in cycles, Table 4.3. The *fatigue-strength reduction factor K*, listed in the table, can be ascertained on the basis of tests or from finite-element analysis. The data will be scattered (in general $K > 0.3$) and considerable variance requires the stress analyst to use a statistically acceptable value. The reversed stress σ_{cr} is computed applying the relations of Table 4.2, as required.

Table 4.3

Fatigue Criterion	Ductile steels ($\sigma_u \leqslant 1750$ MPa)				Brittle (hard) steels ($\sigma_u > 1750$ MPa)			
	σ_f	N_f	σ_e	N_e	σ_f	N_f	σ_e	N_e
Modified Goodman	$0.9\sigma_u$	10^3	$\frac{1}{2}K\sigma_u$	10^6	$0.9\sigma_u$	10^3	$\frac{1}{3}K\sigma_u$	10^8
Soderberg	$0.9\sigma_u$	10^3	$\frac{1}{2}K\sigma_u$	10^6	$0.9\sigma_u$	10^3	$\frac{1}{3}K\sigma_u$	10^8
Gerber	$0.9\sigma_u$	10^3	$\frac{1}{2}K\sigma_u$	10^6	$0.9\sigma_u$	10^3	$\frac{1}{3}K\sigma_u$	10^8
SAE	$\sigma_u + 350\times 10^6$	1	$\frac{1}{2}K\sigma_u$	10^6	σ_u	1	$\frac{1}{3}K\sigma_u$	10^8

The end points header reads: End points for S–N diagram (Fig. 4.2)

*Sullivan, J. L., Fatigue life under combined stress, *Machine Design* (January 25, 1979). See also Juvinall, R. C., *Stress, Strain, and Strength*, New York: McGraw-Hill, 1967.

Alternatively, the fatigue life may be determined graphically from the S–N diagram (Fig. 4.2) constructed by connecting points with coordinates (σ_f, N_f) and (σ_e, N_e). Interestingly, b represents the *slope* of the diagram. Following is a solution of a triaxial stress problem illustrating the use of the preceeding approach.

Example 4.6. A rotating hub and shaft assembly is subjected to bending moment, axial trust, *bidirectional* torque, and a uniform shrink fit pressure so that the following stress levels (in megapascals) occur at an outer critical point of the shaft:

$$\begin{bmatrix} 700 & 14 & 0 \\ 14 & -350 & 0 \\ 0 & 0 & -350 \end{bmatrix}, \quad \begin{bmatrix} -660 & -7 & 0 \\ -7 & -350 & 0 \\ 0 & 0 & -350 \end{bmatrix}$$

These matrices represent the maximum and minimum stress components, respectively. Determine the fatigue life, using the maximum energy of distortion theory of failure together with (a) the SAE fatigue criterion and (b) the Gerber criterion. The material properties are $\sigma_u = 2400$ MPa and $K = 1$.

SOLUTION. From Table 4.3 we have $\sigma_e = 1(2400 \times 10^6)/3 = 800$ MPa, $N_f = 1$ cycle for SAE, $N_f = 10^3$ cycles for Gerber, and $N_e = 10^8$ cycles. The alternating and mean values of the stress components are

$$\sigma_{xa} = \tfrac{1}{2}(700 + 660) = 680 \text{ MPa}$$

$$\sigma_{xm} = \tfrac{1}{2}(700 - 660) = 20 \text{ MPa}$$

$$\sigma_{ya} = \sigma_{za} = \tfrac{1}{2}(-350 + 350) = 0$$

$$\sigma_{ym} = \sigma_{zm} = \tfrac{1}{2}(-350 - 350) = -350 \text{ MPa}$$

$$\tau_{xya} = \tfrac{1}{2}(14 + 7) = 10.5 \text{ MPa}$$

$$\tau_{xym} = \tfrac{1}{2}(14 - 7) = 3.5 \text{ MPa}$$

Upon application of Eq. (4.14a), the equivalent alternating and mean stresses are found to be

$$\sigma_{ea} = \left\{ \tfrac{1}{2}\left[(680-0)^2 + (0-680)^2 + 6(10.5)^2 \right] \right\}^{1/2} = 680.24 \text{ MPa}$$

$$\sigma_{em} = \left\{ \tfrac{1}{2}\left[(20+350)^2 + (-350-20)^2 + 6(3.5^2) \right] \right\}^{1/2} = 370.05 \text{ MPa}$$

(a) The fatigue strength for complete reversal of stress, referring to Table 4.2, is

$$\sigma_{cr} = \frac{\sigma_{ea}}{1 - (\sigma_{em}/\sigma_f)} = \frac{680.24 \times 10^6}{1 - (370.05 \times 10^6/2400 \times 10^6)} = 804.24 \text{ MPa}$$

Equation (4.17) yields

$$b = \frac{\ln(2400/800)}{\ln(1/10^8)} = -0.0596$$

The fatigue life, from Eq. (4.16), is thus

$$N_{cr} = 1\left(\frac{804.24}{2400} \right)^{-(1/0.0596)} = 92.6 \times 10^6 \text{ cycles.}$$

(b) From Table 4.2, we now apply

$$\sigma_{cr} = \frac{\sigma_{ea}}{1 - (\sigma_{em}/\sigma_u)^2} = \frac{680.24 \times 10^6}{1 - (370.05/2400)^2} = 696.81 \text{ MPa}$$

and

$$b = \frac{\ln(0.9 \times 2400/800)}{\ln(10^3/10^8)} = -0.0863$$

It follows that

$$N_{cr} = 10^3 \left(\frac{696.81}{0.9 \times 2400} \right)^{-11.587} = 493.34 \times 10^6 \text{ cycles}$$

Upon comparison of the results of (a) and (b) we observe that the Gerber criterion overestimates the fatigue life.

4.15 Impact or Dynamic Loads

Forces suddenly applied to structures and machines are termed *shock* or *impact* loads and result in *dynamic loading*. Examples include *rapidly* moving loads such as those caused by a railroad train passing over a bridge, or a high speed rocket propelled test sled moving on a track, or direct *impact* loads such as result from a drop hammer. In machine service, impact loads are due to gradually increasing clearances which develop between mating parts with progressive wear, e.g., steering gears and axle journals of automobiles.

A dynamic force acts to modify the static stress and strain fields as well as the resistance properties of a material. Shock loading is usually produced by a sudden application of force or motion to a member, whereas impact loading results from the collision of bodies. When the time of application of a load is *equal to* or *smaller* than the *largest natural period* of vibration of the structural element, shock or impact loading is produced.

While following a shock or impact loading, vibrations commence, our concern here will be only with the influence of impact forces upon the maximum stress and deformation of the body. It is important to observe that the design of engineering structures subject to suddenly applied loads is complicated by a number of factors, and theoretical considerations generally serve only qualitatively to guide the design.* The use of static material properties in the design of members under impact loading is regarded as conservative and satisfactory. Details concerning the behavior of materials under impact loading are presented in the next section.

---

*See R. T. Magner, Simple procedures to follow to design for shock resistance, *Product Engineering* (December 1962); H. A. Rothbart, Ed., *Mechanical Design and Systems Handbook*, New York: McGraw-Hill (1964), Chapter 16.

The impact problem will be analyzed using the elementary theory together with the following assumptions:

1. The displacement is proportional to the applied forces, static and dynamic.
2. The inertia of a member subjected to impact loading may be neglected.
3. The material behaves elastically. In addition, it is assumed that there is no energy loss associated with the local inelastic deformation occurring at the point of impact or at the supports. Energy is thus conserved within the system.

To idealize an elastic system subjected to an impact force, consider Fig. 4.10, in which are shown a weight W, which falls through a distance h, striking the end of a free standing spring. As the velocity of the weight is zero initially, and is again zero at the *instant* of maximum deflection of the spring (δ_{max}), the change in kinetic energy of the system is zero, and likewise the work done on the system. The total work consists of the work done by gravity on the mass as it falls and the resisting work done by the spring:

$$W(h+\delta_{max}) - \tfrac{1}{2}k\delta_{max}^2 = 0 \qquad (a)$$

where k is known as the *spring constant*.

It is noted that the weight is assumed to remain in contact with the spring. The deflection corresponding to a static force equal to the weight of the body is simply W/k. This is termed the *static deflection*, δ_{st}. Then the general expression of maximum dynamic deflection is, from Eq. (a), therefore

$$\delta_{max} = \delta_{st} + \sqrt{(\delta_{st})^2 + 2\delta_{st}h} \qquad (4.18a)$$

or, by rearrangement,

$$\delta_{max} = \delta_{st}\left[1 + \sqrt{1 + \frac{2h}{\delta_{st}}}\,\right] \qquad (4.18b)$$

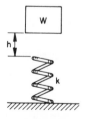

Figure 4.10

The *impact factor*, the ratio of the maximum dynamic deflection to the static deflection, is given by

$$\frac{\delta_{max}}{\delta_{st}} = 1 + \sqrt{1 + \frac{2h}{\delta_{st}}} \tag{4.19}$$

Multiplication of the impact factor by W yields an equivalent or so-called *dynamic load*:

$$P_{dyn} = W\left[1 + \sqrt{1 + \frac{2h}{\delta_{st}}}\right] \tag{4.20}$$

To compute the maximum stress and deflection resulting from impact loading, the above load may be used in the relationships derived for static loading.

Two extreme cases are clearly of particular interest. When $h \gg \delta_{max}$, the work term, $W\delta_{max}$, in Eq. (a) may be neglected, reducing the expression to $\delta_{max} = \sqrt{2\delta_{st}h}$. On the other hand, when $h = 0$, the load is suddenly applied, and Eq. (a) becomes $\delta_{max} = 2\delta_{st}$.

The expressions derived may readily be applied to analyze the dynamic effects produced by a falling weight causing axial, flexural, or torsional loading. Where bending is concerned, the results obtained are acceptable for the deflections, but poor in accuracy for predictions of maximum stress, with the error increasing as h/δ_{st} becomes larger or $h \gg \delta_{st}$. This departure is attributable to the variation in the shape of the actual static deflection curve. Thus, the curvature of the beam axis, and in turn, the maximum stress at the location of the impact, may differ considerably from that obtained through application of the strength of materials approach.

An analysis similar to the above may be employed to derive expressions for the case of a weight W in *horizontal* motion with a velocity v, arrested by an elastic body. In this instance, the kinetic energy $Wv^2/2g$ replaces $W(h + \delta_{max})$, the work done by W, in Eq. (a). Here g is the gravitational acceleration. By so doing, the maximum dynamic load and deflection are found to be, respectively,

$$P_{dyn} = W\sqrt{\frac{v^2}{g\delta_{st}}}, \qquad \delta_{max} = \delta_{st}\sqrt{\frac{v^2}{g\delta_{st}}} \tag{4.21}$$

where δ_{st} is the static deflection caused by a horizontal force W.

Example 4.7. A weight $W = 180$ N is dropped a height $h = 0.1$ m, striking at midspan a simply supported beam of length $L = 1.16$ m. The beam is of rectangular cross section: 0.025 m width and 0.075 m depth. For a material with modulus of elasticity $E = 200$ GPa, determine the instantaneous maximum deflection and maximum stress for the following cases: (a) the beam is rigidly supported, (b) the beam is supported at each end by springs of stiffness $k = 180$ kN/m.

SOLUTION. The deflection of a point at midspan, owing to a statically applied load, is

$$\delta_{st} = \frac{WL^3}{48EI} = \frac{180(1.16)^3(12)}{48(200 \times 10^9)(0.025)(0.075)^3} = 0.033 \times 10^{-3} \text{ m}$$

The maximum static stress, also occuring at midspan, is calculated from

$$\sigma_{st,max} = \frac{Mc}{I} = \frac{180(1.16)(0.0375)(12)}{4(0.025)(0.075)^3} = 2.23 \text{ MPa}$$

(a) The impact factor is, from Eq. (4.14),

$$1 + \sqrt{1 + \frac{2(0.1)}{0.033 \times 10^{-3}}} = 78.86$$

We thus have

$$\delta_{max} = (0.033 \times 10^{-3})(78.86) = 2.602 \times 10^{-3} \text{ m}$$
$$\sigma_{max} = 2.23 \times 78.86 = 175.86 \text{ MPa}$$

(b) The static deflection of the beam due to its own bending and the deformation of the spring is

$$\delta_{st} = 0.033 \times 10^{-3} + \frac{90}{180,000} = 0.533 \times 10^{-3} \text{ m}$$

The impact factor is thus

$$1 + \sqrt{1 + \frac{2(0.1)}{0.533 \times 10^{-3}}} = 20.40$$

Hence,

$$\delta_{max} = (0.533 \times 10^{-3})(20.40) = 0.011 \text{ m}$$
$$\sigma_{max} = 2.23 \times 20.40 = 45.49 \text{ MPa}$$

It is observed from a comparison of the results that dynamic loading increases the value of deflection and stress considerably. Also noted is a reduction in stress with increased flexibility attributable to the springs added to the supports. The values calculated for the dynamic stress are probably somewhat high, as $h \gg \delta_{st}$ in both cases.

4.16 Dynamic and Thermal Effects

We now explore the conditions under which metals may manifest a change from ductile to brittle behavior, and vice versa. The matter of ductile–brittle transition has important application where the operating environment includes a wide variation in temperature or where the rate of loading changes.

Let us, to begin with, identify two tensile stresses. The first, σ_f, leads to brittle fracture, i.e., failure by cleavage or separation. The second, σ_y, corresponds to failure by yielding or permanent deformation. These stresses are shown in Fig. 4.11a as functions of material temperature. Referring to the figure, the point of intersection of the two stress curves defines the critical temperature, T_{cr}. If, at a given temperature above T_{cr}, the stress is progressively increased, failure will occur by yielding and the fracture curve will never be encountered. Similarly, for a test conducted at $T < T_{cr}$, the yield curve is not intercepted, inasmuch as failure occurs by fracture. The principal factors governing whether failure will occur by fracture or yielding are summarized below:

Temperature: If the temperature of the specimen exceeds T_{cr}, resistance to yielding is less than resistance to fracture ($\sigma_y < \sigma_f$), and the specimen yields. If the temperature is less than T_{cr}, then $\sigma_f < \sigma_{cr}$, and the specimen fractures without yielding. Note that σ_f exhibits only a small decrease with increasing temperature.

Loading Rate: Increasing the rate at which the load is applied increases a material's ability to resist yielding while leaving comparatively unaffected its resistance to fracture. The increased loading rate thus results in a shift to the position occupied by the dashed curve. Point C moves to C', meaning that accompanying the increasing loading rate, an increase occurs in the critical temperature. In impact tests, brittle fractures are thus observed to occur at higher temperatures than in static tests.

Triaxiality: The effect on the transition of a three-dimensional stress condition or *triaxiality* is similar to that of loading rate. This phenomenon may be illustrated by comparing the tendency to yield in a uniform cylindrical tensile specimen with that of a specimen containing a circumferential groove. The unstressed region above and below the groove tends to resist the deformation associated with the tensile loading of the central region, therefore contributing to a radial stress field in addition to the longitudinal stress. This state of triaxial stress is thus indicative of a tendency to resist yielding (become less ductile), the material behaving in a more brittle fashion.

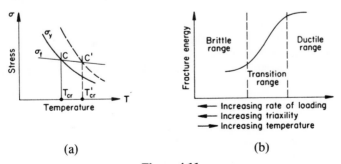

(a)   (b)

Figure 4.11

Referring once more to Fig. 4.11a, in the region to the right of T_{cr} the material behaves in a ductile manner, while to the left of T_{cr} it is brittle. At temperatures close to T_{cr}, the material generally exhibits some yielding prior to a partially brittle fracture. The width of the temperature range over which the transition from brittle to ductile failure occurs is material dependent.

Transition phenomena may also be examined from the viewpoint of the energy required to fracture the material, the *toughness* rather than the stress (Fig. 4.11b). Notches and grooves serve to reduce the energy required to cause fracture and to shift the transition temperature, normally very low, to the range of normal temperatures. This is one reason that experiments are normally performed on notched specimens.

Chapter 4—Problems

Secs. 4.1 to 4.11

4.1. A steel circular cylindrical bar of 0.1-m diameter is subject to compound bending and tension at its ends. The material yield strength is 221 MPa. Assume failure to occur by yielding and take the value of the applied moment to be $M = 17$ kN·m. Determine, using the octahedral shear stress theory, the limiting value of P which can be applied to the bar without causing permanent deformation.

4.2. At a point in a structural member, yielding occured under a state of stress given by

$$\begin{bmatrix} 0 & 4 & 0 \\ 40 & 50 & -60 \\ 0 & -6 & 0 \end{bmatrix} \text{MPa}$$

Determine the uniaxial tensile yield strength of the material according to (a) maximum principal strain theory, (b) maximum shear stress theory, and (c) octahedral shear stress theory.

4.3. A circular shaft of 0.12-m diameter is subjected to end loads $P = 45$ kN, $M = 4$ kN·m, and $M_t = 11.2$ kN·m. Let $\sigma_{yp} = 280$ MPa. What is the factor of safety, assuming failure to occur in accordance with the octahedral shear stress theory?

4.4. Determine the width t of the cantilever of height $2t$ and length 0.25 m subjected to a 450-N concentrated force at its free end. Apply the maximum energy of distortion theory. The tensile and compressive strengths of the material are both 280 MPa.

4.5. Determine the required diameter of a steel transmission shaft 10 m in length and of yield strength 350 MPa in order to resist a torque of up to 500 N·m. The shaft is supported by frictionless bearings at its ends. Design the shaft according to the maximum shear stress theory, selecting a factor of safety of

1.5: (a) neglecting the shaft weight and (b) including the effect of shaft weight. Use $\gamma = 77$ kN/m^3 as the weight per unit volume of steel (Table 2.1).

4.6. A simply supported nonmetallic beam of 0.25-m height, 0.1-m width, and 1.5-m span is subjected to a uniform loading of 6 kN/m. Determine the factor of safety for this loading according to (a) the maximum distortion energy theory with $\sigma_{yp} = 28$ MPa and (b) the maximum principal strain theory with $\varepsilon_{yp} = 1/3{,}000$. Take $E = 14$ GPa and $\nu = 0.3$.

4.7. The state of stress at a point in a machine element of irregular shape, subjected to combined loading, is given by

$$\begin{bmatrix} 3 & 4 & 6 \\ 4 & 2 & 5 \\ 6 & 5 & 1 \end{bmatrix} \text{MPa}$$

A torsion test performed on a specimen made of the same material shows that yielding occurs at a shearing stress of 9 MPa. Assuming the same ratios are maintained between the stress components, predict the values of the normal stresses σ_y and σ_x at which yielding occurs at the point. Use (a) maximum distortion energy theory, (b) Mohr's theory, (c) maximum principal stress theory, and (d) maximum principal strain theory.

4.8. A thin-walled cylindrical pressure vessel of diameter $d = 0.5$ m and wall thickness $t = 5$ mm is fabricated of a material with 280-MPa tensile yield strength and 350-MPa compressive yield strength. Determine the internal pressure p required according to the following theories of failure: (a) Coulomb–Mohr, (b) maximum distortion energy, and (c) maximum shear stress. *Note:* The maximum stress in the tangential direction is given by $\sigma_\theta = pd/2t$ (refer to Sec. 8.2).

4.9. The state of stress at a point is given by

$$\begin{bmatrix} 2.8 & 7 & 14 \\ 7 & 21 & 35 \\ 14 & 35 & 28 \end{bmatrix} \text{MPa}$$

Taking $\sigma_{yp} = 82$ MPa, $\nu = 0.3$, and a factor of safety of 1.2, determine whether failure takes place at the point, using (a) the maximum principal strain theory and (b) the maximum distortion energy theory.

Secs. 4.12 to 4.16

4.10. Simple tension and compression tests on a ductile material reveal that failure occurs by yielding at $\sigma'_{yp} = 260$ MPa and $\sigma''_{yp} = 420$ MPa, respectively. In an actual application, the material is subjected to perpendicular tensile and compressive stresses, σ_1 and σ_2, respectively, such that $\sigma_1/\sigma_2 = -\frac{1}{4}$. Determine the limiting values of σ_1 and σ_2 according to (a) the Mohr theory for a yield stress in torsion of $\tau_{yp} = 175$ MPa and (b) the Coulomb–Mohr theory. [Hint: For case (a), the circle representing the given loading is drawn by a trial-and-error procedure.]

4.11. A thin-walled tube having internal and external diameters of 0.25 m and 0.26 m is subjected to an internal pressure $p_i = 2.8$ MPa, a twisting moment of 31.36 kN·m, and an axial end thrust (tension) $P = 45$ kN. The ultimate strengths in tension and compression are 210 MPa and 500 MPa, respectively. Apply the following theories to evaluate the ability of the tube to resist failure by fracture: (a) maximum shear stress, (b) Coulomb–Mohr, and (c) maximum principal stress.

4.12. The ultimate strengths in tension and compression of a material are 420 MPa and 900 MPa, respectively. If the stress at a point within a member made of this material is

$$\begin{bmatrix} 200 & 150 \\ 150 & 20 \end{bmatrix} \text{MPa}$$

determine the factor of safety according to the following theories of failure: (a) maximum shear stress and (b) Coulomb–Mohr.

4.13. A plate, t meters thick, is fabricated of a material having ultimate strengths in tension and compression of σ_u' and σ_u'' pascals, respectively. Calculate the force P required to punch a hole of d meters in diameter through the plate, Fig. P4.13. Employ (a) the maximum shear stress theory and (b) the Mohr–Coulomb theory. Assume that the shear force is uniformly distributed through the thickness of the plate.

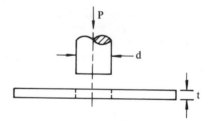

Figure P4.13

4.14. Redo Example 4.5 employing the maximum shear stress theory together with the Soderberg criterion.

4.15. A bolt is subjected to an alternating axial load of maximum and minimum values F_{max} and F_{min}. The static tensile yield and fatigue strength for completely reversed stress of the material are σ_{yp} and σ_{cr}. Verify that, according to the modified Goodman relation, the expression

$$A = \frac{1}{\sigma_{cr}} \left[F_{\text{max}} - \tfrac{1}{2}(F_{\text{max}} + F_{\text{min}})\left(1 - \frac{\sigma_{cr}}{\sigma_{yp}}\right) \right] \tag{P4.15}$$

represents the required cross-sectional area of the bolt.

4.16. A circular rotating shaft is subjected to a static bending moment M and a torque that varies from a value of zero to M_t. Apply the energy of distortion theory together with Soderberg's relation to obtain the following expression for the required shaft radius:

$$r = \left[\frac{M_t}{\pi \sigma_{cr}} \left(\frac{\sigma_{cr}}{\sigma_{yp}} \sqrt{16\left(\frac{M}{M_t}\right)^2 + 3} + \sqrt{3} \right) \right]^{1/3} \tag{P4.16}$$

4.17. Compute the fatigue life of the rotating hub and shaft assembly described in Example 4.6 if at a critical point in the shaft the state of stress is described by $\sigma_{x,max} = 1000$ MPa, $\sigma_{x,min} = -800$ MPa, $\tau_{xy,max} = 300$ MPa, $\tau_{xy,min} = -100$ MPa, and $\sigma_y = \sigma_z = \tau_{xz} = \tau_{yz} = 0$. Employ the maximum shear stress theory of failure together with the four criteria given in Table 4.2. Take $\sigma_{yp} = 1600$ MPa, $\sigma_u = 2400$ MPa, and $K = 1$.

4.18. A weight W is dropped from a height $h = 0.75$ m onto the free end of a cantilever beam of length $L = 1.2$ m. The beam is of 50 mm by 50 mm square cross section. Determine the value of W required to result in yielding. Omit the weight of the beam. Let $\sigma_{yp} = 280$ MPa and $E = 200$ GPa.

4.19. A 0.125-m diameter and 1.5-m long circular shaft has a flywheel at one end and rotates at 240 rpm. The shaft is suddenly stopped at the free end. Determine the maximum shear stress and the maximum angle of twist produced by the impact. It is given that the shaft is made of steel with $G = 80.5$ GPa, $\nu = 0.3$, the weight of the flywheel is 1.09 kN, and the flywheel's radius of gyration is 0.35 m.

Chapter 5

Bending of Beams

5.1 Introduction

We are here concerned with the bending of straight as well as curved *beams*, i.e., structural elements possessing one dimension significantly greater than the other two, usually loaded in a direction normal to the longitudinal axis.

Except in the case of very simple shapes and loading systems, the theory of elasticity yields beam solutions only with considerable difficulty. Practical considerations often lead to assumptions with regard to stress and deformation which results in mechanics of materials or elementary theory solutions. The theory of elasticity can sometimes be applied to test the validity of such assumptions. The role of the theory of elasticity is then threefold. It can serve to place limitations on the use of the elementary theory, it can be used as the basis of approximate solutions through numerical analysis, and it can provide exact solutions where configurations of loading and shape are simple.

PART 1—Exact Solutions

5.2 Pure Bending of Beams of Symmetrical Cross Section

The simplest case of bending is that of a beam possessing a vertical axis of symmetry, subjected to equal and opposite end couples (Fig. 5.1a). The semi-inverse method is now applied to analyze this problem. We shall assume that the normal stress over the cross section varies linearly with y and that the remaining stress components are zero:

$$\sigma_x = ky, \qquad \sigma_y = \sigma_z = \tau_{xy} = \tau_{xz} = \tau_{yz} = 0 \qquad (5.1)$$

Here k is a constant, and $y = 0$ contains the *neutral* surface, i.e., the surface along which $\sigma_x = 0$. The intersection of the neutral surface and the cross section locates the neutral axis. In Fig. 5.1b is shown the linear stress field in a section located an arbitrary distance a from the left end.

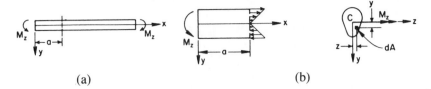

Figure 5.1

Since Eqs. (5.1) indicate that the lateral surfaces are free of stress, we need only be assured that the stresses are consistent with the boundary conditions at the ends. These conditions require that the resultant of the internal forces be zero and that the moments of the internal forces about the neutral axis equal the applied moment M_z:

$$\int_A \sigma_x \, dA = 0, \qquad -\int_A y\sigma_x \, dA = M_z \qquad (5.2)$$

where A is the cross-sectional area. It should be noted that the zero stress components τ_{xy}, τ_{xz} in Eqs. (5.1) satisfy the conditions that no y- and z-directed forces exist at the end faces, and because of the y symmetry of the section, $\sigma_x = ky$ produces no moment about the y axis. The negative sign in the second expression is consistent with the following *sign convention*: a *positive* moment M_z is one which results in *compressive* (negative) stress at points of *positive y*. Substitution of Eqs. (5.1) into Eqs. (5.2) yields

$$k\int_A y \, dA = 0, \qquad -k\int_A y^2 \, dA = M_z \qquad (5.3a,b)$$

Inasmuch as $k \neq 0$, Eq. (5.3a) indicates that the first moment of cross-sectional area about the neutral axis is zero. This requires that *the neutral and centroidal axes of the cross section coincide*. Neglecting body forces, it is clear that the equations of equilibrium (3.4), are satisfied by Eqs. (5.1). It may readily be verified also that the equations (5.1) together with Hooke's law fulfill the compatibility conditions, Eq. (2.8). Thus, Eqs. (5.1) represent an exact solution.

The integral in Eq. (5.3b) defines the moment of inertia I_z of the cross section about the z axis of the beam cross section; therefore

$$k = -\frac{M_z}{I_z} \qquad (a)$$

An expression for normal stress can now be written by combining Eqs. (5.1) and (a):

$$\sigma_x = -\frac{M_z y}{I_z} \qquad (5.4)$$

This is the familiar *flexure formula* applicable to straight beams.

Since, at a given section, M and I are constant, the maximum stress is obtained from Eq. (5.4) by taking $|y|_{max} = c$:

$$\sigma = \frac{Mc}{I} = \frac{M}{I/c} = \frac{M}{Z} \qquad (5.5)$$

Here Z is the *elastic section modulus*. Formula (5.5) is widely employed in practice because of its simplicity. To facilitate its use, section moduli for numerous common sections are tabulated in various handbooks. The stress in extreme fibers, computed from Eq. (5.5) for experimentally obtained *ultimate* bending moment (Sec. 12.5), is termed the *modulus of rupture* of the material in bending. This, $\sigma = M_u/Z$, is frequently used as *a measure of the bending strength* of materials.

Kinematic Relationships

In order to gain further insight into the beam problem, consideration is now given to the geometry of deformation, i.e., beam kinematics. Funda-mental to this discussion is the hypothesis that sections originally plane remain so subsequent to bending. For a beam of symmetrical cross section, Hooke's law and Eq. (5.4) lead to

$$\varepsilon_x = -\frac{M_z y}{EI_z}, \qquad \varepsilon_y = \varepsilon_z = \nu \frac{M_z y}{EI_z}$$

$$\gamma_{xy} = \gamma_{xz} = \gamma_{yz} = 0 \qquad (5.6)$$

where EI_z is the *flexural rigidity*.

Let us examine the deflection of the neutral axis, the axial deformation of which is zero. In Fig. 5.2a is shown an element of an initially straight beam, now in a deformed state. Because the beam is subjected to pure bending, uniform throughout, each element of infinitesimal length experi-ences identical deformation, with the result that the beam curvature is everywhere the same. The deflected axis of the beam or the deflection curve is thus shown deformed, with radius of curvature r_z. The curvature of the beam axis in the xy plane in terms of the y deflection v is

$$\frac{1}{r_z} = \frac{d^2v/dx^2}{\left[1 + (dv/dx)^2\right]^{3/2}} \approx \frac{d^2v}{dx^2} \qquad (5.7)$$

where the approximate form is valid for small deformations ($dv/dx \ll 1$). The sign convention for curvature of the beam axis is such that it is *positive* when the beam is bent concave *downward* as shown in the figure. From the geometry of Fig. 5.2b, the shaded sectors are similar. Hence, the radius of curvature and the strain are related as follows:

$$d\theta = \frac{ds}{r_z} = -\frac{\varepsilon_x \, ds}{y} \qquad (5.8)$$

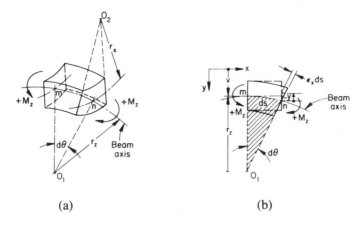

(a)   (b)

Figure 5.2

Here ds is the arc length mn along the longitudinal axis of the beam. For small displacement, $ds \approx dx$, and θ represents the *slope*, dv/dx, of the beam axis. Clearly, for the positive curvature shown, θ increases as we move from left to right along the beam axis. On the basis of the above equation and Eqs. (5.6),

$$\frac{1}{r_z} = -\frac{\varepsilon_x}{y} = \frac{M_z}{EI_z} \tag{5.9a}$$

Following a similar procedure and noting that $\varepsilon_z = -\nu\varepsilon_x$, we may also obtain the curvature in the yz plane as

$$\frac{1}{r_x} = -\frac{\varepsilon_z}{y} = -\frac{\nu M_z}{EI_z} \tag{5.9b}$$

The basic beam equation is obtained by combining Eqs. (5.7) and (5.9a) as follows:

$$\frac{1}{r_z} = \frac{d^2v}{dx^2} = \frac{M_z}{EI_z} \tag{5.10}$$

This expression, relating the beam curvature to the bending moment, is known as the Bernoulli–Euler law of elementary bending theory. It is observed from Fig. 5.2 and Eq. (5.10) that a positive moment produces positive curvature. If the sign convention adopted in this section for either moment or deflection (and curvature) should be reversed, the plus sign in Eq. (5.10) should likewise be reversed.

Reference to Fig. 5.2a reveals that the top and bottom lateral surfaces have been deformed into saddle shaped or *anticlastic* surfaces of curvature $1/r_x$. The vertical sides have been simultaneously rotated as a result of bending. Examining Eq. (5.9b) suggests a method for determining Poisson's ratio. For a given beam and bending moment, a measurement of $1/r_x$ leads directly to ν. The effect of anticlastic curvature is small when the beam depth is comparable to its width.

5.3 Pure Bending of Beams of Asymmetrical Cross Section

The development of the previous section is now extended to the more general case in which a beam of arbitrary cross section is subjected to end couples M_y and M_z about the y and z axes, respectively (Fig. 5.3). Following a procedure similar to that of Sec. 5.2, plane sections are again taken to remain plane. Assume that the normal stress σ_x acting at a point within dA is a linear function of the y and z coordinates of the point; assume further that the remaining stresses are zero. The stress field is thus

$$\sigma_x = c_1 + c_2 y + c_3 z$$
$$\sigma_y = \sigma_z = \tau_{xy} = \tau_{xz} = \tau_{yz} = 0 \tag{5.11}$$

where c_1, c_2, c_3 are constants to be evaluated.

The conditions at the beam ends, as before, relate to the force and bending moment:

$$\int_A \sigma_x \, dA = 0 \tag{a}$$

$$\int_A z\sigma_x \, dA = M_y, \qquad -\int_A y\sigma_x \, dA = M_z \tag{b,c}$$

Substitution of σ_x, as given by Eq. (5.11), into Eqs. (a), (b), and (c) results in the following expressions:

$$c_1 \int_A dA + c_2 \int_A y \, dA + c_3 \int_A z \, dA = 0 \tag{d}$$

$$c_1 \int_A z \, dA + c_2 \int_A yz \, dA + c_3 \int_A z^2 \, dA = M_y \tag{e}$$

$$c_1 \int_A y \, dA + c_2 \int_A y^2 \, dA + c_3 \int_A yz \, dA = -M_z \tag{f}$$

For the origin of the y and z axes to be coincident with the centroid of the section, it is required that

$$\int_A y \, dA = \int_A z \, dA = 0 \tag{g}$$

We conclude, therefore, from Eq. (d) that $c_1 = 0$, and from Eqs. (5.11) that $\sigma_x = 0$ at the origin. The neutral axis is thus observed to pass through the centroid, as in the beam of symmetrical section. It may be verified that the field of stress described by Eqs. (5.11) satisfies the equations of equilibrium and compatibility, and that the lateral surfaces are free of stress. Now

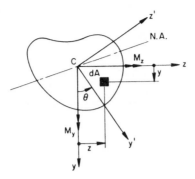

Figure 5.3

consider the defining relationships

$$I_y = \int_A z^2 \, dA, \qquad I_z = \int_A y^2 \, dA, \qquad I_{yz} = \int_A yz \, dA \qquad (5.12)$$

where I_y and I_z are the moments of inertia about the y and z axes, respectively, and I_{yz} is the product of inertia about the y and z axes. From Eqs. (e) and (f), together with Eqs. (5.12), we obtain expressions for c_2 and c_3. Substitution of the constants into Eqs. (5.11) results in the following *generalized flexure formula*:

$$\sigma_x = \frac{(M_y I_z + M_z I_{yz})z - (M_y I_{yz} + M_z I_y)y}{I_y I_z - I_{yz}^2} \qquad (5.13)$$

The equation of the neutral axis is found by equating the above expression to zero:

$$(M_y I_z + M_z I_{yz})z - (M_y I_{yz} + M_z I_y)y = 0 \qquad (5.14)$$

The result indicates a straight line.

There is a specific orientation of the y, z axes for which the product of inertia I_{yz} vanishes. Labeling the axes so oriented y', z', we have $I_{y'z'} = 0$. The flexure formula under these circumstances becomes

$$\sigma_x = \frac{M_{y'} z'}{I_{y'}} - \frac{M_{z'} y'}{I_{z'}} \qquad (5.15)$$

The y', z' axes now coincide with the *principal* axes of inertia of the cross section. The stresses at any point can now be ascertained by applying Eq. (5.13) or (5.15).

The kinematic relationships discussed in Sec. 5.2 are valid for beams of asymmetrical section provided that y and z represent principal axes.

Recall that the two-dimensional stress (or strain) and the moment of inertia of an area are second-order tensors (Appendix A). Thus, *the* *transformation equations for stress and moment of inertia are analogous.* The Mohr's circle analysis and all conclusions drawn for stress therefore apply to the moment of inertia. It can readily be shown* that by replacing $\sigma_x, \sigma_y, \tau_{xy}, \sigma_{x'}$, and $\tau_{x'y'}$ by $I_x, I_y, -I_{xy}, I_{x'}$, and $I_{x'y'}$, respectively, in the appropriate expressions of Chapter 1, relationships for the moment of inertia will be obtained. With reference to the coordinate axes shown in Fig. 5.3, applying Eq. (1.7a), the moment of inertia about the y' axis is found to be

$$I_{y'} = \frac{I_y + I_z}{2} + \frac{I_y - I_z}{2} \cos 2\theta - I_{yz} \sin 2\theta \qquad (5.16)$$

From Eq. (1.8) the orientation of the principal axes is given by

$$\tan 2\theta_p = -\frac{2 I_{yz}}{I_y - I_z} \qquad (5.17)$$

The principal moment of inertia, I_1 and I_2, from Eq. (1.9) are

$$I_{1,2} = \frac{I_y + I_z}{2} \pm \sqrt{\left(\frac{I_y - I_z}{2} \right)^2 + I_{yz}^2} \qquad (5.18)$$

Subscripts 1 and 2 refer to the maximum and minimum values, respectively.

Determination of the moments of inertia and stresses in an asymmetrical section will now be illustrated.

Example 5.1. A 0.15 by 0.15 m slender angle of 0.02 m thickness is subjected to oppositely directed end couples $M_z = 11,000$ N·m, at the centroid of the cross section. What bending stresses exist at points A and B on a section away from the ends (Fig. 5.4a)? Determine the orientation of the neutral axis.

SOLUTION. Equation (5.15) will be applied to ascertain the normal stress. This requires first the determination of a number of section properties, through the use of familiar expressions of mechanics.

Location of the centroid C: Let $\bar{y}$ and $\bar{z}$ represent the distances from C to arbitrary reference lines (denoted Z and Y):

$$\bar{z} = \frac{\Sigma A_i z_i}{\Sigma A_i} = \frac{A_1 z_1 + A_2 z_2}{A_1 + A_2} = \frac{0.13 \times 0.02 \times 0.01 + 0.15 \times 0.02 \times 0.075}{0.13 \times 0.02 + 0.15 \times 0.02} = 0.045 \text{ m}$$

Here z_i represents the z distance from the Y reference line to the centroid of each subarea composing the total cross section. Since the section is symmetrical, $\bar{z} = \bar{y}$.

---

*See, for example, I. Shames, *Engineering Mechanics*, Englewood Cliffs, NJ: Prentice-Hall, 1980, Chapters 8 and 9.

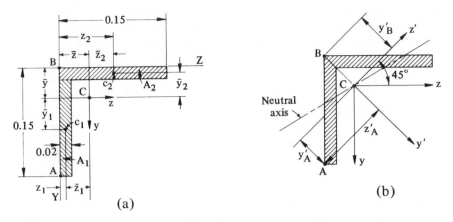

Figure 5.4

Calculation of the moments and products of inertia: Applying the parallel axis theorem,

$$I_z = \int_A y^2\, dA + A_1 \bar{y}_1^2 + \int_A y^2\, dA + A_2 \bar{y}_2^2$$

where I_z represents the moment of inertia of the entire cross section about the z axis, and $\bar{y}_1$ and $\bar{y}_2$ are the distances from the centroid of each subarea (A_1 and A_2) to C. The integrals define the area moments of inertia of A_1 and A_2 about the centroid of each subarea. For a rectangular section of depth h and width b, the moment of inertia about the neutral z axis is $I_z = bh^3/12$. Referring to Fig. 5.4a, we thus write

$$I_z = I_y = \tfrac{1}{12} \times 0.02 \times (0.13)^3 + 0.13 \times 0.02 \times (0.04)^2$$

$$+ \tfrac{1}{12} \times 0.15 \times (0.02)^3 + 0.15 \times 0.02 \times (0.035)^2$$

$$= 11.596 \times 10^{-6}\ \mathrm{m}^4.$$

The parallel axis theorem for a product of inertia yields

$$I_{yz} = \int_{A_1} yz\, dA + A_1 \bar{y}_1 \bar{z}_1 + \int_{A_2} yz\, dA + A_2 \bar{y}_2 \bar{z}_2$$

$$= 0 + 0.13 \times 0.02 \times 0.04 \times (-0.035) + 0 + 0.15 \times 0.02 \times (-0.035) \times 0.03$$

$$= -6.79 \times 10^{-6}\ \mathrm{m}^4$$

Note that at this stage, the normal stresses may be calculated by direct application of Eq. (5.13).

Determination of the directions of the principal axes and calculation of the principal moments of inertia: Employing Eq. (5.17), we have

$$\tan 2\theta_p = \frac{-2(-6.79)}{11.596 - 11.596} = \infty, \qquad 2\theta_p = 90° \text{ and } 270°$$

Therefore the two values of θ_p are $45°$ and $135°$. Substituting the first of these values into Eq. (5.16), we obtain $I_{y'} = [11.596 + 6.79\sin 90°]10^{-6} = 18.386 \times 10^{-6}$ m^4. Since the principal moments of inertia are, by application of Eq. (5.18),

$$I_{1,2} = \left[11.596 \pm \sqrt{0 + 6.79^2} \right]10^{-6} = [11.596 \pm 6.79]10^{-6}$$

it is observed that $I_1 = I_{y'} = 18.386 \times 10^{-6}$ m^4 and $I_2 = I_{z'} = 4.806 \times 10^{-6}$ m^4.

Calculation of stress: The components of bending moment about the principal axes are

$$M_{y'} = 11{,}000\sin 45° = 7778 \text{ N} \cdot \text{m}$$

$$M_{z'} = 11{,}000\cos 45° = 7778 \text{ N} \cdot \text{m}$$

Equation (5.15) is now applied, referring to Fig. 5.4b, with $y'_A = 0.043$ m, $z'_A = -0.106$ m, $y'_B = -0.063$ m, and $z'_B = 0$, determined from geometrical considerations:

$$(\sigma_x)_A = \frac{7778(-0.106)}{18.386 \times 10^{-6}} - \frac{7778(0.043)}{4.806 \times 10^{-6}} = -144.43 \text{ MPa}$$

$$(\sigma_x)_B = 0 - \frac{7778(-0.063)}{4.806 \times 10^{-6}} = 102 \text{ MPa}$$

Determination of the direction of the neutral axis: From Eq. (5.14), with $M_y = 0$,

$$(M_z I_y)y - (M_z I_{yz})z = 0, \qquad \text{from which} \quad 11.596y + 6.79z = 0$$

or

$$z = -1.71y$$

This result is plotted in Fig. 5.4b.

5.4 Bending of a Cantilever of Narrow Section

Consider a narrow cantilever beam of rectangular cross section, loaded at its free end by a concentrated force of such magnitude that the beam weight may be neglected (Fig. 5.5). The situation described may be regarded as a case of plane stress provided that the beam thickness t is small relative to beam depth $2h$. The distribution of stress in the beam, as we have already found in Example 3.1, is given by

$$\sigma_x = -\left(\frac{Px}{I}\right)y, \qquad \sigma_y = 0, \qquad \tau_{xy} = -\frac{P}{2I}(h^2 - y^2) \qquad (5.19)$$

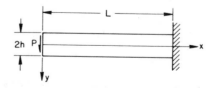

Figure 5.5

In order to derive expressions for the beam displacement, it is necessary to relate stress, described by Eq. (5.19), to strain. This is accomplished through the use of the strain-displacement relations and Hooke's law:

$$\frac{\partial u}{\partial x} = -\frac{Pxy}{EI}, \qquad \frac{\partial v}{\partial y} = \frac{\nu Pxy}{EI} \tag{a,b}$$

$$\frac{\partial u}{\partial y} + \frac{\partial v}{\partial x} = \frac{2(1+\nu)\tau_{xy}}{E} = -\frac{(1+\nu)P}{EI}(h^2-y^2) \tag{c}$$

Integration of Eqs. (a) and (b) yields

$$u = -\frac{Px^2y}{2EI} + u_1(y) \tag{d}$$

$$v = \frac{\nu Pxy^2}{2EI} + v_1(x) \tag{e}$$

Differentiating Eqs. (d) and (e) with respect to y and x, respectively, and substituting into Eq. (c), we have

$$\frac{du_1}{dy} - \frac{P}{2PEI}(2+\nu)y^2 = -\frac{dv_1}{dx} + \frac{P}{2EI}x^2 - \frac{(1+\nu)Ph^2}{EI}$$

In the above expression note that the left and right hand sides depend only upon y and x, respectively. These variables are independent of one another, and it is therefore concluded that the equation can be valid only if each side is equal to the same constant:

$$\frac{du_1}{dy} - \frac{P}{2EI}(2+\nu)y^2 = a_1, \qquad \frac{dv_1}{dx} - \frac{Px^2}{2EI} + \frac{(1+\nu)Ph^2}{EI} = -a_1$$

These are integrated to yield

$$u_1(y) = \frac{P}{6EI}(2+\nu)y^3 + a_1 y + a_2$$

$$v_1(x) = \frac{Px^3}{6EI} - \frac{(1+\nu)Pxh^2}{EI} - a_1 x + a_3$$

in which a_2 and a_3 are constants of integration. The displacements may now be written

$$u = -\frac{Px^2y}{2EI} + \frac{P}{6EI}(2+\nu)y^3 + a_1 y + a_2 \tag{5.20}$$

$$v = \frac{\nu Pxy^2}{2EI} + \frac{Px^3}{6EI} - \frac{(1+\nu)Pxh^2}{EI} - a_1 x + a_3 \tag{5.21}$$

The constants a_1, a_2, and a_3 depend upon known conditions. If, for example, the situation at the fixed end is such that

$$\frac{\partial u}{\partial y} = 0, \qquad v = u = 0 \qquad (x=L, y=0)$$

then, from Eqs. (5.20) and (5.21),

$$a_1 = \frac{PL^2}{2EI}, \qquad a_2 = 0, \qquad a_3 = \frac{PL^3}{3EI} + \frac{PLh^2(1+\nu)}{EI}$$

The beam displacement is therefore

$$u = \frac{P}{2EI}(L^2 - x^2)y + \frac{(2+\nu)Py^3}{6EI} \tag{5.22}$$

$$v = \frac{P}{EI}\left[\frac{x^3}{6} + \frac{L^3}{3} + \frac{x}{2}(\nu y^2 - L^2) + h^2(1+\nu)(L-x) \right] \tag{5.23}$$

It is clear upon examining the above equations that u and v do not obey a simple linear relationship with y and x. We conclude, therefore, that plane sections do not, as assumed in elementary theory, remain plane subsequent to bending.

The vertical displacement of the beam axis is obtained by substituting $y = 0$ into Eq. (5.23):

$$(v)_{y=0} = \frac{Px^3}{6EI} - \frac{PL^2x}{2EI} + \frac{PL^3}{3EI} + \frac{Ph^2(1+\nu)}{EI}(L-x) \tag{5.24}$$

It is now a simple matter to compare the total vertical deflection at the free end $(x=0)$ with the deflection derived in elementary theory. Substituting $x=0$ into Eq. (5.24), the total deflection is

$$(v)_{x=y=0} = \frac{PL^3}{3EI} + \frac{Ph^2(1+\nu)L}{EI} = \frac{PL^3}{3EI} + \frac{Ph^2L}{2GI} \tag{5.25}$$

wherein the deflection associated with shear is clearly $Ph^2L/2GI = 3PL/2GA$. The ratio of the shear deflection to the bending deflection at $x=0$ provides a measure of beam slenderness:

$$\frac{Ph^2L/2GI}{PL^3/3EI} = \frac{3}{2}\frac{h^2E}{L^2G} = \frac{3}{4}(1+\nu)\left(\frac{2h}{L}\right)^2 \approx \left(\frac{2h}{L}\right)^2$$

If, for example, $L = 10 (2h)$, the above quotient is only $\frac{1}{100}$. For a slender beam, $2h \ll L$, and it is clear that the deflection is mainly due to bending. It should be mentioned here, however, that in vibration at higher modes, and in wave propagation, the effect of shear is of great importance in slender as well as in other beams.

In the case of the *wide beams* $(t \gg 2h)$, Eq. (5.25) must be modified by replacing E and ν as indicated in Table 3.1.

5.5 Bending of a Simply Supported, Narrow Beam

Consideration is now given the stress distribution in a narrow beam of thickness t and depth $2h$ subjected to a uniformly distributed loading (Fig. 5.6). The situation as described is one of plane stress, subject to the following boundary conditions, consistent with the origin of an x, y

coordinate system located at midspan and midheight of the beam, as shown:

$$(\tau_{xy})_{y=\pm h}=0, \qquad (\sigma_y)_{y=+h}=0, \qquad (\sigma_y)_{y=-h}=-p/t \qquad \text{(a)}$$

Since at the ends no longitudinal load is applied, it would appear reasonable to state that $\sigma_x=0$ at $x=\pm L$. However, this boundary condition leads to a complicated solution, and a less severe statement is instead used:

$$\int_{-h}^{h} \sigma_x t\, dy = 0 \qquad \text{(b)}$$

The corresponding conditior for bending couples at $x=\pm L$ is

$$\int_{-h}^{h} \sigma_x ty\, dy = 0 \qquad \text{(c)}$$

For y equilibrium, it is required that

$$\int_{-h}^{h} \tau_{xy} t\, dy = \pm pL \qquad (x=\pm L) \qquad \text{(d)}$$

The problem is treated by superimposing the solutions ϕ_2, ϕ_3, and ϕ_5 (Sec. 3.5), with

$$c_2=b_2=a_3=c_3=a_5=b_5=c_5=e_5=0$$

We then have

$$\phi=\phi_2+\phi_3+\phi_5=\frac{a_2}{2}x^2+\frac{b_3}{2}x^2y+\frac{d_3}{6}y^3+\frac{d_5}{6}x^2y^3-\frac{2d_5}{60}y^5$$

The stresses are

$$\sigma_x=d_3y+d_5\left(x^2y-\tfrac{2}{3}y^3\right)$$

$$\sigma_y=a_2+b_3y+\frac{d_5}{3}y^3 \qquad \text{(e)}$$

$$\tau_{xy}=-b_3x-d_5xy^2$$

The conditions (a) are

$$-b_3-d_5h^2=0$$

$$a_2+b_3h+\frac{d_5}{3}h^3=0$$

$$a_2-b_3-\frac{d_5}{3}h^3=-\frac{p}{t}$$

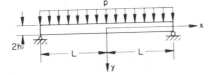

Figure 5.6

and the solution is

$$a_2 = -p/2t, \qquad b_3 = 3p/4th, \qquad d_5 = -3p/4th^3$$

The constant d_3 is obtained from condition (c) as follows

$$\int_{-h}^{h} \left[d_3 y + d_5 \left(L^2 y - \frac{2}{3} y^3 \right) \right] yt \, dy = 0$$

or

$$d_3 = -d_5 \left(L^2 - \frac{2}{5} h^2 \right) = \frac{3p}{4th} \left(\frac{L^2}{h^2} - \frac{2}{5} \right)$$

Expressions (e) together with the values obtained for the constants also fulfill conditions (b) and (d).

The state of stress is thus represented by

$$\sigma_x = \frac{py}{2I}(L^2 - x^2) + \frac{py}{I} \left(\frac{y^2}{3} - \frac{h^2}{5} \right) \tag{5.26}$$

$$\sigma_y = \frac{-p}{2I} \left(\frac{y^3}{3} - h^2 y + \frac{2h^3}{3} \right) \tag{5.27}$$

$$\tau_{xy} = \frac{-px}{2I}(h^2 - y^2) \tag{5.28}$$

Here $I = \frac{2}{3} th^3$ is the area moment of inertia taken about a line through the centroid, parallel to the z axis. Although the solutions given by Eqs. (5.26), (5.27), and (5.28) satisfy the equations of elasticity and the boundary conditions, they are nevertheless not exact. This is indicated by substituting $x = \pm L$ into Eq. (5.26) to obtain the following expression for the normal distributed forces per unit area at the ends:

$$T_x = \frac{py}{I} \left(\frac{y^2}{3} - \frac{h^2}{5} \right)$$

which cannot exist, as no forces act at the ends. From St. Venant's principle we may conclude, however, that the solutions do predict the correct stresses throughout the beam, except near the supports.

Recall that the longitudinal normal stress derived from elementary beam theory is $\sigma_x = -My/I$; this is equivalent to the first term of Eq. (5.26). The second term is then the difference between the longitudinal stress results given by the two approaches. To gauge the magnitude of the deviation, consider the ratio of the second term of Eq. (5.26) to the result of elementary theory at $x = 0$. At this point, the bending moment is a maximum. Substituting $y = h$ for the condition of maximum stress, we obtain

$$\frac{\Delta \sigma_x}{(\sigma_x)_{\text{elem. theory}}} = \frac{(ph/I)(h^2/3 - h^2/5)}{phL^2/2I} = \frac{4}{15} \left(\frac{h}{L} \right)^2$$

For a beam of length 10 times its depth, the above ratio is small, $\frac{1}{1500}$. For beams of ordinary proportions, we can conclude that elementary theory provides a result of sufficient accuracy for σ_x. As for σ_y, this stress is not found in the elementary theory. The result for τ_{xy} is, on the other hand, the same as that of elementary beam theory.

The displacement of the beam may be determined in a manner similar to that described for a cantilever beam (Sec. 5.4).

PART 2—Approximate Solutions

5.6 Elementary Theory of Bending

We may conclude, on the basis of the previous sections, that exact solutions are difficult to obtain. It was also observed that for a slender beam the results of the exact theory do not differ markedly from that of the mechanics of materials or elementary approach provided that solutions close to the ends are not required. The bending deflection was found to be very much larger than the shear deflection. Thus the stress associated with the former predominates. We deduce therefore that the normal strain ε_y resulting from transverse loading may be neglected. Because it is more easily applied, the elementary approach is usually preferred in engineering practice. The exact and elementary theories should be regarded as complementary rather than competitive approaches, enabling the analyst to obtain the degree of accuracy required in the context of the specific problem at hand.

The basic assumptions of the elementary theory are

$$\varepsilon_y = \frac{\partial v}{\partial y} = 0, \qquad \gamma_{xy} = \frac{\partial u}{\partial y} + \frac{\partial v}{\partial x} = 0$$

$$\varepsilon_x = \frac{\sigma_x}{E} \quad \text{(independent of } z\text{)}$$

$$(5.29)$$

$$\varepsilon_z = 0, \qquad \gamma_{yz} = \gamma_{xz} = 0 \tag{5.30}$$

The first equation of (5.29) is equivalent to the assertion $v = v(x)$. Thus, all points in a beam at a given longitudinal location x experience identical deformation. The second equation of (5.29), together with $v = v(x)$, yields, after integration,

$$u = -y \frac{dv}{dx} + u_0(x) \tag{a}$$

The third equation of (5.29) and Eqs. (5.30) imply that the beam is considered *narrow*.

At $y = 0$, the bending deformation should vanish. Referring to Eq. (a), it is clear, therefore, that $u_0(x)$ must represent axial deformation. The term dv/dx is the slope θ of the beam axis, as shown in Fig. 5.7a, and is very

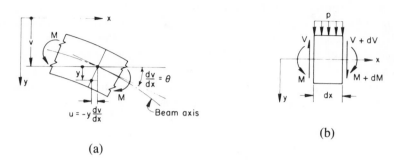

Figure 5.7

much smaller than unity. Therefore, for the case of pure bending,

$$u = -y\frac{dv}{dx} = -y\theta$$

Since u is a linear function of y, the above serves to restate the kinematic hypothesis of the elementary theory of bending: *plane sections perpendicular to the longitudinal axis of the beam remain plane subsequent to bending.* This assumption is confirmed by the exact theory *only* in the case of *pure bending*.

In the next section, we will obtain the stress distribution in a beam according to the elementary theory. We now derive some useful relations involving the shear force V, the bending moment M, the load per unit length p, the slope θ, and the deflection v. Consider a beam element of length dx subjected to a distributed loading (Fig. 5.7b). Note that as dx is small, the variation in the load per unit length p is omitted. In the free body diagram, all the forces and the moments are positive. The shear force obeys the sign convention discussed in Sec. 1.3; the bending moment is in agreement with the convention adopted in Sec. 5.2. In general, the shear force and bending moment vary with the distance x, and it thus follows that these quantities will have different values on each face of the element. The increments in shear force and bending moment are denoted by dV and dM, respectively. Equilibrium of forces in the vertical direction is governed by $V - (V + dV) - p\, dx = 0$, or

$$\frac{dV}{dx} = -p \qquad (5.31)$$

That is, the rate of change of shear force with respect to x is equal to algebraic value of the distributed loading. Equilibrium of the moments about a z axis through the left end of the element, neglecting the higher-order infinitesimals, leads to

$$\frac{dM}{dx} = -V \qquad (5.32)$$

The above relation states that the rate of change of bending moment is equal to the algebraic value of the shear force, valid only if a distributed load or no load acts on the beam segment. Combining Eqs. (5.31) and (5.32), we have

$$\frac{d^2M}{dx^2} = p \tag{5.33}$$

The basic equation of bending of a beam, Eq. (5.10), combined with Eq. (5.33), may now be written

$$\frac{d^2}{dx^2}\left(EI\frac{d^2v}{dx^2}\right) = p \tag{5.34}$$

For a beam of constant section the beam equations derived above may be expressed as

$$EI\frac{d^4v}{dx^4} = p$$

$$EI\frac{d^3v}{dx^3} = -V$$

$$EI\frac{d^2v}{dx^2} = M \tag{5.35}$$

$$EI\frac{dv}{dx} = EI\theta = \int M\,dx$$

These relationships also apply to *wide beams* provided that $E/(1-v^2)$ is substituted for E (Table 3.1).

In many problems of practical importance, the deflection due to transverse loading of a beam may be obtained through successive integration of the beam equation:

$$EIv^{\text{IV}} = EI\frac{d^4v}{dx^4} = p$$

$$EIv''' = \int_0^x p\,dx + c_1$$

$$EIv'' = \int_0^x dx \int_0^x p\,dx + c_1 x + c_2 \tag{5.36}$$

$$EIv' = \int_0^x dx \int_0^x dx \int_0^x p\,dx + \tfrac{1}{2}c_1 x^2 + c_2 x + c_3$$

$$EIv = \int_0^x dx \int_0^x dx \int_0^x dx \int_0^x p\,dx + \tfrac{1}{6}c_1 x^3 + \tfrac{1}{2}c_2 x^2 + c_3 x + c_4$$

Alternately, one could begin with $EIv'' = M(x)$ and integrate twice to obtain

$$EIv = \int_0^x dx \int_0^x M\,dx + c_3 x + c_4 \tag{5.37}$$

In either case, the constants c_1, c_2, c_3, c_4, which correspond to the homogeneous solution of the differential equations, may be evaluated from the boundary conditions. The constants $c_1, c_2, c_3/EI$, and c_4/EI represent the values at the origin of V, M, θ, and v, respectively. In the method of successive integration there is no need to distinguish between statically determinate and statically indeterminate systems (Sec. 5.9), because the equilibrium equations represent only two of the boundary conditions (on the first two integrals), and because the *total* number of boundary conditions is always equal to the total number of unknowns.

5.7 The Normal and Shear Stresses

When a beam is bent by transverse loads, there usually will be both a bending moment M and a shear force V acting on each cross section. The distribution of the normal stress associated with the bending moment is given by the flexure formula, Eq. (5.4):

$$\sigma_x = -\frac{My}{I} \tag{5.38}$$

where M and I are taken with respect to the z axis (Fig. 5.7).

In accordance with the assumptions of elementary bending, Eqs. (5.29) and (5.30), the contribution of the shear strains to beam deformation is omitted. However, shear stresses do exist, and the shearing forces are the resultant of the stresses. The shearing stress τ acting at section mn, assumed uniformly distributed over the area $b \cdot dx$, can be determined on the basis of the equilibrium of forces acting on the shaded part of the beam element (Fig. 5.8). Here b is the width of the beam a distance y_0 from the neutral axis and dx is the length of the element. The distribution of normal stresses produced by M and $M + dM$ is indicated in the figure. The normal force distributed over the left hand face mr on the shaded area A^* is equal to

$$\int_{-b/2}^{b/2}\int_{y_0}^{h_1} \sigma_x \, dy \, dz = \int_{A^*} -\frac{My}{I} \, dA \tag{a}$$

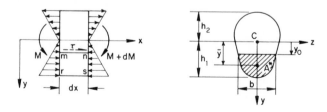

Figure 5.8

Similarly, an expression for the normal force on the right hand face *ns* may be written in terms of $M + dM$. The equilibrium of x directed forces acting on the beam element is governed by

$$-\int_{A^*} \frac{(M+dM)y}{I}\, dA - \int_{A^*} -\frac{My}{I}\, dA = \tau b\, dx$$

from which we have

$$\tau = -\frac{1}{Ib} \int_{A^*} \frac{dM}{dx} y\, dA$$

Upon substitution of Eq. (5.32), the shear stress assumes the form

$$\tau = \frac{V}{Ib} \int_{A^*} y\, dA = \frac{VQ}{Ib} \tag{5.39}$$

The integral represented by Q is the *first moment of the shaded area* A^* with respect to the neutral axis z:

$$Q = \int_{A^*} y\, dA = A^* \bar{y} \tag{5.40}$$

By definition, $\bar{y}$ is the distance from the neutral axis to the centroid of A^*. In the case of sections of regular geometry, $A^*\bar{y}$ provides a convenient means of calculating Q.

For example, in the case of a *rectangular cross section* of width b and depth $2h$, the shear stress at y_0 is

$$\tau = \frac{V}{Ib} \int_{-b/2}^{b/2} \int_{y_0}^{h} y\, dy\, dz = \frac{V}{2I}(h^2 - y_0^2) \tag{5.41}$$

This shows that the shear stress varies parabolically with y_0; it is zero when $y_0 = \pm h$, and has its maximum value at the neutral axis, $y_0 = 0$:

$$\tau_{max} = \frac{Vh^2}{2I} = \frac{3}{2}\frac{V}{2bh} \tag{5.42}$$

In the above, $2bh$ is the area of the rectangular cross section. It is observed that the maximum shear stress (either horizontal or vertical: $\tau = \tau_{xy} = \tau_{yx}$) is 1.5 times larger than the average shear stress V/A. As observed in Sec. 5.4, for a *thin* rectangular beam the above is the exact distribution of shear stress. However, in general, for wide rectangular sections and for other sections, Eq. (5.39) yields only approximate values of the shearing stress.

It should be pointed out that the maximum shear stress does not always occur at the neutral axis. For instance, in the case of a cross section having nonparallel sides, such as a triangular section, the maximum value of Q/b (and thus τ) takes place at midheight, $h/2$, while the neutral axis is located at a distance $h/3$ from the base.

The following sample problem illustrates the application of the shear stress formula.

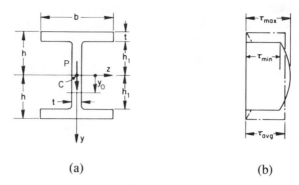

(a)   (b)

Figure 5.9

Example 5.2. A cantilever I-beam is loaded by a force P at the free end acting through the centroid of the section. The beam is of constant thickness t (Fig. 5.9a). Determine the shear stress distribution in the section.

SOLUTION. The vertical shear force at every section is P. It is assumed that the shear stress τ is uniformly distributed over the web thickness. Then, in the web, for $0 \leqslant y_0 \leqslant h_1$, applying Eq. (5.39),

$$\tau = \frac{V}{Ib}A^*\bar{y} = \frac{P}{It}\left[b(h-h_1)\left(h_1 + \frac{h-h_1}{2}\right) + t(h_1-y_0)\left(y_0 + \frac{h_1-y_0}{2}\right) \right]$$

The above may be written

$$\tau = \frac{P}{It}\left[\frac{b}{2}(h^2 - h_1^2) + \frac{t}{2}(h_1^2 - y_0^2) \right] \tag{b}$$

The shearing stress thus varies parabolically in the web (Fig. 5.9b). The extreme values of τ found at $y_0 = 0$ and $y_0 = h_1$ are, from Eq. (b), as follows:

$$\tau_{\max} = \frac{P}{2It}(bh^2 - bh_1^2 + th_1^2), \qquad \tau_{\min} = \frac{P}{2It}(bh^2 - bh_1^2)$$

Note that it is usual that $t \ll b$, and therefore the maximum and minimum stresses do not differ appreciably, as is seen in the figure. Similarly, the shear stress in the flange, for $h_1 < y_0 \leqslant h$, is

$$\tau = \frac{P}{Ib}\left[b(h-y_0)\left(y_0 + \frac{h-y_0}{2}\right) \right] = \frac{P}{2I}(h^2 - y_0^2) \tag{c}$$

This is the parabolic equation for the variation of stress in the flange, shown by the dashed lines in the figure. Clearly, for a thin flange, the shear stress is very small as compared with the shear stress in the web. It is concluded that the approximate average value of shear stress in the beam may be found by dividing P by the web cross section: $\tau_{\text{avg}} = P/2th_1$. This is indicated by the dotted lines in the figure. The distribution of stress given by Eq. (c) is fictitious, because the inner planes of the flanges must be free of shearing stress, as they are load-free boundaries of the beam. The above contradiction cannot be resolved by the elementary theory; the theory of elasticity must be applied to obtain the correct solution. Fortunately, this defect of the shearing stress formula does not lead to serious error, since as pointed out previously, the web carries almost all the shear force. In order to reduce the stress concentration at the juncture of the web and the flange, the sharp corners should be rounded.

5.8   The Shear Center

Given any cross-sectional configuration, one point may be found in the plane of the cross section through which passes the resultant of the transverse shearing forces. A load acting on the beam must act through this point, called the *shear center* or *flexural center*, if no twisting is to occur.* The center of shear is sometimes defined as the point in the end section of a cantilever beam at which an applied load results in bending only. When the load does not act through the shear center, a twisting action results, and torsion of the beam takes place (Sec. 6.1). The location of the shear center is independent of the direction and magnitude of the transverse forces. For symmetrical sections, the shear center is found on the axis of symmetry, while for a beam with two axes of symmetry, the shear center coincides with their point of intersection (also the centroid). It is not necessary, in general, for the shear center to lie on a principal axis, and it may be located outside of the cross section of the beam.

For thin-walled sections, the shearing stresses are taken to be distributed uniformly over the thickness of the wall and directed so as to parallel the boundary of the cross section. If the shear center S for the typical section of Fig. 5.10a is required, we begin by calculating the shear stresses by means of Eq. (5.39). The moment M_x of these stresses about arbitrary point A is then obtained. Inasmuch as the external moment attributable to V_y about A is $V_y e$, the distance between A and the shear center is given by

$$e = \frac{M_x}{V_y} \qquad (5.43)$$

If the force is parallel to the z axis rather than the y axis, the position of the line of action may be established in the manner discussed above. In the event that both V_y and V_z exist, the intersection of the two lines of action locates the shear center.

The determination of M_x is simplified by propitious selection of point A, such as in Fig. 5.10b. Here it is observed that the moment M_x of the shear forces about A is zero; point A is also the shear center. For all sections consisting of two intersecting rectangular elements, the same situation exists.

---

*For a detailed discussion, see I. S. Sokolnikoff, *Mathematical Theory of Elasticity*, New York: McGraw-Hill, 1956, Sec. 53.

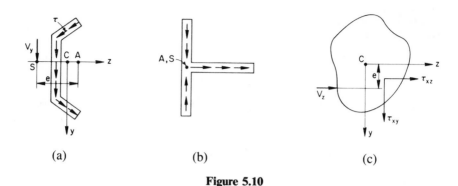

Figure 5.10

The above considerations can be extended to beams of arbitrary solid cross section, in which the shearing stress varies with both cross-sectional coordinates, y and z. For these sections, the exact theory can, in some cases, be successfully applied to locate the shear center. Examine the section of Fig. 5.10c, subjected to the shear force V_z, which produces the stresses indicated. Denote y and z as the principal directions. The moment about the x axis is

$$M_x = \int \int (x_y z - \tau_{xz} y)\, dz\, dy \tag{5.44}$$

V_z must be located a distance e from the z axis, where $e = M_x/V_z$.

In the following examples, the determination of the shear center of an open, thin-walled section is illustrated in the solution for two typical situations. The first refers to a section having only one axis of symmetry, the second to an asymmetrical section.

Example 5.3. Locate the shear center of the channel section loaded as a cantilever (Fig. 5.11a). Assume that the flange thicknesses are small when compared with the depth and width of the section.

SOLUTION.   The shearing stress in the upper flange at any section nn will be found first. This section is located a distance s from the free edge m, as shown in the figure. At m the shearing stress is zero. The first moment of area st_1 about the z axis is $Q_z = st_1 h$. The shear stress at nn, from Eq. (5.39), is thus

$$\tau = \frac{V_y Q_z}{I_z b} = P \frac{sh}{I_z} \tag{a}$$

The direction of τ along the flange can be determined from the equilibrium of the forces acting on an element of length dx and width s (Fig. 5.11b). Here the normal force $N = t_1 s \sigma_x$, owing to the bending of the beam, increases with dx by dN. Hence, the x equilibrium of the element requires that $\tau t_1 \cdot dx$ must be directed as shown. As a consequence this flange force is directed to the left, as the shear forces must intersect at the corner of the element.

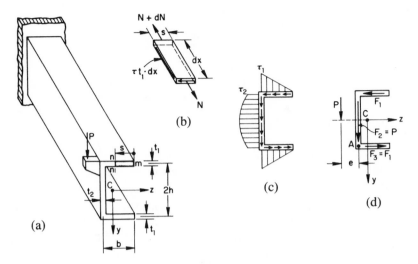

Figure 5.11

The distribution of the shear stress on the flange, as Eq. (a) indicates, is linear with s. Its maximum value occurs at $s = b$:

$$\tau_1 = P \frac{bh}{I_z} \tag{b}$$

Similarly, the stress τ_2 at the top of the web is

$$\tau_2 = P \frac{bt_1 h}{t_2 I_z} \tag{c}$$

The stress varies parabolically over the web, and its maximum value is found at the neutral axis. A sketch of the shear stress distribution in the channel is shown in Fig. 5.11c. As the shear stress is linearly distributed across the flange length, from Eq. (b), the flange force is expressed by

$$F_1 = \frac{1}{2} \tau_1 b t_1 = P \frac{b^2 h t_1}{2 I_z} \tag{d}$$

Symmetry of the section dictates that $F_1 = F_3$ (Fig. 5.11d). We shall assume that the web force $F_2 = P$, since the vertical shearing force transmitted by the flange is negligibly small, as shown in Example 5.2. The shearing force components acting in the section must be statically equivalent to the resultant shear load P. Thus, the principle of the moments for the system of forces in Fig. 5.11d, applied at A, yields $Pe = 2F_1 h$. Upon substituting F_1 from Eq. (d) into this expression, we obtain

$$e = \frac{b^2 h^2 t_1}{I_z}$$

where

$$I_z = \tfrac{2}{3} t_2 h^3 + 2 b t_1 h^2$$

The shear center is thus located by the expression

$$e = \frac{3}{2} \frac{b^2 t_1}{h t_2 + 3 b t_1} \tag{e}$$

Note that e is dependent upon only section dimensions. Examining the above reveals that e may vary from a minimum of zero to a maximum of $b/2$. A zero or near zero value of e corresponds to either a flangeless beam ($b=0, e=0$) or an especially deep beam ($h \gg b$). The extreme case, $e=b/2$, is obtained for an infinitely wide beam.

Example 5.4. Locate the shear center S for the asymmetrical channel section shown in Fig. 5.12a. All dimensions are in millimeters. Assume that the beam thickness $t = 1.25$ mm is constant.

SOLUTION. The centroid C of the section is located by $\bar{y}$ and $\bar{z}$ with respect to nonprincipal axes z and y. By performing the procedure given in Example 5.1, we obtain $\bar{y} = 15.87$ mm, $\bar{z} = 5.28$ mm, $I_y = 4765.62$ mm^4, $I_z = 21{,}054.69$ mm^4, and $I_{yz} = 3984.37$ mm^4. Equation (5.17) then yields the direction of the principal axis x', y' as $\theta_p = 13.05°$, and Eq. (5.18), the principal moments of inertia $I_{y'} = 3828.12$ mm^4, $I_{z'} = 21{,}953.12$ mm^4. (Fig. 5.12a).

Let us now assume that a shear load $V_{y'}$ is applied in the y', z' plane (Fig. 5.12b). This force may be considered the resultant of force components, F_1, F_2, and F_3 acting in the flanges, and web in the directions indicated in the figure. The algebra will be minimized if we choose point A, where F_2 and F_3 intersect, in finding the line of action of $V_{y'}$ by applying the principle of moments. In so doing, we need to determine the value of F_1 acting in the upper flange. The shear stress τ in this flange, from Eq. (5.39), is

$$\tau = \frac{V_{y'} Q_{z'}}{I_{z'} b} = \frac{V_{y'}}{I_{z'} t} \left[st \left(19.55 + \frac{1}{2} s \sin 13.05° \right) \right] \tag{f}$$

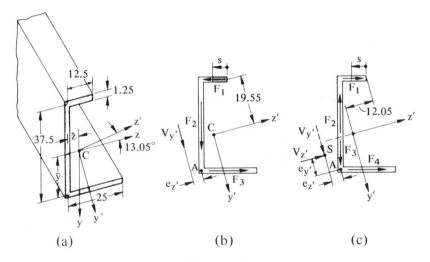

Figure 5.12

where s is measured from right to left along the flange. Note that $Q_{z'}$, the bracketed expression, is the first moment of the shaded flange element area with respect to the z' axis. The constant 19.55 is obtained from the geometry of the section. Upon substituting the numerical values and integrating Eq. (f), the total shear force in the upper flange is found to be

$$F_1 = \int_0^s \tau t \, ds = \frac{V_{y'} t}{I_{z'}} \int_0^{12.5} s \left(19.55 + \frac{1}{2} s \sin 13.05° \right) ds = 0.0912 V_{y'} \qquad \text{(g)}$$

Application of the principle of moments at A gives $V_{y'} e_{z'} = 37.5 F_1$. Introducing F_1 from Eq. (g) into the above, the distance $e_{z'}$, which locates the line of action of $V_{y'}$ from A, is

$$e_{z'} = 3.42 \text{ mm} \qquad \text{(h)}$$

Next, assume that the shear loading $V_{z'}$ acts on the beam (Fig. 5.12c). The distance $e_{y'}$ may be obtained as in the situation described above. Because of $V_{z'}$, the force components F_1 to F_4 will be produced in the section. The shear stress in the upper flange is given by

$$\tau = \frac{V_{z'} Q_{y'}}{I_{y'} b} = \frac{V_{z'}}{I_{y'} t} \left[st \left(12.05 - \frac{1}{2} s \cos 13.05° \right) \right] \qquad \text{(i)}$$

Here $Q_{y'}$ represents the first moment of the flange segment area with respect to the y' axis, and 12.05 is found from the geometry of the section. The total force F_1 in the flange is

$$F_1 = \frac{V_{z'}}{I_{y'}} \int_0^{12.05} st \left(12.05 - \frac{1}{2} s \cos 13.05° \right) ds = 0.204 V_{z'}.$$

The principle of moments applied at A, $V_{z'} e_{y'} = 37.5 F_1 = 7.65 V_{z'}$, leads to

$$e_{y'} = 7.65 \text{ mm} \qquad \text{(j)}$$

Thus, the intersection of the lines of action of $V_{y'}$ and $V_{z'}$, $e_{z'}$ and $e_{y'}$, locates the shear center S of the asymmetrical channel section.

5.9 Statically Indeterminate Systems

A large class of problems of considerable practical interest relates to structural systems for which the equations of statics are not sufficient (though necessary) for determination of the reactions or other unknown forces. Such systems are *statically indeterminate*, requiring for solution supplementary information. Additional equations usually describe certain geometrical conditions associated with strain. These equations, also referred to as the *conditions of continuity*, state that the strain owing to deflection or rotation must be such as to preserve continuity. With this additional information, the solution proceeds in essentially the same manner as for statically determinate systems.

Several methods are available to analyze statically indeterminate structures. The principle of superposition, briefly discussed next, offers for many cases an effective approach. In Sec. 5.6, and in Chapters 7 and 10, several methods are discussed for the solution of indeterminate beam, frame, and truss problems.

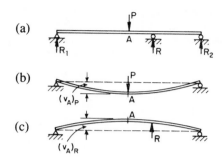

Figure 5.13

The Method of Superposition

In the event of complicated load configurations, the method of *superposition* may be used to good advantage to simplify the analysis. Consider, for example, the continuous beam of Fig. 5.13a, replaced by the beams shown in Fig. 5.13b and c. At point A, the beam now experiences the deflections $(v_A)_P$ and $(v_A)_R$, due respectively to P and R. Subject to the restrictions imposed by small deformation theory and a material obeying Hooke's law, the deflections and stresses are linear functions of transverse loading, and superposition is valid:

$$v_A = (v_A)_P + (v_A)_R$$

$$\sigma_A = (\sigma_A)_P + (\sigma_A)_R$$

The procedure may in principle be extended to situations involving any degree of indeterminacy.

5.10   Strain Energy in Beams. Castigliano's Theorem

Strain energy methods are frequently employed to analyze the deflections of beams and other structural elements. Of the many approaches available, *Castigliano's second theorem* is one of the most widely used. In applying this theory, the strain energy must be represented as a function of loading. Detailed discussions of energy techniques are found in Chapter 10. In this section we limit ourselves to a simple example to illustrate how the strain energy in a beam is evaluated and how the deflection is obtained by the use of Castigliano's theorem (Sec. 10.4).

In determining the strain energy in a beam in *bending alone*, consider that for this loading, only the axial stress $\sigma_x = \sigma$ exists. The strain energy density at a point in the beam is, from Eq. (2.22), $U_0 = \sigma^2/2E$. Substituting $\sigma = My/I$, one has $U_0 = M^2 y^2/2EI^2$. Integrating U_0 over the volume of the beam, we obtain

$$U = \int \frac{1}{2} \frac{M^2}{EI^2} \left[\int \int y^2 \, dy \, dz \right] dx \tag{a}$$

Here the integral in the bracket is the moment of inertia I and $M = EI(d^2v/dx^2)$. The strain energy stored in a beam under flexural stress only is thus

$$U = \int \frac{M^2 \, dx}{2EI} \quad \text{or} \quad U = \int \frac{EI}{2} \left(\frac{d^2v}{dx^2} \right)^2 dx \tag{5.45}$$

where the integrations are carried out over the beam length.

We next determine the strain energy stored in a beam, *only* due to the *shear* loading V. As we have described in Sec. 5.7, this force produces shear stress τ at every point in the beam. The strain energy density is, from Eq. (2.24), $U_0 = \tau^2/2G$. Substituting τ as expressed by Eq. (5.39), we have $U_0 = V^2Q^2/2GI^2b^2$. Integrating this expression over the volume of the beam of cross-sectional area A, we obtain

$$U = \int \frac{V^2}{2GI^2} \left[\int \frac{Q^2}{b^2} \, dA \right] dx \tag{b}$$

Let us denote

$$f_s = \frac{A}{I^2} \int \frac{Q^2}{b^2} \, dA \tag{5.46}$$

This is termed the *form factor for shear*, which when substituted in Eq. (b) yields

$$U = \int \frac{f_s V^2 \, dx}{2GA} \tag{5.47}$$

where the integration is carried over the beam length. The form factor is a dimensionless quantity specific to a given cross section geometry. For example, for a *rectangular* cross section of width b and height $2h$, the first moment Q, from Eq. (5.41), is $Q = (b/2)(h^2 - y_0^2)$. As $A/I^2 = 9/2bh^5$, Eq. (5.46) provides the following result:

$$f_s = \frac{9}{2bh^5} \int_{-h}^{h} \frac{1}{4} \left(h^2 - y_0^2 \right)^2 b \, dy_0 = \frac{6}{5}$$

In a like manner, the form factor for other cross sections can be determined. Table 5.1 lists several typical cases. Following the determination of f_s, the strain energy is evaluated by applying Eq. (5.47).

Table 5.1

Cross-section	f_s
A	$\dfrac{6}{5}$
B	$\dfrac{A}{A_{web}}$
C	$\dfrac{10}{9}$
D	2

For a linearly elastic beam, Castigliano's theorem, from Eq. (10.6), is expressed by

$$e = \frac{\partial U}{\partial P} \tag{5.48}$$

where P is a load acting on the beam and e is the displacement of the point of application in the direction of P.

As an illustration, consider the bending of a cantilever beam of rectangular cross section and length L, subjected to a concentrated force P at the free end (Fig. 5.5). The bending moment at any section is $M = Px$, and the shear force V is equal in magnitude to P. Upon substituting these together with $f_s = \frac{6}{5}$ into Eqs. (5.45) and (5.47) and integrating, the strain energy stored in the cantilever is found to be

$$U = \frac{P^2 L^3}{6EI} + \frac{3P^2 L}{5GA}$$

The displacement of the free end owing to bending and shear is, by application of Castigliano's theorem, therefore

$$e = \frac{PL^3}{3EI} + \frac{6PL}{5GA}$$

The exact solution is given by Eq. (5.25).

PART 3—Curved Beams

5.11   Exact Solution

A curved bar or beam is a structural element for which the locus of the centroids of the cross sections is a curved line. This section concerns itself with an application of the theory of elasticity. We deal here with a bar

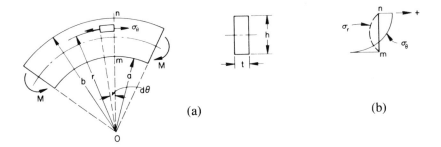

Figure 5.14

characterized by a constant narrow rectangular cross section and a circular axis. The axis of symmetry of the cross section lies in a single plane throughout the length of the member.

Consider a beam subjected to equal end couples M such that bending takes place in the plane of curvature, as shown in Fig. 5.14a. Inasmuch as the bending moment is constant along the length of the bar, the stress distribution should be identical in any radial cross section. Stated differently, we seek a distribution of stress displaying θ independence. It is clear that the appropriate expression of equilibrium is Eq. (8.2),

$$\frac{d\sigma_r}{dr} + \frac{\sigma_r - \sigma_\theta}{r} = 0 \qquad (a)$$

and that the condition of compatability for plane stress, Eq. (3.25),

$$\frac{d^2(\sigma_r + \sigma_\theta)}{dr^2} + \frac{1}{r}\frac{d(\sigma_r + \sigma_\theta)}{dr} = 0$$

must also be satisfied. The latter is an equidimensional equation, reducible to a second-order equation with constant coefficients by substituting $r = e^t$ or $t = \ln r$. Direct integration then leads to $\sigma_r + \sigma_\theta = c'' + c'\ln r$, which may be written in the form $\sigma_r + \sigma_\theta = c''' + c'\ln(r/a)$. Solving the above expression together with Eq. (a) results in the following equations for the radial and tangential stress:

$$\sigma_r = c_1 + c_2\ln\frac{r}{a} + \frac{c_3}{r^2}$$

$$\sigma_\theta = c_1 + c_2\left(1 + \ln\frac{r}{a}\right) - \frac{c_3}{r^2} \qquad (5.49)$$

To evaluate the constants of integration, the boundary conditions are applied as follows:

1. No normal forces act along the curved boundaries at $r = a$ and $r = b$, and therefore

$$(\sigma_r)_{r=a} = (\sigma_r)_{r=b} = 0 \qquad (b)$$

2. Because there is no force acting at the ends, the normal stresses acting at the straight edges of the bar must be distributed to yield a zero resultant:

$$t\int_a^b \sigma_\theta\, dr = 0 \tag{c}$$

where t represents the beam thickness.

3. The normal stresses at the ends must produce a couple M:

$$t\int_a^b r\sigma_\theta\, dr = M \tag{d}$$

The conditions (c) and (d) apply not only at the ends, but because of θ independence, at any θ. In addition, shearing stresses have been assumed zero throughout the beam, and $\tau_{r\theta} = 0$ is thus satisfied at the boundaries, where no tangential forces exist.

Combining the first equation of (5.49) with the conditions (b) above, we find that

$$c_3 = -a^2 c_1, \qquad c_1\left(\frac{a^2}{b^2} - 1\right) = c_2 \ln\frac{b}{a}$$

These constants together with the second of Eqs. (5.49) satisfy the condition (c). From the above we have

$$c_1 = \frac{b^2 \ln(b/a)}{a^2 - b^2} c_2, \qquad c_3 = \frac{a^2 b^2 \ln(b/a)}{b^2 - a^2} c_2 \tag{e}$$

Finally, substitution of the second of Eqs. (5.49) and (e) into (d) provides

$$c_2 = \frac{M}{N}\frac{4(b^2 - a^2)}{tb^4} \tag{f}$$

where

$$N = \left(1 - \frac{a^2}{b^2}\right)^2 - 4\frac{a^2}{b^2}\ln^2\frac{b}{a} \tag{5.50}$$

When the expressions for constants c_1, c_2, and c_3 are inserted into Eq. (5.49), the following equations are obtained for the radial and tangential stress:

$$\sigma_r = \frac{4M}{tb^2 N}\left[\left(1 - \frac{a^2}{b^2}\right)\ln\frac{r}{a} - \left(1 - \frac{a^2}{r^2}\right)\ln\frac{b}{a}\right]$$

$$\sigma_\theta = \frac{4M}{tb^2 N}\left[\left(1 - \frac{a^2}{b^2}\right)\left(1 + \ln\frac{r}{a}\right) - \left(1 + \frac{a^2}{r^2}\right)\ln\frac{b}{a}\right] \tag{5.51}$$

If the end moments are applied so that the force couples producing them are distributed in the manner indicated by Eq. (5.51), then these equations are applicable throughout the bar. If the distribution of applied stress (to produce M) differs from Eq. (5.51), the results may be regarded as valid in regions away from the ends, in accordance with St. Venant's principle. The foregoing results, when applied to a beam with radius a, large relative to its depth h, yield an interesting comparison between straight and curved beam theory. For $h \ll a$, σ_r in Eq. (5.51) becomes negligible, and σ_θ is approximately the same as that obtained from My/I.

The bending moment is taken as positive when it tends to decrease the radius of curvature of the beam, as in Fig. 5.14a. Employing this sign convention, σ_r as determined from Eq. (5.51) is always negative, indicating that it is compressive. Similarly, when σ_θ is found to be positive, it is tensile; otherwise, compressive. In Fig. 5.14b, a plot of the stresses at section mn is presented. Note that the maximum stress magnitude is found at the extreme fiber of the concave side.

Substitution of σ_r and σ_θ from Eq. (5.51) into Hooke's law provides expressions for the strains ε_θ, ε_r, and $\gamma_{r\theta}$. The displacements u and v then follow, upon integration, from the strain-displacement relationships, Eqs. (3.20). The resulting displacements indicate that plane sections of the curved beam subjected to pure bending remain plane subsequent to bending. Castigliano's theorem (Sec. 5.10) is particularly attractive for determining the deflection of curved members. For beams in which the *depth of the member* is *small relative to the radius of curvature* or, as is usually assumed, $R/c > 4$, the initial curvature may be neglected in evaluating the strain energy. Here R represents the radius to the centroid, and c is the distance from the centroid to the extreme fiber on the concave side (Fig. 5.15). Thus, the strain energy due to the bending of a straight beam [eq. (5.45)] is a good approximation also for curved, slender beams.

5.12 Winkler's Theory

The approach to curved beams now explored is due to Winkler and relies upon the following basic assumptions:

1. All cross sections possess a vertical axis of symmetry lying in the plane of the centroidal axis passing through C (Fig. 5.15a).

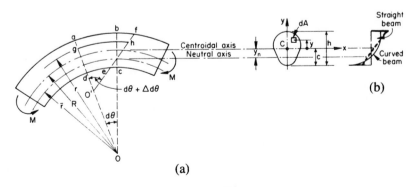

(a)

(b)

Figure 5.15

2. The beam is subjected to end couples M. The bending moment vector is everywhere normal to the plane of symmetry of the beam.
3. Sections originally plane and perpendicular to the centroidal beam axis remain so subsequent to bending. (The influence of transverse shear upon beam deformation is not taken into account.)

Referring to assumption (3), note the relationship in Fig. 5.15a between lines bc and ef representing plane sections before and after the bending of an initially curved beam. Note also that the initial length of a beam fiber such as gh is dependent upon the distance r from the center of curvature O. On the basis of plane sections remaining plane, we can state that the total deformation of a beam fiber obeys a linear law, as the beam element rotates through small angle $\Delta\,d\theta$. The tangential strain ε_θ *does not* follow a linear relationship, however. The deformation of arbitrary fiber gh is $\varepsilon_c R\,d\theta + y\Delta\,d\theta$, where ε_c denotes the strain of the centroidal fiber. Since the original length of gh is $(R+y)\,d\theta$, the tangential strain of this fiber is given by $\varepsilon_\theta = (\varepsilon_c R\,d\theta + y\Delta\,d\theta)/(R+y)\,d\theta$. Through introduction of Hooke's law, the tangential stress acting on area dA is then

$$\sigma_\theta = \frac{\varepsilon_c R + y(\Delta\,d\theta/d\theta)}{R+y}\,E$$

Denoting the angular strain $\Delta\,d\theta/d\theta$ by λ and adding and subtracting $\varepsilon_c y$ in the numerator, we put the above expression into a more convenient form:

$$\sigma_\theta = \left[\varepsilon_c + (\lambda - \varepsilon_c)\frac{y}{R+y}\right]E \qquad (a)$$

The beam section must, of course, satisfy the conditions of static equilibrium, $F_z = 0$ and $M_x = 0$, respectively:

$$\int \sigma_\theta\,dA = 0, \qquad \int \sigma_\theta y\,dA = M \qquad (b)$$

When the tangential stress of Eq. (a) is inserted into Eq. (b), we obtain

$$\varepsilon_c \int dA = -(\lambda - \varepsilon_c)\int \frac{y}{R+y}\,dA$$

$$M = \left[\varepsilon_c \int y\,dA + (\lambda - \varepsilon_c)\int \frac{y^2}{R+y}\,dA\right]E \qquad (c)$$

Note that $\int dA = A$, and since y is measured from the centroidal axis, $\int y\,dA = 0$. We now introduce the notation for a *property of the area*

$$m = -\frac{1}{A}\int \frac{y}{R+y}\,dA \qquad (5.52)$$

It follows that

$$\int \frac{y^2}{R+y}\,dA = \int \left(y - \frac{Ry}{R+y}\right)dA = -R\int \frac{y}{R+y}\,dA = mAR \qquad (d)$$

Equations (c) are thus written $\varepsilon_c = (\lambda - \varepsilon_c) m$ and $M = E(\lambda - \varepsilon_c) \cdot mAR$. From these,

$$\varepsilon_c = \frac{M}{AER}, \qquad \lambda = \frac{1}{AE}\left(\frac{M}{R} + \frac{M}{mR}\right) \qquad \text{(e)}$$

Substitution of Eqs. (e) into Eq. (a) provides an expression for the tangential stress in a curved beam subject to pure bending:

$$\sigma_\theta = \frac{M}{AR}\left[1 + \frac{y}{m(R+y)}\right] \qquad (5.53)$$

The variation of stress over the cross section is thus *hyperbolic*, as sketched in Fig. 5.15b. The sign convention applied to bending moment is the same as that used in Sec 5.11. The bending moment is positive when directed toward the concave side of the beam, as shown in the figure. If Eq. (5.53) results in a positive value, it is indicative of a tensile stress.

The distance between the centroidal axis ($y = 0$) and the neutral axis is found by setting equal to zero the tangential stress in Eq. (5.53):

$$1 + \frac{y_n}{m(R+y_n)} = 0 \qquad \text{(f)}$$

where y_n denotes the distance between axes, as indicated in Fig. 5.15. From the above,

$$y_n = -\frac{mR}{m+1} \qquad (5.54)$$

This expression is valid for the case of pure bending only.

The tangential stress given by Eq. (5.53) may be added to the stress produced by a normal load P acting through the centroid of cross-sectional area A. For this simple case of superposition,

$$\sigma_\theta = \frac{P}{A} + \frac{M}{AR}\left[1 + \frac{y}{m(R+y)}\right] \qquad (5.55)$$

As before, a negative sign would be associated with a compressive load P.

It now proves useful to compare the results of the various theories. To do this, consider a curved beam of rectangular cross section and unit thickness experiencing pure bending. The tangential stress predicted by the elementary theory (based upon a linear distribution of stress) is My/I. The Winkler approach, leading to a hyperbolic distribution, is given by Eq. (5.53), while the exact theory results in Eqs. (5.51). In each case, the maximum and minimum values of stress are expressible by

$$\sigma_\theta = B\frac{M}{a^2}$$

In Table 5.2 values of B are listed as a function of b/a for the three cases cited. It is observed that there is good agreement between the exact and Winkler results. On this basis as well as more extensive comparisons, it may be concluded that the Winkler approach is adequate for practical

Table 5.2. Values of _B_

b/a	Elementary theory	Winkler's theory		Exact theory	
		r=a	r=b	r=a	r=b
1.5	±24	−26.971	20.647	−27.858	21.275
2.0	± 6.00	− 7.725	4.863	− 7.755	4.917
3.0	± 1.50	− 2.285	1.095	− 2.292	1.130

applications. Its advantage lies in the relative ease with which it may be applied to _any_ symmetric section.

The agreement between the Winkler and exact analyses is not as good in situations of combined loading as for the case of pure bending. As might be expected, for beams of only slight curvature, the simple flexure formula provides good results while requiring only simple computation. The _linear and hyperbolic stress distributions are approximately the same for $R/c > 20$._

Finally, it is noted that where I-, T-, or thin-walled tubular curved beams are involved, the stresses predicted by the approaches developed in this chapter will be in error. This is attributable to high stresses existing in certain sections such as the flanges, which cause significant beam distortion. A modified Winkler's equation finds application in such situations if more accurate results are required.

Example 5.5. A load P of 70 kN is applied to the circular steel frame shown in Fig. 5.16a. The rectangular cross section (Fig. 5.16b) is 0.1-m wide and 0.05-m thick. Determine (a) the area property m of the cross section, and (b) the tangential stress at points 1 and 2.

SOLUTION

(a) Applying Eq. (5.52) with $c_1 = c_2 = c$,

$$m = -\frac{1}{2tc}\int_{-c}^{c}\frac{y}{R+y}\,t\,dy = -\frac{1}{2c}\int_{-c}^{c}\frac{y}{R+y}\,dy \qquad (g)$$

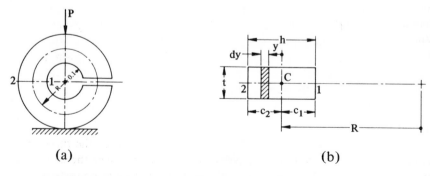

(a)   (b)

Figure 5.16

This expression may in general be evaluated through direct integration by use of the binomial expansion and integration, or by numerical techniques. Through direct integration, Eq. (g) yields

$$m = -\frac{1}{2c} \int_{-c}^{c} \left(1 - \frac{R}{R+y}\right) dy = -1 + \frac{R}{2c} \ln\left(\frac{R+c}{R-c}\right) \tag{h}$$

Alternatively, expanding in a binomial series,

$$\frac{1}{R+y} = (R+y)^{-1} = \frac{1}{R}\left(1 + \frac{y}{R}\right)^{-1}$$

Substituting this expression in Eq. (g), we have

$$m = -\frac{1}{2c} \int_{-c}^{c} \left(\frac{y}{R} - \frac{y^2}{R^2} + \frac{y^3}{R^3} - \frac{y^4}{R^4} + \cdots \right) dy$$

$$= \frac{1}{3}\left(\frac{c}{R}\right)^2 + \frac{1}{5}\left(\frac{c}{R}\right)^4 + \frac{1}{7}\left(\frac{c}{R}\right)^6 + \cdots \tag{i}$$

Employing similar methods, expressions for the area property m for other sections may be found. Table 5.3 lists some commonly encountered examples.

(b) From Eq. (h) with $R = 0.1 + 0.05 = 0.15$, $c_1 = c_2 = c = 0.05$ m (Fig. 5.16b), it is found that

$$m = -1 + \tfrac{3}{2} \ln 2 = 0.0397$$

Table 5.3

Cross section	Formula for m
A	$m = -1 + 2\left(\frac{R}{c}\right)^2 - 2\left(\frac{R}{c}\right)\sqrt{\left(\frac{R}{c}\right)^2 - 1}$
B	$m = -1 + \frac{2R}{c^2 - c_1^2}\left[\sqrt{R^2 - c_1^2} - \sqrt{R^2 - c^2}\right]$
C	$m = -1 + \frac{R}{Ah}\left\{\left[b_1 h + (R + c_1)(b - b_1)\right] \cdot \ln\frac{R + c_1}{R - c} - (b - b_1)h\right\}$ For rectangular section: $c = c_1$, $b = b_1$ For triangular section: $b_1 = 0$
D	$m = -1 + \frac{R}{A}\left[t \cdot \ln(R + c_1) + (b - t) \cdot \ln(R - c_2) - b \cdot \ln(R - c)\right]$
E	$m = -1 + \frac{R}{A}\left[b_1 \cdot \ln(R + c_1) + (t - b_1) \cdot \ln(R + c_3)\right.$ $\left. + (b - t) \cdot \ln(R - c_2) - b \cdot \ln(R - c)\right]$

The stresses at the inner and outer edges of section 1–2, with $M = PR$, are thus

$$\sigma_{\theta 1} = -\frac{P}{A} + \frac{M}{AR}\left[1 + \frac{-c_1}{m(R-c_1)}\right] = -\frac{Pc_1}{mA(R-c_1)}$$

$$= -\frac{70,000 \times 0.05}{0.0397 \times 0.005(0.15 - 0.05)} = -176 \text{ MPa}$$

$$\sigma_{\theta 2} = -\frac{P}{A} + \frac{M}{AR}\left[1 + \frac{c_2}{m(R+c_2)}\right] = \frac{Pc_2}{mA(R+c_2)}$$

$$= \frac{70,000 \times 0.05}{0.0397 \times 0.005(0.15 + 0.05)} = 88 \text{ MPa}$$

Example 5.6. A steel ring of 0.35-m mean diameter and of uniform rectangular section 0.06-m wide and 0.012-m thick is shown in Fig. 5.17a. A rigid bar is fitted across diameter AB, and a tensile force P applied to the ring as shown. Assuming an allowable stress of 140 MPa, determine the maximum tensile force that can be carried by the ring.

SOLUTION. Let the thrust induced in bar AB be denoted by $2F$. The moment at any section mn (Fig. 5.17b) is then

$$M_\theta = -FR\sin\theta + M_B + \frac{PR}{2}(1 - \cos\theta) \qquad (j)$$

Note that before and after deformation, the relative slope between B and C remains unchanged. Therefore the relative angular rotation between B and C is zero. Applying Eq. (5.35), we therefore obtain

$$EI\theta = 0 = \int_B^C M_\theta\, dx = R\int_0^{\pi/2} M_\theta\, d\theta$$

where $dx = ds = R\, d\theta$ is the length of beam segment corresponding to $d\theta$. Upon substitution of Eq. (j) the above becomes, after integrating,

$$\frac{1}{2}\pi M_B + \frac{1}{2}PR\left(\frac{\pi}{2} - 1\right) - FR = 0 \qquad (k)$$

This expression involves two unknowns, M_B and F. Another expression in terms of M_B and F is found by recognizing that the deflection at B is zero. By application of

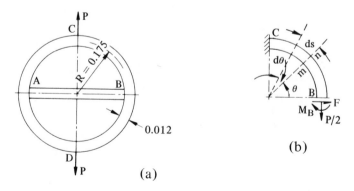

(a)

(b)

Figure 5.17

Castigliano's theorem, Eq. (5.48),

$$e_B = \frac{\partial U}{\partial F} = \frac{1}{EI} \int_0^{\pi/2} M_\theta \frac{\partial M_\theta}{\partial F} (R\,d\theta) = 0$$

where U is the strain energy of the segment. The above expression, upon introduction of Eq. (j), takes the form

$$\int_0^{\pi/2} \left\{ -FR\sin\theta + M_B + \frac{PR}{2}(1-\cos\theta) \right\} \sin\theta\,d\theta = 0$$

After integration,

$$-\tfrac{1}{4}\pi FR + \tfrac{1}{4}PR + M_B = 0 \tag{l}$$

Solution of Eqs. (k) and (l) yields $M_B = 0.1131PR$ and $FR = 0.4625PR$. Substituting the above, Eq. (j) gives, for $\theta = 90°$,

$$M_C = -FR + M_B + \tfrac{1}{2}PR = 0.1506PR$$

Thus, $M_C > M_B$. Since $R/c = 0.175/0.006 = 29$, the simple flexure formula offers the most efficient means of computation. The maximum stress is found at points A and B:

$$(\sigma_\theta)_{A,B} = \frac{P/2}{A} + \frac{M_B c}{I} = 694P + 13,809P = 14,503P$$

Similarly, at C and D,

$$(\sigma_\theta)_{C,D} = \frac{M_C c}{I} = 18,387P$$

Hence $\sigma_{\theta C} > \sigma_{\theta B}$. Since $\sigma_{max} = 140$ MPa, $140 \times 10^6 = 18,387P$. The maximum tensile load is therefore $P = 7641$ N.

Chapter 5—Problems

Secs. 5.1 to 5.9

5.1. A simply supported beam constructed of a $0.15 \times 0.15 \times 0.015$ m angle is loaded by concentrated force $P = 22.5$ kN at its midspan (Fig. P5.1). Calculate stresses σ_x and σ_y at A, and the orientation of the neutral axis. Neglect the effect of shear in bending and assume that beam twisting is prevented.

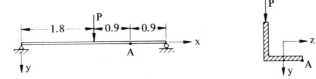

Figure P5.1

5.2. For the thin cantilever of Fig. P5.2, the stress function is given by

$$\phi = -c_1 xy + c_2 \frac{x^3}{6} - c_3 \frac{x^3 y}{6} - c_4 \frac{xy^3}{6} - c_5 \frac{x^3 y^3}{9} - c_6 \frac{xy^5}{20}$$

a. Determine the stresses σ_x, σ_y, and τ_{xy} by using the elasticity method.
b. Determine the stress σ_x by using the elementary method.
c. Compare the values of maximum stress obtained by the above approaches, for $L = 10h$.

Figure P5.2

5.3. Consider a cantilever beam of constant unit thickness, subjected to a uniform load of $p = 2000$ kN per unit length (Fig. P5.3). Determine the maximum stress in the beam:

a. based upon a stress function

$$\phi = \frac{p}{0.43}\left[-x^2 + xy + (x^2 + y^2)\left(0.78 - \tan^{-1}\frac{y}{x}\right)\right]$$

b. based on the elementary theory. Compare the results of (a) and (b).

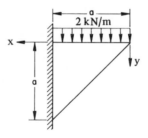

Figure P5.3

5.4. A beam is constructed of half a hollow tube of mean radius R and wall thickness t (Fig. P5.4). Assuming $t \ll R$, locate the shear center S. The moment of inertia of the section about the z axis is $I_z = \pi R^3 t / 2$.

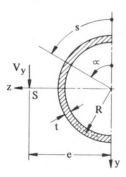

Figure P5.4

5.5. An H-section beam with unequal flanges is subjected to a vertical load P (Fig. P5.5). The following assumptions are applicable:
a. The total resisting shear occurs in the flanges.
b. The rotation of a plane section during bending occurs about the symmetry axis so that the radii of curvature of both flanges are equal.
Determine the location of the shear center S.

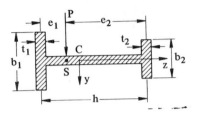

Figure P5.5

5.6. Determine the shear center S of the section shown in Fig. P5.6. All dimensions are in millimeters.

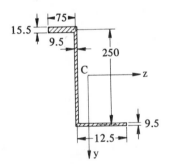

Figure P5.6

5.7. A wooden, simply supported beam of length L is subjected to a uniform load p. Determine the beam length and the loading necessary to develop simultaneously $\sigma_{max} = 8.4$ MPa and $\tau_{max} = 0.7$ MPa. Take thickness $t = 0.05$ m and depth $h = 0.15$ m.

5.8. A box beam supports the loading shown in Fig. P5.8. Determine the maximum value of P such that a flexural stress $\sigma = 7$ MPa or a shearing stress $\tau = 0.7$ MPa will not be exceeded.

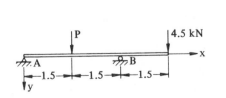
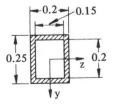

Figure P5.8

5.9. The slope at the wall of a built-in beam (Fig. P5.9a) is as shown in Fig. P5.9b and is given by $pL^3/96EI$. Determine the force acting at the simple support, expressed in terms of p, L, E, and I.

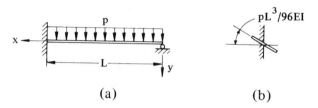

(a)   (b)

Figure P5.9

5.10. A fixed-ended beam of length L is subjected to a concentrated force P at a distance c away from the left end. Derive the equations of the elastic curve.

5.11. A welded bimetallic strip, Fig. P5.11, is initially straight. A temperature increment ΔT causes the element to curve. The coefficients of thermal expansion of the constituent metals are α_1 and α_2. Assuming elastic deformation and $\alpha_2 > \alpha_1$, determine (a) the radius of curvature to which the strip bends, (b) the maximum stress occurring at the interface, and (c) the temperature increase that would result in the simultaneous yielding of both elements.

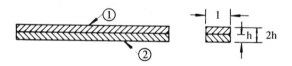

Figure P5.11

Secs. 5.10 to 5.12

5.12. Verify the values of f_s in Figs. B, C, and D of Table 5.1.

5.13. Consider a curved bar subjected to pure bending, Fig. 5.14. Assume the stress function

$$\phi = A \ln r + Br^2 \ln r + Cr^2 + D$$

to rederive the stress field in the bar given by Eqs. (5.51).

5.14. The cross section of a curved beam is shown in Fig. P5.14. Derive the property m by means of integration. Compare the result with that given for Fig. D in Table 5.3.

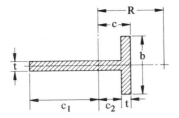

Figure P5.14

5.15. Verify the value of m in Fig. B of Table 5.3.

5.16. For the hook of circular cross section shown in Fig. P5.16, determine (a) the maximum load P that may be supported without exceeding a stress of 161 MPa at point 1 and (b) the tangential stress at point 2 of section 1–2 for the load obtained in (a).

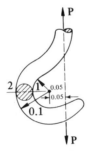

Figure P5.16

5.17. For curved members of solid section such as in Fig. 5.16a, Castigliano's theorem may be applied to provide an approximate expression for the deflection:

$$e_p = \int \left(\frac{N}{EA} \frac{\partial N}{\partial P} + \frac{M}{EI} \frac{\partial M}{\partial P} + \frac{f_s V}{GA} \frac{\partial V}{\partial P} \right) ds \qquad (P5.17)$$

where M, N, and V denote the moment, normal and shear force, respectively, at a section.

Calculate the approximate deflection along the line of action of the load P in the frame of Fig. 5.16a. Take $E = \frac{5}{2} G$.

5.18. A ring of mean radius R and constant rectangular section is subjected to a concentrated load (Fig. P5.18). Derive the following general expression for the tangential stress at any section of the ring:

$$\sigma_\theta = -\frac{(P/2)\cos\theta}{A} + \frac{M_\theta}{AR}\left[1 + \frac{y}{m(R+y)}\right] \qquad \text{(P5.18)}$$

where

$$M_\theta = 0.182\, PR - \tfrac{1}{2}PR(1 - \cos\theta)$$

Use Castigliano's theorem.

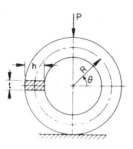

Figure P5.18

5.19. The ring shown in Fig. P5.18 has the following dimensions: $R=0.15$ m, $t=0.05$ m, and $h=0.1$ m. Taking $E=\tfrac{5}{2}G$, determine (a) the tangential stress on the inner fiber at $\theta=\pi/4$ and (b) the deflection along the line of action of the load P.

Chapter 6

Torsion of Prismatic Bars

6.1 Introduction

In this chapter, consideration is given stresses and deformations in prismatic members subject to equal and opposite end torques. In general, the bars are assumed free of end constraint. The reader will recall from an earlier study of the mechanics of solids two important expressions relevant to the torsion of *circular* bars:

$$\tau = \frac{M_t r}{J} \tag{a}$$

$$\theta = \frac{1}{L} \int_L \frac{M_t \, dz}{GJ} \tag{b}$$

Here τ represents the shear stress, M_t the applied torque, r the radius at which the stress is required, G the shear modulus, θ the angle of twist per unit longitudinal length, L the length, and z the axial coordinate. The polar moment of inertia J, defined by $\int_A r^2 \, dA$, is $\pi a^4/2$ for a circular cross section, where a is the radius.

The usual assumptions associated with an elementary approach to the derivation of Eqs. (a) and (b) are as follows:

1. The material is homogeneous and obeys Hooke's law.
2. All plane sections perpendicular to the longitudinal axis remain plane following the application of a torque, i.e., points in a given cross-sectional plane remain in that plane after twisting.
3. Subsequent to twisting, cross sections are undistorted in their individual planes, i.e., the shearing strain varies linearly with the distance from the central axis.
4. The angle of twist per unit length is constant.

Usually members that transmit torque, such as propeller shafts and torque tubes of power equipment, are circular or tubular in cross section, but in some situations, slender members with other than circular cross sections are used (Fig. 6.1). In treating noncircular prismatic bars, initially

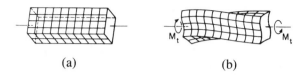

Figure 6.1

plane cross sections (Fig. 6.1a) experience out-of-plane deformation or *warping* (Fig. 6.1b), and assumptions (2) and (3) are no longer appropriate. Consequently, a different analytic approach is employed, that of the theory of elasticity, discussed next.

6.2 General Solution of the Torsion Problem

Consider a prismatic bar of constant arbitrary cross section subjected to equal and opposite twisting moments applied at the ends, as in Fig. 6.2a. The origin of x, y, z in the figure is located at the *center of twist* of the cross section, about which the cross section rotates during twisting. It is sometimes defined as the point at rest in every cross section of a bar in which one end is fixed and the other twisted by a couple. At this point, u and v, the x and y displacements, are thus zero. The location of the center of twist is a function of the shape of the cross section. It is to be noted that while the center of twist is referred to in the derivations of the basic relationships, it is not dealt with explicitly in the solution of torsion problems (see Problem 6.3). The z passes through the centers of twist of all cross sections.

In general, the cross sections warp, as already noted. We now explore the problem of torsion with free warping, applying the St. Venant semi-inverse method. As a fundamental assumption, the warping deformation is taken to be independent of axial location, i.e., identical for any cross section:

$$w = f(x, y) \tag{a}$$

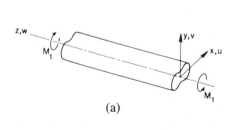

(a)

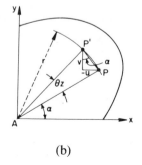

(b)

Figure 6.2

It is also assumed that the projection on the xy plane of any warped cross section rotates as a rigid body, and that the angle of twist per unit length is constant.

We refer now to Fig. 6.2b, which shows the partial end view of the bar (and could represent any section). An arbitrary point on the cross section, point $P(x, y)$, located a distance r from center of twist A, has moved to $P'(x-u, y+v)$ as a result of torsion. Assuming that no rotation occurs at end $z=0$ and that θ is *small*, the x and y displacements of P are, respectively,

$$u = -(r\theta z)\sin\alpha = -y\theta z$$
$$v = (r\theta z)\cos\alpha = x\theta z \tag{b}$$

where the angular displacement of AP at a distance z from the right end is θz, then x, y and z are the coordinates of point P, and α is the angle between AP and the x axis. Clearly, Eqs. (b) specify the rigid body rotation of any cross section through a small angle θz. By substituting Eqs. (a) and (b) into Eq. (2.3), the following are obtained:

$$\varepsilon_x = \gamma_{xy} = \varepsilon_y = \varepsilon_z = 0$$
$$\gamma_{zx} = \frac{\partial w}{\partial x} - y\theta, \qquad \gamma_{zy} = \frac{\partial w}{\partial y} + x\theta \tag{c}$$

Equation (2.17) together with the above expressions leads to the following:

$$\sigma_x = \sigma_y = \sigma_z = \tau_{xy} = 0 \tag{d}$$

$$\tau_{zx} = G\left(\frac{\partial w}{\partial x} - y\theta\right)$$
$$\tau_{zy} = G\left(\frac{\partial w}{\partial y} + x\theta\right) \tag{e}$$

By now substituting Eq. (d) into the equations of equilibrium (1.5), assuming negligible body forces, we obtain

$$\frac{\partial \tau_{zx}}{\partial z} = 0, \qquad \frac{\partial \tau_{zy}}{\partial z} = 0, \qquad \frac{\partial \tau_{zx}}{\partial x} + \frac{\partial \tau_{zy}}{\partial y} = 0 \tag{6.1}$$

Differentiating the first equation of (e) with respect to y and the second with respect to x, and subtracting the second from the first, we obtain an equation of compatibility:

$$\frac{\partial \tau_{zx}}{\partial y} - \frac{\partial \tau_{zy}}{\partial x} = H \tag{6.2}$$

where

$$H = -2G\theta \tag{6.3}$$

The stress in a bar of arbitrary section may thus be determined by solving Eqs. (6.1) and (6.2) along with the given boundary conditions.

Stress Function

As in the case of beams, the torsion problem formulated above is commonly solved by introducing a single stress function. If a function $\phi(x, y)$, the so-called *Prandtl stress function*, is assumed to exist, such that

$$\tau_{zx} = \frac{\partial \phi}{\partial y}, \qquad \tau_{zy} = -\frac{\partial \phi}{\partial x} \tag{6.4}$$

then the equations of equilibrium (6.1) are satisfied. The equation of compatibility (6.2) becomes, upon substitution of Eq. (6.4),

$$\frac{\partial^2 \phi}{\partial x^2} + \frac{\partial^2 \phi}{\partial y^2} = H \tag{6.5}$$

The stress function ϕ must therefore satisfy *Poisson's equation* if the compatibility requirement is to be satisfied.

We are now prepared to consider the boundary conditions, treating first the *load-free* lateral surface. Recall from Sec. 1.3 that τ_{xz} is a z-directed shearing stress acting on a plane whose normal is parallel to the x axis, i.e., the yz plane. Similarly, τ_{zx} acts on the xy plane and is x-directed. By virtue of the symmetry of the stress tensor, we have $\tau_{xz} = \tau_{zx}$ and $\tau_{yz} = \tau_{zy}$. Therefore, the stresses given by Eq. (e) may be indicated on the xy plane near the boundary as shown in Fig. 6.3. The boundary element is associated with arc length ds. Note that ds increases in the counterclockwise direction. When ds is zero, the element represents a point at the boundary. Then, referring to Fig. 6.3 together with Eq. (1.28), which relates the surface forces to the internal stress, and noting that the cosine of the angle between z and a unit normal $\mathbf{n}$ to the surface is zero [i.e., $\cos(n, z) = 0$], we have

$$\tau_{zx} l + \tau_{zy} m = 0 \tag{f}$$

According to Eq. (f), the resultant shear stress τ must be tangent to the boundary (Fig. 6.3). From the figure, it is clear that

$$l = \cos(n, x) = \frac{dy}{ds}, \qquad m = \cos(n, y) = -\frac{dx}{ds} \tag{g}$$

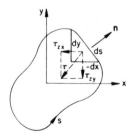

Figure 6.3

Note that as one proceeds in the direction of increasing s, x decreases and y increases. This accounts for the algebraic sign in front of dx and dy in Fig. 6.3. Substitution of Eqs. (6.4) and (g) into Eq. (f) yields

$$\frac{\partial \phi}{\partial y} \frac{dy}{ds} + \frac{\partial \phi}{\partial x} \frac{dx}{ds} = \frac{d\phi}{ds} = 0 \qquad \text{(on the boundary)} \qquad (6.6)$$

This expression states that the directional deviation along a boundary curve is zero. Thus the function $\phi(x, y)$ must be an *arbitrary* constant on the lateral surface of the prism. Examination of Eq. (6.4) indicates that the stresses remain the same regardless of additive constants, i.e., if $\phi + $ constant is substituted for ϕ, the stresses will not change. For solid cross sections, we are therefore free to set ϕ equal to zero at the boundary. In the case of multiply connected cross sections, such as hollow or tubular members, an arbitrary value may be assigned at the boundary of *only one* of the contours $s_0, s_1, \ldots, s_n$. For such members it is necessary to extend the mathematical formulation presented in this section. Solutions for thin-walled, multiply connected cross sections are treated in Sec. 6.5 by use of the membrane analogy.

Returning to the member of solid cross section, we complete discussion of the boundary conditions by considering the ends, at which the normals are parallel to the z axis and therefore $\cos(n, z) = n = \pm 1, l = m = 0$. Equation (1.28) now gives, for $T_z = 0$,

$$T_x = \pm \tau_{zx}, \qquad T_y = \pm \tau_{zy} \qquad (h)$$

where the algebraic sign depends upon the relationship between the outer normal and the positive z direction. For example, it is negative for the end face at the origin in Fig. 6.2a.

We now confirm the fact that the summation of forces over the ends of the bar is zero:

$$\int \int T_x \, dx \, dy = \int \int \tau_{zx} \, dx \, dy = \int \int \frac{\partial \phi}{\partial y} \, dx \, dy$$

$$= \int dx \int_{y_1}^{y_2} \frac{\partial \phi}{\partial y} \, dy = \int \phi |_{y_1}^{y_2} dx = \int (\phi_2 - \phi_1) \, dx = 0$$

Here y_1 and y_2 represent the y coordinates of points located on the surface. Inasmuch as $\phi = $ constant on the surface of the bar, the values of ϕ corresponding to y_1 and y_2 must be equal to a constant, $\phi_1 = \phi_2 = $ constant. Similarly it may be shown that

$$\int \int \tau_{zy} \, dx \, dy = 0$$

The end forces, while adding to zero, must nevertheless provide the required twisting moment or externally applied torque about the z axis:

$$M_t = \int\int(x\tau_{zy} - y\tau_{zx})\,dx\,dy = -\int\int x\frac{\partial\phi}{\partial x}\,dx\,dy - \int\int y\frac{\partial\phi}{\partial y}\,dx\,dy$$

$$= -\int dy \int x\frac{\partial\phi}{\partial x}\,dx - \int dx \int y\frac{\partial\phi}{\partial y}\,dy$$

Integrating by parts,

$$M_t = -\int x\phi\big|_{x_1}^{x_2}\,dy + \int\int\phi\,dx\,dy - \int y\phi\big|_{y_1}^{y_2}\,dx + \int\int\phi\,dx\,dy$$

Since $\phi=$ constant at the boundary and x_1, x_2, y_1, y_2 denote points on the lateral surface, it follows that

$$M_t = 2\int\int\phi\,dx\,dy \qquad\qquad (6.7)$$

Inasmuch as $\phi(x, y)$ has a value at each point on the cross section, it is clear that Eq. (6.7) represents twice the volume beneath the ϕ surface.

What has resulted from the foregoing development is a ϕ of equations satisfying all the conditions of the prescribed torsion problem. Equilibrium is governed by Eqs. (6.4), compatibility by Eq. (6.5), and the boundary condition by Eq. (6.6). Torque is related to stress by Eq. (6.7). To ascertain the distribution of stress, it is necessary to determine a stress function which satisfies Eqs. (6.5) and (6.6), as is demonstrated in the following example.

Example 6.1. Consider a solid bar of elliptical cross section (Fig. 6.4a). Determine the maximum shearing stress and the angle of twist per unit length. Also derive an expression for the warping $w(x, y)$. Take $M_t = 1200\pi$ N·m, $a = 0.05$ m, $b = 0.025$ m, and $G = 80$ GPa.

SOLUTION. Equations (6.5) and (6.6) are satisfied by selecting the stress function

$$\phi = k\left(\frac{x^2}{a^2} + \frac{y^2}{b^2} - 1\right)$$

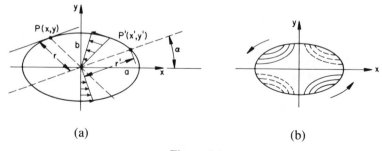

(a)   (b)

Figure 6.4

where k is a constant. Substituting the above into Eq. (6.5), we obtain

$$k = \frac{a^2 b^2}{2(a^2+b^2)} H$$

Hence

$$\phi = \frac{a^2 b^2 H}{2(a^2+b^2)} \left(\frac{x^2}{a^2} + \frac{y^2}{b^2} - 1 \right) \tag{i}$$

and Eq. (6.7) yields

$$M_t = \frac{a^2 b^2 H}{a^2+b^2} \left(\frac{1}{a^2} \int\int x^2\, dx\, dy + \frac{1}{b^2} \int\int y^2\, dx\, dy - \int\int dx\, dy \right)$$

$$= \frac{a^2 b^2 H}{a^2+b^2} \left(\frac{1}{a^2} I_y + \frac{1}{b^2} I_x - A \right)$$

where A is the cross-sectional area. Inserting expressions for I_x, I_y, and A results in

$$M_t = \frac{a^2 b^2 H}{a^2+b^2} \left(\frac{1}{a^2} \frac{\pi b a^3}{4} + \frac{1}{b^2} \frac{\pi a b^3}{4} - \pi a b \right) = \frac{-\pi a^3 b^3 H}{2(a^2+b^2)} \tag{j}$$

from which

$$H = - \frac{2 M_t (a^2+b^2)}{\pi a^3 b^3} \tag{k}$$

The stress function is now expressed as

$$\phi = - \frac{M_t}{\pi a b} \left(\frac{x^2}{a^2} + \frac{y^2}{b^2} - 1 \right)$$

and the shearing stresses are found readily from Eq. (6.4):

$$\tau_{zx} = \frac{\partial \phi}{\partial y} = - \frac{2 M_t y}{\pi a b^3} = - \frac{M_t y}{2 I_x}$$

$$\tau_{zy} = - \frac{\partial \phi}{\partial x} = \frac{2 M_t x}{\pi a^3 b} = \frac{M_t x}{2 I_y} \tag{l}$$

The ratio of these stress components is proportional to y/x and thus constant along any radius of the ellipse:

$$\frac{\tau_{zx}}{\tau_{zy}} = - \frac{y}{x} \frac{a^2}{b^2} = - \frac{y}{x} \frac{I_y}{I_x}$$

The resultant shearing stress,

$$\tau_{za} = \left(\tau_{zx}^2 + \tau_{zy}^2 \right)^{1/2} = \frac{2 M_t}{\pi a b} \left(\frac{x^2}{a^4} + \frac{y^2}{b^4} \right)^{1/2} \tag{m}$$

has a direction parallel to a tangent drawn at the boundary at its point of intersection with the radius containing the point under consideration. Note that α represents an arbitrary angle (Fig. 6.4a). To determine the location of the maximum resultant shear, which from Eq. (1) is somewhere on the boundary, consider a point $P'(x', y')$ located on a diameter conjugate to that containing $P(x, y)$ (Fig. 6.4a). Note that OP' is parallel to the tangent line at P. The coordinates of P and

P' are related by

$$x = \frac{a}{b}y', \qquad y = \frac{b}{a}x'$$

When these expressions are substituted into Eq. (m) we have

$$\tau_{z\alpha} = \frac{2M_t}{\pi a^2 b^2}(x'^2 + y'^2)^{1/2} = \frac{2M_t}{\pi a^2 b^2}r' \qquad (n)$$

Clearly, $\tau_{z\alpha}$ will have its maximum value corresponding to the largest value of the conjugate semidiameter r'. This occurs where $r' = a$, or $r = b$. The maximum resultant shearing stress thus occurs at $P(x, y)$ corresponding to the extremities of the minor axis as follows: $x = 0, y = \pm b$. From Eq. (n),

$$\tau_{max} = \frac{2M_t}{\pi a b^2} = \frac{2(1200\,\pi)}{\pi(0.05)(0.025)^2} = 76.8 \text{ MPa}$$

The angle of twist per unit length is obtained by substituting Eq. (k) into Eq. (6.3):

$$\theta = \frac{(a^2 + b^2)M_t}{\pi a^3 b^3 G} = \frac{(0.05^2 + 0.025^2)(1200\,\pi)}{\pi(0.05)^3(0.025)^3(80 \times 10^9)} = 0.024 \text{ rad/m} \qquad (o)$$

We note that the factor by which the twisting moment is divided to determine the twist per unit length is called the *torsional rigidity* or *torsional stiffness*, commonly denoted C. That is,

$$C = \frac{M_t}{\theta} \qquad (6.8)$$

The torsional rigidity, for an elliptical cross section, from Eq. (o), is thus

$$C = \frac{\pi a^3 b^3}{a^2 + b^2}G$$

The components of displacement u and v are then found from Eq. (b). To obtain the warpage $w(x, y)$, consider Eq. (e) into which have been substituted the previously derived relations for τ_{zx}, τ_{zy}, and θ:

$$\tau_{zx} = -\frac{2M_t y}{\pi a b^3} = G\left[\frac{\partial w}{\partial x} - \frac{y(a^2 + b^2)M_t}{\pi a^3 b^3 G}\right]$$

$$\tau_{zy} = \frac{2M_t x}{\pi a^3 b} = G\left[\frac{\partial w}{\partial y} + \frac{x(a^2 + b^2)M_t}{\pi a^3 b^3 G}\right]$$

Integration of these equations leads to identical expressions for $w(x, y)$, except that the first also yields an arbitrary function of y, $f(y)$, and the second an arbitrary function of x, $f(x)$. Since $w(x, y)$ must give the same value for a given $P(x, y)$, we conclude that $f(x) = f(y) = 0$; what remains is

$$w(x, y) = \frac{M_t}{G}\frac{(b^2 - a^2)xy}{\pi a^3 b^3}$$

The contour lines, obtained by setting $w = $ constant, are the hyperbolas shown in Fig. 6.4b. The solid lines indicate the portions of the section that become convex, and the dashed lines indicate the portions of the section that become concave, when the bar is subjected to a torque in the direction shown.

The results obtained in this example for an elliptical section may readily be reduced to the case of circular section by setting b equal to a.

6.3 Prandtl's Membrane Analogy

It is demonstrated below that the differential equation for the stress function, Eq. (6.5), is of the same form as the equation describing the deflection of a membrane or soap film subject to pressure. Hence, an analogy exists between the torsion and membrane problems, serving as the basis of a number of experimental techniques. Consider an edge supported homogeneous membrane, given its boundary contour by a hole cut in a plate (Fig. 6.5a). The *shape* of hole is the same as that of the twisted bar to be studied; the *sizes* need not be identical.

The equation describing the z deflection of the membrane is derived from considerations of equilibrium applied to the isolated element *abcd*. Let the tensile forces per unit membrane length be denoted by S. From a small z deflection, the inclination of S acting on side *ab* may be expressed

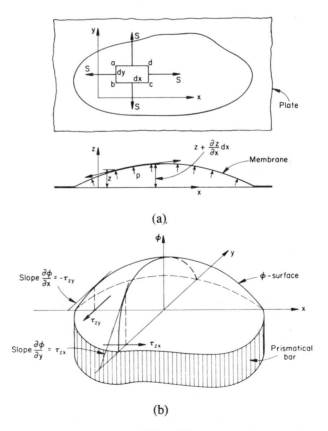

(a)

(b)

Figure 6.5

as $\beta \approx \partial z / \partial x$. Since z varies from point to point, the angle at which S is inclined on side dc is

$$\beta + \frac{\partial \beta}{\partial x} dx \approx \frac{\partial z}{\partial x} + \frac{\partial^2 z}{\partial x^2} dx$$

Similarly, on sides ad and bc, the angles of inclination for the tensile forces are $\partial z / \partial y$ and $\partial z / \partial y + (\partial^2 z / \partial y^2) dy$, respectively. In the development which follows, S is regarded as a constant, and the weight of the membrane is ignored. For a uniform lateral pressure p, the equation of vertical equilibrium is then

$$-(S\,dy)\frac{\partial z}{\partial x} + S\,dy\left(\frac{\partial z}{\partial x} + \frac{\partial^2 z}{\partial x^2}dx\right) - (S\,dx)\frac{\partial z}{\partial y}$$

$$+ (S\,dx)\left(\frac{\partial z}{\partial y} + \frac{\partial^2 z}{\partial y^2}dy\right) + p\,dx\,dy = 0$$

leading to

$$\frac{\partial^2 z}{\partial x^2} + \frac{\partial^2 z}{\partial y^2} = -\frac{p}{S} \tag{6.9}$$

This is again Poisson's equation. Upon comparison of Eq. (6.9) with Eqs. (6.5) and (6.4) the following quantities are observed to be analogous:

Membrane problem	Torsion problem
z	ϕ
$\dfrac{1}{S}$	G
p	2θ
$-\dfrac{\partial z}{\partial x},\ \dfrac{\partial z}{\partial y}$	τ_{zy},τ_{zx}
$2\cdot$(volume beneath membrane)	M_t

The membrane, subject to the conditions outlined, thus represents the ϕ surface (Fig. 6.5b). In view of the derivation, the restriction with regard to smallness of slope must be borne in mind.

Below is outlined one method by which the foregoing theory can be reduced to a useful experiment. In two thin stiff plates, bolted together, are cut two adjacent holes; one conforms to the outline of the irregular cross section, the other is circular. The plates are then separated and a thin sheet of rubber stretched across the holes (with approximately uniform and equal tension). The assembly is then bolted together. Subjecting one side of the membrane to a uniform pressure p causes a different distribution of deformation for each cross section, with the circular hole providing calibration data. The measured geometric quantities associated with the circular hole, together with the known solution, provide the needed proportionalities between pressure and angle of twist, slope and stress, volume and torque. These are then applied to the irregular cross section, for which the measured slopes and volume yield τ and M_t. The need for precise information concerning the membrane stress is thus obviated.

The membrane analogy provides more than a useful *experimental* technique. As is demonstrated in the next section, it also serves as the basis for obtaining approximate *analytical* solutions for bars of narrow cross section as well as for members of open thin-walled section.

For reference purposes, Table 6.1 presents the shear and angle of twist for a number of commonly encountered shapes.

Table 6.1

Cross section	Shear stress (τ)		Angle of twist per unit length (θ)
$2a$, $2b$, A For circular bar: $a = b$	$\tau_A = \dfrac{2M_t}{\pi ab^2}$		$\dfrac{a^2 + b^2}{\pi a^3 b^3} \cdot \dfrac{M_t}{G}$
A, a Equilateral triangle	$\tau_A = \dfrac{20M_t}{a^3}$		$\dfrac{46.2}{a^4} \cdot \dfrac{M_t}{G}$
a, b, A	$\tau_A = \dfrac{M_t}{\alpha ab^2}$		$\dfrac{1}{\beta ab^3} \cdot \dfrac{M_t}{G}$
	a/b	β	α
	1.0	0.141	0.208
	1.5	0.196	0.231
	2.0	0.229	0.246
	2.5	0.249	0.256
	3.0	0.263	0.267
	4.0	0.281	0.282
	5.0	0.291	0.292
	10.0	0.312	0.312
	∞	0.333	0.333
t_1, A, B, t, b, a	$\tau_A = \dfrac{M_t}{2abt_1}$ $\tau_B = \dfrac{M_t}{2abt}$		$\theta = \dfrac{at + bt_1}{2tt_1 a^2 b^2} \cdot \dfrac{M_t}{G}$
t, A, $2b$, $2a$ For circular tube: $a = b$	$\tau_A = \dfrac{M_t}{2\pi abt}$		$\theta = \dfrac{\sqrt{2(a^2 + b^2)}}{4\pi a^2 b^2 t} \cdot \dfrac{M_t}{G}$

6.4   Torsion of Thin-Walled Members of Open Cross Section

In applying the analogy to a bar of narrow rectangular cross section, it is usual to assume a constant cylindrical membrane shape over the entire dimension b (Fig. 6.6). Subject to this approximation, $\partial z/\partial y=0$, and Eq. (6.9) reduces to $d^2z/dx^2=-p/S$, which is twice integrated to yield the parabolic deflection

$$z=\frac{1}{2}\frac{p}{S}\left[\left(\frac{t}{2}\right)^2-x^2\right] \tag{a}$$

To arrive at Eq. (a), the boundary conditions that $dz/dx=0$ at $x=0$ and $z=0$ at $x=t/2$ have been employed. The volume bounded by the parabolic cylindrical membrane and the xy plane is given by $V=pbt^3/12S$. According to the analogy, p is replaced by 2θ and $1/S$ by G, and consequently $M_t=2V=\frac{1}{3}bt^3G\theta$. The torsional rigidity for a thin rectangular section is therefore

$$C=\frac{M_t}{\theta}=\frac{1}{3}bt^3G=JG$$

Here J represents the polar moment of inertia of the section. The analogy also requires that

$$\tau_{zy}=-\frac{\partial z}{\partial x}=2G\theta x \tag{b}$$

The angle of twist per unit length is, from the expression for torque,

$$\theta=\frac{3M_t}{bt^3G} \tag{6.10}$$

Maximum shear occurs at $\pm t/2$:

$$\tau_{max}=G\theta t=\frac{3M_t}{bt^2} \tag{6.11}$$

or

$$M_t=\frac{1}{3}bt^2\tau_{max} \tag{6.12}$$

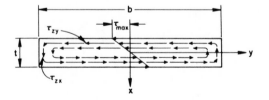 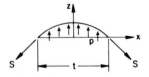

Figure 6.6

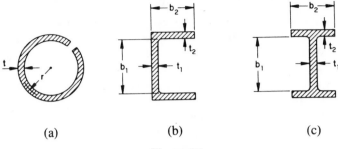

Figure 6.7

According to Eq. (b) the shearing stress is linear in x, as in Fig. 6.6, producing a twisting moment M_t about z given by

$$M_t = 2\left(\frac{1}{2}\tau_{max}\frac{t}{2}\right)\left(\frac{t}{3}\right)(b) = \frac{1}{6}bt^2\tau_{max}$$

This is exactly one-half the torque given by Eq. (6.12). The remaining applied torque is evidently resisted by the shearing stresses τ_{zx}, neglected in the original analysis in which the membrane is taken as cylindrical. The membrane slope at $y = \pm b/2$ is smaller than that at $x = \pm t/2$, or equivalently, $(\tau_{zx})_{max} < (\tau_{zy})_{max}$. It is clear, therefore, that Eq. (6.11) represents the maximum shearing stress in the bar, of a magnitude unaffected by the original approximation. That the lower τ_{zx} stresses can provide a resisting torque equal to that of the τ_{zy} stresses is explained on the basis of the longer moment arm for the stresses near $y = \pm b/2$.

Equations (6.10) and (6.11) are also applicable to thin-walled sections such as those shown in Fig. 6.7. Because the foregoing expressions neglect stress concentration, the points of interest should be reasonably distant from the corners of the section (Figs. 6.7b and c). The validity of the foregoing approach depends upon the degree of similarity between the membrane shape of Fig. 6.6 and that of the geometry of the component section. Consider, for example, the I-section of Fig. 6.7c. Summing the torsional rigidity of the three rectangular components, we obtain

$$\theta = \frac{M_t}{G\left(\frac{1}{3}b_1t_1^3 + \frac{2}{3}b_2t_2^3\right)} = \frac{3M_t}{G}\frac{1}{b_1t_1^3 + 2b_2t_2^3} \tag{6.13}$$

$$\tau_{max} = G\theta t_i = \frac{3M_t t_i}{b_1t_1^3 + 2b_2t_2^3} \tag{6.14}$$

where t_i is the larger of t_1 and t_2. The effect of the stress concentrations at the corners will be examined in Sec. 6.6.

6.5 Torsion of Multiply Connected Thin-Walled Sections

The membrane analogy may be applied to good advantage to analyze the torsion of thin tubular members, provided that some care is taken. Consider the deformation which would occur if a membrane subject to pressure were to span a hollow tube of arbitrary section (Fig. 6.8a). Since the membrane surface is to describe the stress function (and its slope, the stress at any point), arc ab cannot represent a meaningful stress function. This is simply because in the region ab, the stress must be zero, as no material exists there. If the curved surface ab is now replaced by a plane representing constant ϕ, the zero-stress requirement is satisfied. For bars containing multiply connected regions, each boundary is also a line of constant ϕ, of different value. The absolute value of ϕ is meaningless, and therefore at one boundary, ϕ may arbitrarily be equated to zero, the others adjusting accordingly.

Based upon the foregoing considerations, the membrane analogy is extended to a thin tubular member (Fig. 6.8b), in which the fixed plate to which the membrane is attached has the same contour as the outer boundary of the tube. The membrane is also attached to a "weightless" horizontal plate having the same shape as the inner boundary nn of the tube, thus bridging the inner and outer contours over a distance t. The inner horizontal plate, made "weightless" by a counterbalance system, is permitted to seek its own vertical position, but is guided so as not to experience sideward motion. As we have assumed the tube to be thin-walled, the membrane curvature may be disregarded; i.e., lines nn may be considered straight. We are thus led to conclude that the slope is constant over a given thickness t, and consequently the shearing stress is likewise constant, given by

$$\tau = \frac{h}{t} \qquad \text{or} \qquad h = \tau t \qquad \text{(a)}$$

where h is the membrane deflection and t the tube thickness. Note that the tube thickness may vary circumferentially.

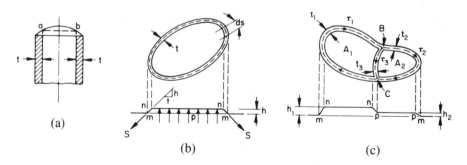

(a)   (b)   (c)

Figure 6.8

The dashed line in Fig. 6.8b indicates the mean perimeter, which may be used to determine the volume bounded by the membrane. Letting A represent the *area enclosed by the mean perimeter*, the volume *mnnm* is simply Ah, and the analogy gives

$$M_t = 2Ah \quad \text{or} \quad h = \frac{M_t}{2A} \tag{b}$$

Combining Eqs. (a) and (b), we have

$$\tau = \frac{M_t}{2At} \tag{6.15}$$

The application of the above equation is limited to thin-walled members displaying no abrupt variations in thickness and no reentrant corners in their cross sections.

To develop a relationship for the angle of twist from the membrane analogy, we again consider Fig. 6.8b, in which $h \ll t$ and consequently $\tan(h/t) \approx h/t = \tau$. Vertical equilibrium therefore yields

$$pA = \oint \left(\frac{h}{t} S\right) ds = \oint \tau S \, ds$$

Here s is the *length of the mean perimeter* of the tube. Since the membrane tension is constant, h is independent of S. The above is then written

$$\frac{p}{S} = \frac{1}{A} \oint \tau \, ds = \frac{h}{A} \oint \frac{ds}{t} = 2G\theta$$

where the last term follows from the analogy. The angle of twist per unit length is now found directly:

$$\theta = \frac{h}{2GA} \oint \frac{ds}{t} = \frac{1}{2GA} \oint \tau \, ds \tag{6.16}$$

Eqs. (6.15) and (6.16) are known as *Bredt's formulas*.

In Eqs. (a), (b), and (6.16), the quantity h possesses the dimensions of force per unit length, representing the resisting force per unit length along the tube perimeter. For this reason, h is referred to as the *shear flow*.

Example 6.2. A hollow aluminum tube of rectangular cross section (Fig. 6.9a, with the dimensions given in meters) is subjected to a torque of 56.5 kN·m along its longitudinal axis. Determine the shearing stresses and the angle of twist. Assume $G = 28$ GPa.

SOLUTION. Referring to Fig. 6.9b, which shows the membrane surface *mnnm* (representing ϕ), the applied torque is, according to Eq. (b),

$$M_t = 2Ah = 2(0.125h) = 56,500 \text{ N·m}$$

from which $h = 226,000$ N/m. The shearing stresses are found from Eq. (a) as

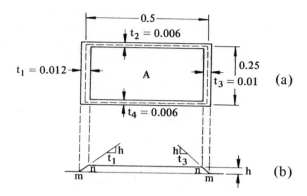

Figure 6.9

follows:

$$\tau_1 = \frac{h}{t_1} = \frac{226,000}{0.012} = 18.833 \text{ MPa}$$

$$\tau_2 = \frac{h}{t_2} = \frac{h}{t_4} = \frac{226,000}{0.006} = 37.667 \text{ MPa}$$

$$\tau_3 = \frac{h}{t_3} = \frac{226,000}{0.01} = 22.6 \text{ MPa}$$

Applying Eq. (6.16), the angle of twist per unit length is

$$\theta = \frac{h}{2GA} \oint \frac{ds}{t} = \frac{226,000}{2 \times 28 \times 10^9 \times 0.125} \left(\frac{0.25}{0.012} + 2\frac{0.5}{0.006} + \frac{0.25}{0.01} \right)$$
$$= 0.00686 \text{ rad/m}$$

If multiply connected regions exist within a tubular member, as in Fig. 6.8c, the foregoing techniques are again appropriate. As before, the thicknesses are assumed small, so that lines such as *mn*, *np*, and *pm* are regarded as straight. The stress function is then represented by the membrane surface *mnnppm*. As in the case of a simple hollow tube, lines *nn* and *pp* are straight by virtue of flat, weightless plates with contours corresponding to the inner openings. Referring to the figure, the shearing stresses are

$$\tau_1 = \frac{h_1}{t_1}, \qquad \tau_2 = \frac{h_2}{t_2} \tag{c}$$

$$\tau_3 = \frac{h_1 - h_2}{t_3} = \frac{t_1\tau_1 - t_2\tau_2}{t_3} \tag{d}$$

The stresses are produced by a torque equal to twice the volume beneath surface *mnnppm*,

$$M_t = 2A_1 h_1 + 2A_2 h_2 \tag{e}$$

or, upon substitution of Eqs. (c),

$$M_t = 2A_1 t_1 \tau_1 + 2A_2 t_2 \tau_2 \tag{f}$$

Assuming the thicknesses t_1, t_2, and t_3 constant, application of Eq. (6.16) yields

$$\tau_1 s_1 + \tau_3 s_3 = 2G\theta A_1 \tag{g}$$

$$\tau_2 s_2 - \tau_3 s_3 = 2G\theta A_2 \tag{h}$$

where s_1, s_2, and s_3 represent the paths of integration indicated by the dashed lines. Note the relationship between the algebraic sign, the assumed direction of stress, and the direction in which integration proceeds. There are thus four equations [(d), (f), (g), and (h)], containing four unknowns: τ_1, τ_2, τ_3, and θ.

If now Eq. (d) is written in the form

$$\tau_1 t_1 = \tau_2 t_2 + \tau_3 t_3 \tag{i}$$

it is observed that the shear flow $h = \tau t$ is constant and distributes itself in a manner analogous to a liquid circulating through a channel of shape identical with that of the tubular bar. This analogy proves very useful in writing expressions for shear flow in tubular sections of considerably greater complexity.

Example 6.3. A multiply connected hollow steel tube (Fig. 6.10) resists a torque of 12 kN·m. The wall thicknesses are $t_1 = t_2 = t_3 = 6$ mm, and $t_4 = t_5 = 3$ mm. Determine the maximum shearing stresses and the angle of twist per unit length. Let $G = 80$ GPa.

SOLUTION. Assuming the shearing stresses directed as shown, consideration of shear flow yields

$$\tau_1 t_1 = \tau_2 t_2 + \tau_4 t_4, \qquad \tau_2 t_2 = \tau_5 t_5 + \tau_3 t_3 \tag{j}$$

The torque associated with the shearing stresses must resist the externally applied torque, and an expression similar to Eq. (f) is obtained:

$$M_t = 2A_1 t_1 \tau_1 + 2A_2 t_2 \tau_2 + 2A_3 t_3 \tau_3 = 12{,}000 \text{ N·m} \tag{k}$$

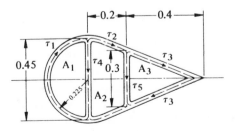

Figure 6.10

Three more equations are available through application of Eq. (6.16) over areas A_1, A_2, and A_3:

$$\tau_1 s_1 + \tau_4 s_4 = 2G\theta A_1$$

$$-\tau_4 s_4 + 2\tau_2 s_2 + \tau_5 s_5 = 2G\theta A_2 \qquad (1)$$

$$-\tau_5 s_5 + 2\tau_3 s_3 = 2G\theta A_3$$

In the above $s_1 = 0.7069$ m, $s_2 = 0.2136$ m, $s_3 = 0.4272$ m, $s_4 = 0.45$ m, $s_5 = 0.3$ m, $A_1 = 0.079522$ m^2, $A_2 = 0.075$ m^2, and $A_3 = 0.06$ m^2. There are six equations in the five unknown stresses and the angle of twist per unit length. Thus, simultaneous solution of Eqs. (j), (k), and (l) leads to the following rounded values: $\tau_1 = 4.902$ MPa, $\tau_2 = 5.088$ MPa, $\tau_3 = 3.809$ MPa, $\tau_4 = -0.373$ MPa, $\tau_5 = 2.558$ MPa, and $\theta = 0.0002591$ rad/m. The positive values obtained for τ_1, τ_2, τ_3, and τ_5 indicate that the directions of these stresses have been correctly assumed in Fig. 6.10. The negative sign of τ_4 means that the direction initially assumed was incorrect, i.e., τ_4 is actually upward directed.

6.6   Fluid Flow Analogy and Stress Concentration

Examination of Eq. 6.4 suggests a similarity between the stress function ϕ and the *stream function* ψ of fluid mechanics:

$$\tau_{zx} = \frac{\partial \phi}{\partial y}, \qquad \tau_{zy} = -\frac{\partial \phi}{\partial x}$$

$$V_x = \frac{\partial \psi}{\partial y}, \qquad V_y = -\frac{\partial \psi}{\partial x} \qquad (6.17)$$

In Eqs. (6.17), V_x and V_y represent the x and y components of the fluid velocity $\mathbf{V}$. Recall that for an incompressible fluid, the equation of continuity may be written

$$\frac{\partial V_x}{\partial x} + \frac{\partial V_y}{\partial y} = 0$$

Continuity is thus satisfied when $\psi(x, y)$ is defined as in Eqs. (6.17). The vorticity $\omega = \frac{1}{2}(\nabla \times \mathbf{V})$, is for two-dimensional flow,

$$\omega = \frac{1}{2}\left(-\frac{\partial V_x}{\partial y} + \frac{\partial V_y}{\partial x}\right)$$

where $\nabla = (\partial/\partial x)\mathbf{i} + (\partial/\partial y)\mathbf{j}$. In terms of the stream function, we obtain

$$\frac{\partial^2 \psi}{\partial y^2} + \frac{\partial^2 \psi}{\partial x^2} = -2\omega \qquad (6.18)$$

This expression is clearly analogous to Eq. (6.5) with -2ω replacing $-2G\theta$. The completeness of the analogy is assured if it can be demonstrated that ψ is constant along a streamline (and hence on a boundary), as ϕ is constant over a boundary. Since the equation of a streamline in two-dimensional flow is

$$\frac{dy}{dx} = \frac{V_y}{V_x} \qquad \text{or} \qquad V_x\,dy - V_y\,dx = 0$$

in terms of the stream function, we have

$$\frac{\partial\psi}{\partial y}\,dy + \frac{\partial\psi}{\partial x}\,dx = 0 \qquad (6.19)$$

This is simply the total differential $d\psi$, and therefore ψ is constant along a streamline.

Based upon the foregoing, experimental techniques have been developed in which the analogy between the motion of an ideal fluid of constant vorticity and the torsion of a bar is successfully exploited. The tube in which the fluid flows and the cross section of the twisted member are identical in these experiments, useful in visualizing stress patterns in torsion. Moreover, a vast body of literature exists which deals with flow patterns around bodies of various shapes, and the results presented are often directly applicable to the torsion problem.

The hydrodynamic analogy is especially valuable in dealing with stress concentration in shear, which we have heretofore neglected. In this regard, consider first the torsion of a circular bar containing a small circular hole (Fig. 6.11a). In Fig. 6.11b is shown the analogous flow pattern produced by a solid cylindrical obstacle placed in a circulating fluid. From hydrodynamic theory, it is found that the maximum velocity (at points a and b) is twice the value in the undisturbed stream at the respective radii. From this, it is concluded that a small hole has the effect of doubling the shearing stress normally found at a given radius.

Of great importance also is the shaft keyway shown in Fig. 6.11c. According to the hydrodynamic analogy, the points a ought to have zero stress, since they are stagnation points of the fluid stream. In this sense, the material in the immediate vicinity of points a is excess. On the other hand, the velocity at the points b is theoretically infinite, and by analogy, so is the stress. It is therefore not surprising that most torsional fatigue failures have their origins at these sharp corners, and the lesson is thus supplied that it is profitable to round such corners.

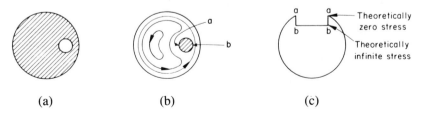

(a)   (b)   (c)

Figure 6.11

6.7 Torsion of Restrained Thin-Walled Members of Open Cross Section

It is a basic premise of previous sections of this chapter that all cross sections of a bar subject to torques applied at the ends suffer free warpage. As a consequence, one must assume that the torque is produced by pure shearing stresses distributed over the ends as well as all other cross sections of the member. In this way, the stress distribution is obtained from Eq. (6.5) and satisfies the boundary conditions, Eq. (6.6).

If any section of the bar is held rigidly, it is clear that the rate of change of the angle of twist as well as the warpage will now vary in the longitudinal direction. The longitudinal fibers are therefore subject to tensile or compressive stresses. Equations (6.5) and (6.6) are, in this instance, applied with satisfactory results in regions away from the re-strained section of the bar. While this restraint has negligible influence upon the torsional resistance of members of solid section such as rectangles and ellipses, it is significant when dealing with open thin-walled sections such as channels or I-beams. Consider, for example, the case of a cantilever I-beam, shown in Fig. 6.12. The applied torque causes each cross section to rotate about the axis of twist (z), thereby resulting in bending of the flanges. According to beam theory, the associated bending stresses in the flanges are zero at the juncture with the web. Consequently, the web does not depart from a state of simple torsion. In resisting the bending of the flanges or the warpage of a cross section, considerable torsional stiffness can, however, be imparted to the beam.

Referring to Fig. 6.12, the applied torque M_t is balanced in part by the action of torsional shearing stresses and in part by the resistance of the flanges to bending. At the representative section AB, consider the in-

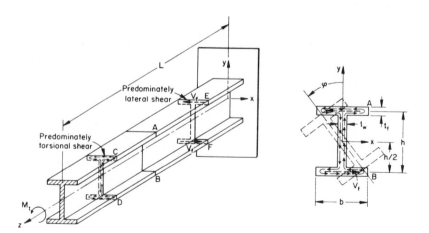

Figure 6.12

fluence of torques M_{t1} and M_{t2}. The former is attributable to pure torsional shearing stresses in the entire cross section, assumed to occur as though each cross section were free to warp. Torque M_{t1} is thus related to the angle of twist of section AB by the expression

$$M_{t1} = C \frac{d\varphi}{dz} \tag{a}$$

in which C is the torsional rigidity of the beam. The right hand rule should be applied to furnish the sign convention for both torque and angle of twist. A pair of lateral shearing forces *owing to bending of the flanges* acting through moment arm h gives rise to torque M_{t2}:

$$M_{t2} = V_f h \tag{b}$$

An expression for V_f may be derived by considering the x displacement, u. Because the beam cross section is symmetrical and the deformation small, we have $u = (h/2)\varphi$, and

$$\frac{du}{dz} = \frac{h}{2} \frac{d\varphi}{dz} \tag{c}$$

Thus, the bending moment M_f and shear V_f in the flange are

$$M_f = EI_f \frac{d^2 u}{dz^2} = \frac{EI_f h}{2} \frac{d^2 \varphi}{dz^2} \tag{d}$$

$$V_f = -EI_f \frac{d^3 u}{dz^3} = -\frac{EI_f h}{2} \frac{d^3 \varphi}{dz^3} \tag{e}$$

where I_f is the moment of inertia of one flange about the y axis. Now Eq. (b) becomes

$$M_{t2} = -\frac{EI_f h^2}{2} \frac{d^3 \varphi}{dz^3} \tag{f}$$

The total torque is therefore

$$M_t = M_{t1} + M_{t2} = C \frac{d\varphi}{dz} - \frac{EI_f h^2}{2} \frac{d^3 \varphi}{dz^3} \tag{6.20}$$

The conditions appropriate to the flange ends are

$$\left(\frac{d\varphi}{dz} \right)_{z=0} = 0, \qquad \left(\frac{d^2 \varphi}{dz^2} \right)_{z=L} = 0$$

indicating that the slope and bending moment are zero at the fixed and free ends, respectively. The solution of Eq. (6.20) is, upon satisfying these conditions,

$$\frac{d\varphi}{dz} = \frac{M_t}{C} \left[1 - \frac{\cosh \alpha (L-z)}{\cosh \alpha L} \right] \tag{g}$$

where

$$\alpha = \left(\frac{2C}{EI_f h^2} \right)^{1/2} \tag{6.21}$$

For a beam of *infinite length*, Eq. (g) reduces to

$$\frac{d\varphi}{dz} = \frac{M_t}{C}(1 - e^{-\alpha z}) \tag{h}$$

By substituting Eq. (h) into Eqs. (a) and (f), the following expressions result:

$$\begin{aligned} M_{t1} &= M_t(1 - e^{-\alpha z}) \\ M_{t2} &= M_t e^{-\alpha z} \end{aligned} \tag{6.22}$$

From the above, it is noted that at the fixed end ($z=0$), $M_{t1}=0$ and $M_{t2}=M_t$. At this end, the applied torque is counterbalanced by the effect of shearing forces only, which from Eq. (b) are given by $V_f = M_{t2}/h = M_t/h$. The torque distribution, Eq. (6.22), indicates that sections such as EF, close to the fixed end, contain predominantly lateral shearing forces (Fig. 6.12). Sections such as CD, near the free end, contain mainly torsional shearing stresses (as Eq. 6.22 indicates for $z \to \infty$).

The flange bending moment, obtained from Eqs. (d) and (h), is a maximum at $z=0$:

$$M_{f,\max} = \frac{EI_f h\alpha}{2C} M_t \tag{i}$$

The maximum bending moment, occurring at the fixed end of the flange, is found by substituting the relations (6.21) into (i):

$$M_{f,\max} = \frac{M_t}{\alpha h} \tag{6.23}$$

An expression for the angle of twist is determined by integrating Eq. (h) and satisfying the condition $\varphi = 0$ at $z=0$:

$$\varphi = \frac{M_t}{C}\left[z + \frac{1}{\alpha}(e^{-\alpha z} - 1) \right] \tag{j}$$

For relatively long beams, for which $e^{-\alpha z}$ may be neglected, the total angle of twist at the free end is, from Eq. (j),

$$\varphi_{z=L} = \frac{M_t}{C}\left(L - \frac{1}{\alpha} \right) \tag{6.24}$$

In this equation the term $1/\alpha$ indicates the influence of flange bending upon the angle of twist. Since for pure torsion, the total angle of twist is given by $\varphi = M_t L/C$, it is clear that end restraint increases the stiffness of the beam in torsion.

Example 6.4. A cantilever I-beam with the idealized cross section shown in Fig. 6.12 is subjected to a torque of 1.2 kN·m. Determine (a) the maximum longitudinal stress, (b) the total angle of twist, φ. Take $G = 80$ GPa and $E = 200$ GPa. Let $t_f = 10$ mm, $t_w = 7$ mm, $b = 0.1$ m, $h = 0.2$ m, and $L = 2.4$ m.

SOLUTION.

(a) The torsional rigidity of the beam is, from Eq. (6.13),

$$C = \frac{M_t}{\theta} = \left(b_1 t_1^3 + 2 b_2 t_2^3 \right) \frac{G}{3} = (0.19 \times 0.007^3 + 2 \times 0.1 \times 0.01^3) \frac{G}{3} = 8.839 \times 10^{-8} G$$

The flexural rigidity of one flange is

$$I_f E = \frac{0.01 \times 0.1^3}{12} E = 8.333 \times 10^{-7} E$$

Hence, from Eq. (6.21) we have

$$\frac{1}{\alpha} = h \sqrt{\frac{EI_f}{2C}} = h \sqrt{\frac{2.5 \times 8.333 \times 10^{-7}}{8.839 \times 10^{-8} \times 2}} = 3.43h$$

From Eq. (6.23), the bending moment in the flange is found to be 3.43 times larger than the applied torque, M_t. Thus the maximum longitudinal bending stress in the flange is

$$\sigma_{f, \max} = \frac{M_{f, \max} x}{I_f} = \frac{3.43 M_t \times 0.05}{8.333 \times 10^{-7}} = 0.2058 \times 10^6 M_t = 246.97 \text{ MPa}$$

(b) Since $e^{-\alpha L} = 0.03$, we can apply Eq. (6.24) to calculate the angle of twist at the free end:

$$\varphi = \frac{M_t}{C} \left(L - \frac{1}{\alpha} \right) = \frac{1200}{8.839 \times 10^{-8} \times 80 \times 10^9} (2.4 - 3.43 \times 0.2) = 0.2908 \text{ rad}$$

It is interesting to note that if the ends of the beam were both free, the total angle of twist would be $\varphi = M_t L / C = 0.4073$ rad, and the beam would experience $\varphi_{\text{free}} / \varphi_{\text{fixed}} = 1.4$ times more twist under the same torque.

6.8 Curved Circular Bars

The assumptions of Sec. 6.1 are also valid for a *curved, circular bar*, provided that the radius r of the bar is small in comparison with the radius of curvature R. When $r/R = \frac{1}{12}$ for example, the maximum stress computed on the basis of the torsion formula, $\tau = M_t r / J$, is approximately 5% too low.* On the other hand, if the radius of the bar is large relative to the radius of curvature, the length differential of the surface elements must be taken into consideration and there is a stress concentration at the inner point of the bar. We are concerned here with the torsion of slender curved members for which $r/R \ll 1$.

---

*See A. M. Wahl, *Mechanical Springs*, New York: McGraw-Hill, 1963.

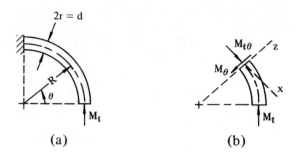

Figure 6.13

Frequently, a curved bar is subjected to loads which at any cross section produce a twisting moment as well as a bending moment. Expressions for the strain energy in torsion and bending have already been developed (Secs. 2.9 and 5.10) and application of Castigliano's theorem (Sec. 10.4) leads readily to the displacements.

Consider the case of a cylindrical rod or bar bent into a quarter circle of radius R as shown in Fig. 6.13a. The rod is fixed at one end and loaded at the free end by a twisting moment M_t. The bending and twisting moments at any section are (Fig. 6.13b)

$$M_\theta = M_t \sin \theta, \qquad M_{t\theta} = -M_t \cos \theta \tag{a}$$

Substituting these quantities into Eqs. (2.45) and (2.30) together with $dx = ds = R \, d\theta$ yield

$$U = \frac{1}{2} \int \left(\frac{M_\theta^2}{EI} + \frac{M_{t\theta}^2}{GJ} \right) ds \tag{b}$$

or

$$U = \frac{1}{2} M_t^2 R \int_0^{\pi/2} \left(\frac{\sin^2 \theta}{EI} + \frac{\cos^2 \theta}{GJ} \right) d\theta$$

where $J = \pi d^4/32 = 2I$. The strain energy in the entire rod is obtained by integrating the above:

$$U = \frac{8(2+\nu)}{d^4 E} M_t^2 \tag{6.25}$$

Upon application of $\varphi = \partial U/\partial M_t$, it is found that

$$\varphi = \frac{16(2+\nu)R}{d^4 E} M_t \tag{6.26}$$

for the *angle of twist* at the free end.

A helical spring, produced by wrapping a wire around a cylinder in such a way that the wire forms a helix of uniformly spaced turns, as typifies a curved bar, is discussed in the following example.

Example 6.5. An *open-coiled helical spring* wound from wire of diameter d, with *pitch angle* α and n *number of coils* of radius R, is extended by an axial load P (Fig. 6.14). (a) Develop expressions for maximum stress and deflection. (b) Redo part (a) for the spring *closely coiled*.

SOLUTION. An element of the spring located between two adjoining sections of the wire may be treated as a straight circular bar in torsion and bending. This is because a tangent to the coil at any point such as A is not perpendicular to the load. At cross section A, components $P\cos\alpha$ and $P\sin\alpha$ produce the following respective torque and moment.

$$M_t = PR\cos\alpha, \qquad M = PR\sin\alpha \qquad \text{(c)}$$

(a) The stresses, from Eq. (d) of Sec. 4.8, are given by

$$\sigma_{1,2} = \frac{16}{\pi d^3}\left(M \pm \sqrt{M^2 + M_t^2}\right), \qquad \tau_{\max} = \frac{16}{\pi d^3}\sqrt{M^2 + M_t^2}$$

The *maximum normal stress* and the *maximum shear stress* are thus

$$\sigma_{\max} = \frac{16PR}{\pi d^3}(1 + \sin\alpha) \qquad \text{(6.27a)}$$

and

$$\tau_{\max} = \frac{16PR}{\pi d^3} \qquad \text{(6.27b)}$$

The deflection is computed by applying Castigliano's theorem (Sec. 10.4) together with Eqs. (b) and (c):

$$e = \int_0^L \left(\frac{PR^2\sin^2\alpha}{EI} + \frac{PR^2\cos^2\alpha}{GJ}\right) ds$$

where the length of the coil $L = 2\pi Rn$. It follows that the relationship

$$e = 2\pi PR^3 n\left(\frac{\sin^2\alpha}{EI} + \frac{\cos^2\alpha}{GJ}\right)$$

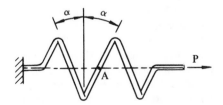

Figure 6.14

or

$$e = \frac{128 PR^3 n}{d^4}\left(\frac{\sin^2 \alpha}{E} + \frac{\cos^2 \alpha}{2G}\right) \tag{6.28}$$

defines the *axial end deflection* of an open-coil helical spring.

(b) For a *closely coiled helical spring* the angle of pitch α of the coil is quite small. The deflection is now produced *entirely* by the torsional stresses induced in the coil. To derive expressions for the stress and deflection, let $\alpha = \sin \alpha = 0$ and $\cos \alpha = 1$ in Eqs. (6.27) and (6.28). In so doing, we obtain

$$\tau_{max} = \sigma_{max} = \frac{16 PR}{\pi d^3} \tag{6.29}$$

and

$$e = \frac{64 PR^3 n}{G d^4} \tag{6.30}$$

The foregoing results are applicable to *both tension and compression* helical springs, the wire diameter of which are small in relation to coil radius.

Chapter 6—Problems

Secs. 6.1 and 6.2

6.1. Consider two bars, one having a circular section of radius b, the other an elliptic section with semiaxes a, b. Determine (a) for equal angles of twist, which bar experiences the larger shearing stress, and (b) for equal allowable shearing stresses, which one resists a larger twisting moment.

6.2. The stress function appropriate to a solid bar subjected to torques at its free ends is given by

$$\phi = k(a^2 - x^2 + by^2)(a^2 + bx^2 - y^2)$$

where a and b are constants. Determine the value of k.

6.3. Show that Eqs. (6.2) through (6.7) are not altered by a shift of the origin of x, y, z from the center of twist to any point within the cross section defined by $x = a, y = b$, where a and b are constants. [Hint: The displacements are now expressed $u = -\theta z(y - b)$, $v = \theta z(x - a)$, and $w = w(x, y)$.]

6.4. Rederive Eq. (6.7) for the case in which the stress function $\phi = c$ on the boundary, where c is a nonzero constant.

6.5. The thin circular ring of cross-sectional radius r, shown in Fig. P6.5, is subjected to a distributed torque per unit length, $M_{t\theta} = M_t \cos^2 \theta$. Determine the angle of twist at sections A and B in terms of M_t, a, and r. Assume that the radius a is large enough to permit the effect of curvature on the torsion formula to be neglected.

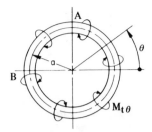

Figure P6.5

6.6. The torsion solution for a cylinder of equilateral triangular section (Fig. P6.6) is derivable from the stress function

$$\phi = k\left(x - \sqrt{3}\,y - \tfrac{2}{3}h\right)\left(x + \sqrt{3}\,y - \tfrac{2}{3}h\right)\left(x + \tfrac{1}{3}h\right)$$

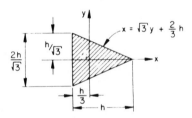

Figure P6.6

Derive expressions for the maximum and minimum shearing stresses and the twisting angle.

6.7. The torsional rigidity of a circle, and ellipse, and an equilateral triangle (Fig. P6.6) are denoted by C_c, C_e, and C_t, respectively. If the cross-sectional areas of these sections are equal, demonstrate that the following relationships exist:

$$C_e = \frac{2ab}{a^2 + b^2}\,C_c, \qquad C_t = \frac{2\pi\sqrt{3}}{15}\,C_c$$

where a and b are the semiaxes of the ellipse in the x and y directions.

6.8. Two thin-walled circular tubes, one having a seamless section, the other (Fig. 6.7a) a split section, are subjected to the action of identical twisting moments. Both tubes have equal outer diameter d_o, inner diameter d_i, and thickness t. Determine the ratio of their angles of twist.

Secs. 6.3 and 6.4

6.9. A steel bar of slender rectangular cross section (5 mm× 125 mm) is subjected to twisting moments of 80 N·m at the ends. Calculate the maximum shearing stress and the angle of twist per unit length. Take $G = 80$ GPa.

6.10. Derive an approximate expression for the twisting moment in terms of G, θ, b, t_0 for the thin triangular section shown in Fig. P6.10. Assume that at any y, the expression for the stress function ϕ corresponds to a parabolic membrane appropriate to the width at that y: $\phi = G\theta[(t/2)^2 - x^2]$.

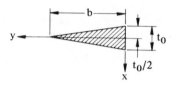

Figure P6.10

6.11. Consider the sections described below: (a) A hollow tube of 50-mm outside diameter and 2.5-mm wall thickness. (b) An equal angle, having the same perimeter and thickness as above. (c) A square box section with 50-mm sides and 2.5-mm wall thickness. Compare the torsional rigidities and the maximum shearing stresses for the same applied torque.

6.12. The cross section of a 3-m-long steel bar is an equilateral triangle with 50-mm sides. The bar is subjected to end twisting moments causing a maximum shearing stress equal to two-thirds of the elastic strength in shear ($\tau_{yp} = 420$ MPa). Determine, using Table 6.1, the angle of twist between the ends. Let $G = 80$ GPa.

Secs. 6.5 to 6.8

6.13. Show that when Eq. (6.8) is applied to a thin-walled tube, it reduces to Eq. (6.15).

6.14. Redo Example 6.2 with a 0.01-m-thick vertical wall at the middle of the section.

6.15. The cross section of a thin-walled aluminum tube is an equilateral triangular section of mean side length 50 mm and wall thickness 3.5 mm. If the tube is subjected to a torque of 40 N·m, what are the maximum shearing stress and angle of twist per unit length? Let $G = 28$ GPa.

6.16. A hollow, multicell aluminum tube (cross section shown in Fig. P6.16) resists a torque of 4 kN·m. The wall thicknesses are $t_1 = t_2 = t_4 = t_5 = 0.5$ mm,

$t_3 = 0.75$ mm. Determine the maximum shearing stresses and the angle of twist per unit length. Let $G = 28$ GPa.

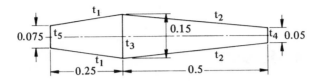

Figure P6.16

6.17. Consider two closely-coiled helical springs, one made of steel, the other of copper, each 0.01 m in diameter, one fitting within the other. Each has an identical number of coils $n = 20$ and ends constrained to deflect the same amount. The steel outer spring, has a diameter of 0.124 m and the copper inner spring, a diameter of 0.1 m. Determine (a) the total axial load the two springs can jointly sustain if the shear stresses in the steel and the copper are not to exceed 500 MPa and 300 MPa, respectively, and (b) the ratio of spring constants. For steel and copper, use shear moduli of elasticity $G_s = 79$ GPa and $G_c = 41$ GPa, respectively.

Chapter 7

Numerical Methods

7.1 Introduction

In this chapter we discuss a number of numerical approaches to the solution of problems in elasticity and the mechanics of materials. The use of numerical methods enables the engineer to expand his ability to solve practical design problems. He may now treat real shapes as distinct from the somewhat limited variety of shapes amenable to simple analytic solution. Similarly, he need no longer force a complex loading system to fit a more regular load configuration in order to conform to the dictates of a purely academic situation. Numerical analysis thus provides a tool with which the engineer may feel freer to undertake the solution of problems as they are found in practice.

Analytical solutions of the type discussed in earlier chapters have much to offer beyond the specific cases for which they have been derived. For example, they enable one to gain insight into the variation of stress and deformation with basic shape and property changes. In addition, they provide the basis for rough approximations in preliminary design even though there is only crude similarity between the analytical model and the actual case. In other situations, analytical methods provide a starting point or guide in numerical solutions.

Numerical analyses lead often to a system of linear algebraic equations. The most appropriate method of solution then depends upon the nature and the number of such equations, as well as the type of computing equipment available. The techniques introduced in this chapter and applied in the chapters following have clear application to computation by means of electronic digital computer.

7.2 An Informal Approach to Numerical Analysis

The factors serving to complicate the analysis of real problems may generally be reduced to irregularities in the shape of the structural element being loaded and nonuniformity in the applied load. By replacing the actual configurations of load and structure with suitable approximations which can be treated analytically, frequently very little sacrifice in accuracy is encountered.

Consider the case of a beam with a variable cross-sectional area. There are a number of ways of determining an equivalent beam which will not compromise the elastic response significantly. One can, for example, replace the actual beam with a series of rigid weightless segments hinged at their points of connection, and given at these points that degree of bending rigidity required to match the actual beam. Whatever the true load configuration may be, it is then reduced to a series of concentrated forces applied at the points of connection. Another approach to this problem is to regard the beam as stepped, i.e., to replace the beam with a series of sections of constant area. Still more improvement results from the use of an equivalent beam whose moment of inertia varies linearly from segment to segment, thus avoiding abrupt changes in cross section.

As to replacing a variable loading, almost any method which permits analytical solution in a piecewise fashion is acceptable. The distributed load may thus be converted to a series of concentrated forces, or to a number of uniformly distributed segments. Refinements, which serve to reduce the number of segments necessary to effect an accurate solution, include the reduction of the true force distribution to a series of linearly or parabolically varying loadings.

Consider a simply supported beam whose depth is a function of axial distance x. It is necessary to find the deflection v described by Eqs. (5.10) and (5.34):

$$EI\frac{d^2v}{dx^2} = M(x), \qquad \frac{d^2}{dx^2}\left(EI\frac{d^2v}{dx^2}\right) = p(x) \qquad (a,b)$$

where $p(x)$ represents the variable distributed load, and $I = I(x)$ is the moment of inertia. An analytical solution is, in general, not available, but the deflection for a particular loading $p(x)$ may be determined as follows. Let the beam be divided into N segments of constant cross section and equal length $\Delta x = L/N$. The distributed load in the interval

$$x_n - \frac{\Delta x}{2} < x < x_n + \frac{\Delta x}{2}$$

is replaced by equivalent concentrated load P_n, as represented in Fig. 7.1a. For simplicity of computation, *the moment M_n at $x = x_n - \Delta x/2$ may, for small values of Δx, be regarded as constant over the nth segment.* Let I_n represent the constant moment of inertia of the nth segment, equal to the moment of inertia at $x = x_n - \Delta x/2$. Referring to Fig. 7.1b, assuming small deflections and small Δx, the following recurrence relations apply:

$$\theta_n = \theta_{n-1} + \Delta\theta_n = \theta_{n-1} + \frac{M_n}{EI_n}\Delta x$$

$$v_n = v_{n-1} + \Delta v_n = v_{n-1} + \theta_{n-1}\Delta x \qquad (c)$$

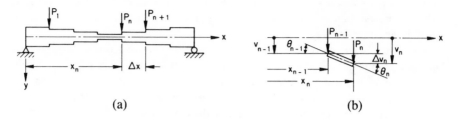

Figure 7.1

Here θ_n and v_n are the slope and the deflection at x_n. The above expressions may be more conveniently written

$$\theta_n = \theta_0 + \sum_{j=1}^{n} \frac{M_j}{EI_j} \Delta x \tag{7.1}$$

$$v_n = v_0 + \sum_{j=1}^{n-1} \theta_j \Delta x \tag{7.2}$$

where v_0 is the deflection and θ_0 the slope at $x=0$. The foregoing express the first and second integrals of Eq. (a). The solution is obtained by applying Eqs. (7.1) and (7.2) while satisfying the conditions at the beam ends. We have, for simplicity, chosen the Δx's to be equal. It is clear that the same procedure may be followed for unequal Δx, this being the more general case.

The optimum choice of interval size is dependent upon the method used to simplify beam shape and loading. For example, it may be possible to increase accuracy without decreasing interval size by selecting a more sophisticated replacement for $I(x)$ and $p(x)$ than simply a series of constant segments. A number of references are available for those seeking a more thorough treatment.*

Example 7.1. Calculate the deflection at the points of the beam indicated in Fig. 7.2a. Use two approaches: (a) Lump the distributed load to form concentrated loads. (b) Replace the given load by segments of constant distributed load. Compare the results with those obtained by means of the direct integration method.

SOLUTION

(a) The actual load at point 1 is

$$\int_0^{L/4} p_0 \sin \frac{\pi x}{2L} \, dx = 0.048 p_0 L$$

---

*See, for example, M. G. Salvadori and M. L. Baron, *Numerical Methods in Engineering*, Englewood Cliffs, NJ: Prentice-Hall, 1959.

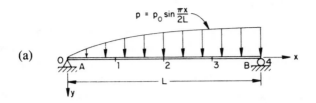

Figure 7.2

The loads at points 2, 3, and 4 are similarly found and the corresponding reactions at A and B determined (Fig. 7.2b). Referring to this figure, values of M_n for each segment are then obtained. Applying Eqs. (7.1) and (7.2) and satisfying the end conditions $v(0)=v(L)=0$, we obtain $v_0=0$ and $\theta_0=0.014p_0L^3/EI$. The deflection at each point is then calculated, and the results, to slide rule accuracy, are given in Table 7.1.

(b) The loading is now replaced by uniform distributions in each segment (Fig. 7.2c). Using these values, the reactions are as given in the figure. Following the same procedure as in (a), the slope at $x=0$ is calculated to be $\theta_0=0.019p_0L^3/EI$, and the deflections are $v_1 = 0.005p_0L^4/EI$, $v_2 = 0.007p_0L^4/EI$, $v_3 = 0.006p_0L^4/EI$, $v_4=0$.

According to the method of direct integration, the midpoint deflection is $v_2=0.008p_0L^4/EI$, while from approximations (a) and (b) we have $v_2=0.006p_0L^4/EI$ and $v_2=0.007p_0L^4/EI$, respectively.

Table 7.1

Units	n	1	2	3	4
N·m	M_n	$-0.0195p_0L^2$	$-0.0525p_0L^2$	$-0.0623p_0L^2$	$-0.0295p_0L^2$
rad	$\theta_n-\theta_0$	$-0.0048p_0L^3/EI$	$-0.0178p_0L^3/EI$	$-0.0333p_0L^3/EI$	$-0.0405p_0L^3/EI$
m	$v_n-n\Delta x\theta_0$	0	$-0.0012p_0L^4/EI$	$-0.0057p_0L^4/EI$	$-0.014p_0L^4/EI$
rad	θ_n	$0.0092p_0L^3/EI$	$-0.0038p_0L^3/EI$	$-0.0193p_0L^3/EI$	$-0.0266p_0L^3/EI$
m	v_n	$0.0035p_0L^4/EI$	$-0.006p_0L^4/EI$	$-0.0048p_0L^4/EI$	0

7.3 Finite Differences

The numerical solution of a differential equation is essentially a table in which values of the required function are listed next to corresponding values of the independent variable(s). In the case of an ordinary differential equation, the unknown function (y) is listed at specific *pivot* or *nodal* points spaced along the x axis. For a two-dimensional partial differential equation, the nodal points will be in the xy plane.

The basic *finite difference* expressions follow logically from the fundamental rules of calculus. Consider the definition of the first derivative with respect to x of a continuous function $y = f(x)$ (Fig. 7.3):

$$\left(\frac{dy}{dx}\right)_n = \lim_{\Delta x \to 0} \frac{y(x_n + \Delta x) - y(x_n)}{\Delta x} = \lim_{\Delta x \to 0} \frac{y_{n+1} - y_n}{\Delta x}$$

The subscript n denotes any point on the curve. If the increment in the independent variable does not become vanishingly small but instead assumes a finite value $\Delta x = h$, the above expression represents an approximation to the derivative:

$$\left(\frac{dy}{dx}\right)_n \approx \frac{\Delta y_n}{h} = \frac{y_{n+1} - y_n}{h}$$

Here Δy_n is termed the first difference of y at point x_n,

$$\Delta y_n = y_{n+1} - y_n \approx h\left(\frac{dy}{dx}\right)_n \tag{7.3}$$

Because the above relationship is expressed in terms of the numerical value of the function at the point in question (n) and a point ahead of it ($n + 1$), the difference is termed a *forward difference*. The *backward difference* at n, denoted ∇y_n, is given by

$$\nabla y_n = y_n - y_{n-1} \tag{7.4}$$

Central differences involve pivot points symmetrically located with respect to x_n, and often result in more accurate approximations than forward or backward differences. The latter are especially useful where, because of geometrical limitations (as near boundaries), central differences cannot be

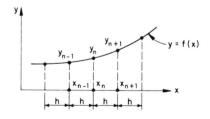

Figure 7.3

employed. In terms of symmetrical pivot points, the derivative of y at x_n is

$$\left(\frac{dy}{dx}\right)_n \approx \frac{y(x_n+h)-y(x_n-h)}{2h} = \frac{1}{2h}(y_{n+1}-y_{n-1}) \tag{7.5}$$

The *first central difference* δy is thus

$$\delta y = \frac{1}{2}(y_{n+1}-y_{n-1}) \approx h\left(\frac{dy}{dx}\right)_n \tag{7.6}$$

A procedure similar to that used above will yield the higher order derivatives. Referring once again to Fig. 7.3, employing Eq. (7.3), we have

$$\left(\frac{d^2y}{dx^2}\right)_n = \frac{d}{dx}\left(\frac{dy}{dx}\right) \approx \frac{1}{h^2}\Delta(\Delta y_n) = \frac{1}{h^2}\Delta(y_{n+1}-y_n)$$

$$= \frac{1}{h^2}(\Delta y_{n+1}-\Delta y_n) = \frac{1}{h^2}(y_{n+2}-y_{n+1}) - \frac{1}{h^2}(y_{n+1}-y_n)$$

The *second forward difference* at x_n is therefore

$$\Delta^2 y_n = y_{n+2}-2y_{n+1}+y_n \approx h^2\left(\frac{d^2y}{dx^2}\right)_n \tag{7.7}$$

The *second backward difference* is found in the same way

$$\nabla^2 y_n = \nabla(\nabla y_n) = \nabla(y_n-y_{n-1}) = \nabla y_n - \nabla y_{n-1}$$

$$= (y_n-y_{n-1}) - (y_{n-1}-y_{n-2})$$

$$\nabla^2 y_n = y_n - 2y_{n-1}+y_{n-2} \approx h^2\left(\frac{d^2y}{dx^2}\right)_n \tag{7.8}$$

It is simple matter to verify the fact that the coefficients of the pivot values in the mth forward and backward differences are the same as the coefficients of the binomial expansion $(a-b)^m$. Using this scheme, higher order forward and backward differences are easily written.

The second central difference at x_n is the difference of the first central differences:

$$\delta^2 y_n = \delta(\delta y_n) = \frac{1}{2}\delta(y_{n+1}-y_{n-1}) = \frac{1}{2}(\delta y_{n+1}-\delta y_{n-1})$$

$$= \frac{1}{4}(y_{n+2}-y_n) - \frac{1}{4}(y_n-y_{n-2})$$

$$= \frac{1}{4}(y_{n+2}-2y_n+y_{n-2})$$

This quantity is thus observed to depend upon values of the function two intervals ahead of and behind the point under consideration. To improve the approximation, the second difference is expressed in terms of the function only one interval ahead of and behind the point. To do this, consider the following:

$$\delta^2 y_n = \Delta(\nabla y_n) \quad \text{or} \quad \delta^2 y_n = \nabla(\Delta y_n)$$

The *second central difference* at x_n is thus

$$\delta^2 y_n = \Delta(\nabla y_n) = \Delta(y_n - y_{n-1}) = \Delta y_n - \Delta y_{n-1}$$
$$= (y_{n+1} - y_n) - (y_n - y_{n-1})$$
$$= y_{n+1} - 2y_n + y_{n-1} \approx h^2 \left(\frac{d^2 y}{dx^2} \right)_n \tag{7.9}$$

In a like manner, the *third* and *fourth central differences* are readily determined:

$$\delta^3 y_n = \delta(\delta^2 y_n) = \delta(y_{n+1} - 2y_n + y_{n-1}) = \delta y_{n+1} - 2\delta y_n + \delta y_{n-1}$$
$$= \tfrac{1}{2}(y_{n+2} - y_n) - (y_{n+1} - y_{n-1}) + \tfrac{1}{2}(y_n - y_{n-2})$$
$$= \tfrac{1}{2}(y_{n+2} - 2y_{n+1} + 2y_{n-1} - y_{n-2}) \approx h^3 \left(\frac{d^3 y}{dx^3} \right)_n \tag{7.10}$$

$$\delta^4 y_n = y_{n+2} - 4y_{n+1} + 6y_n - 4y_{n-1} + y_{n-2} \approx h^4 \left(\frac{d^4 y}{dx^4} \right)_n \tag{7.11}$$

Examination of Eqs. (7.9) and (7.11) reveals that for *even order* derivatives, the coefficients of y_n, y_{n+1}, are equal to the coefficients in the binomial expansion $(a-b)^m$.

Unless otherwise specified, we shall use the term finite differences to refer to central differences.

We now discuss a continuous function $w(x, y)$ of two variables. The *partial derivatives* may be approximated by the following procedures, similar to those discussed in the previous article. For purposes of illustration, consider a rectangular boundary as in Fig. 7.4. By taking $\Delta x = \Delta y = h$, a square net is obtained. Application of Eqs. (7.6) and (7.9) yields

$$\frac{\partial w}{\partial x} \approx \frac{1}{h} \delta_x w, \qquad \frac{\partial w}{\partial y} \approx \frac{1}{h} \delta_y w \tag{7.12}$$

$$\frac{\partial^2 w}{\partial x^2} \approx \frac{1}{h^2} \delta_x^2 w, \qquad \frac{\partial^2 w}{\partial y^2} \approx \frac{1}{h^2} \delta_y^2 w, \qquad \frac{\partial^2 w}{\partial x \partial y} \approx \frac{1}{h} \delta_x \left(\frac{\partial w}{\partial y} \right) \tag{7.13}$$

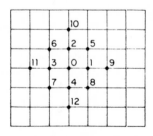

Figure 7.4

The subscripts x and y applied to the δ's indicate the coordinate direction appropriate to the difference being formed. The above expressions written for the point 0 are

$$\frac{\partial w}{\partial x} \approx \frac{1}{2h}\left[w(x+h,y)-w(x-h,y)\right] = \frac{1}{2h}(w_1-w_3)$$

$$\frac{\partial w}{\partial y} \approx \frac{1}{2h}\left[w(x,y+h)-w(x,y-h)\right] = \frac{1}{2h}(w_2-w_4) \tag{7.14}$$

and

$$\frac{\partial^2 w}{\partial x^2} \approx \frac{1}{h^2}\left[w(x+h,y)-2w(x,y)+w(x-h,y)\right]$$

$$= \frac{1}{h^2}(w_1-2w_0+w_3) \tag{7.15}$$

$$\frac{\partial^2 w}{\partial y^2} \approx \frac{1}{h^2}(w_2-2w_0+w_4)$$

$$\frac{\partial^2 w}{\partial x \, \partial y} \approx \frac{1}{2h^2}(\delta_x w_2 - \delta_x w_4) = \frac{1}{4h^2}(w_5-w_6+w_7-w_8) \tag{7.16}$$

Similarly, Eqs. (7.10) and (7.11) lead to expressions for approximating the third- and fourth-order partial derivatives.

7.4 Finite Difference Equations

We are now in a position to transform a differential equation into an algebraic equation. This is accomplished by substituting the appropriate finite difference expressions into the differential equation. At the same time, the boundary conditions must also be converted to finite difference form. The solution of a differential equation thus reduces to the simultaneous solution of a set of linear, algebraic equations, written for every nodal point within the boundary.

Example 7.2. Analyze the torsion of a bar of square section using finite difference techniques.

SOLUTION. The governing partial differential equation is (see Sec. 6.2)

$$\frac{\partial^2 \phi}{\partial x^2} + \frac{\partial^2 \phi}{\partial y^2} = -2G\theta \tag{7.17}$$

where ϕ may be assigned the value of zero at the boundary. Referring to Fig. 7.4, the finite difference equation about the point 0, corresponding to Eq. (7.17), is

$$\phi_1+\phi_2+\phi_3+\phi_4-4\phi_0 = -2G\theta h^2 \tag{7.18}$$

A similar expression is written for every other nodal point within the section. The solution of the problem then requires the determination of those values of ϕ which satisfy the system of algebraic equations.

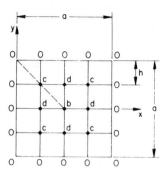

Figure 7.5

The domain is now divided into a number of small squares, 16 for example. In labeling nodal points, it is important to take into account any conditions of symmetry which may exist. This has been done in Fig. 7.5. Note that $\phi = 0$ has been substituted at the boundary. Equation (7.18) is now applied to nodal points b, c, and d, resulting in the following set of expressions:

$$4\phi_d - 4\phi_b = -2G\theta h^2$$

$$2\phi_d - 4\phi_c = -2G\theta h^2$$

$$2\phi_c + \phi_b - 4\phi_d = -2G\theta h^2$$

Simultaneous solution yields

$$\phi_b = 2.25G\theta h^2, \qquad \phi_c = 1.375G\theta h^2, \qquad \phi_d = 1.75G\theta h^2 \tag{a}$$

The results for points b and d are tabulated in the second column of Table 7.2.

To determine the partial derivatives of the stress function, we shall assume a smooth curve containing the values in Eq. (a) to represent the function ϕ. *Newton's interpolation formula*[*], used for fitting such a curve, is

$$\phi(x) = \phi_0 + \frac{x}{h}\Delta\phi_0 + \frac{x(x-h)}{2!h^2}\Delta^2\phi_0 + \frac{x(x-h)(x-2h)}{3!h^3}\Delta^3\phi_0$$

$$+ \frac{x(x-h)(x-2h)(x-3h)}{4!h^4}\Delta^4\phi_0 + \cdots + \frac{x(x-h)\cdots(x-nh+h)}{n!h^n}\Delta^n\phi_0$$

$$\tag{7.19}$$

---

[*]See I. S. Sokolnikoff and R. M. Redheffer, *Mathematics of Physics and Modern Engineering*, New York: McGraw-Hill, 1966, p. 668.

Table 7.2

x	$\phi/G\theta h^2$	$\Delta/G\theta h^2$	$\Delta^2/G\theta h^2$	$\Delta^3/G\theta h^2$	$\Delta^4/G\theta h^2$
0	0	1.75	-1.25	0.25	-0.5
h	1.75	0.5	-1.0	-0.25	
$2h$	2.25	-0.5	-1.25		
$3h$	1.75	-1.75			
$4h$	0				

where the $\Delta\phi_0$'s are the forward differences, calculated at $x=0$ as follows:

$$\Delta\phi_0 = \phi_d - \phi_0 = (1.75-0)G\theta h^2 = 1.75G\theta h^2$$

$$\Delta^2\phi_0 = \phi_b - 2\phi_d + \phi_0 = -1.25G\theta h^2$$

$$\Delta^3\phi_0 = \phi_d - 3\phi_b + 3\phi_d - \phi_0 = 0.25G\theta h^2$$

$$\Delta^4\phi_0 = \phi_0 - 4\phi_d + 6\phi_b - 4\phi_d + \phi_0 = -0.5G\theta h^2$$

(b)

The differences are also calculated at $x=h$, $x=2h$, etc. and are listed in Table 7.2. Note that one can readily obtain the values given in Table 7.2 (for the given ϕ's) by starting at node $x=4h$: $0-1.75=-1.75$, $1.75-2.25=-0.5$, $-1.75-(-0.5)=-1.25$, etc.

The maximum shear stress, which occurs at $x=0$, is obtained from $(\partial\phi/\phi x)_0$. Thus, differentiating Eq. (7.19) with respect to x and then setting $x=0$, the result is

$$\left(\frac{\partial\phi}{\partial x}\right)_0 = \frac{1}{h}\left(\Delta\phi_0 - \frac{1}{2}\Delta^2\phi_0 + \frac{1}{3}\Delta^3\phi_0 - \frac{1}{4}\Delta^4\phi_0 + \cdots\right)$$

(c)

Substituting the values in the first row of Table 7.2 into Eq. (c), we obtain

$$\left(\frac{\partial\phi}{\partial x}\right)_0 = \tau_{max} = \frac{31}{12}G\theta h = 0.646G\theta a$$

The exact value, given in Table 6.1 as $\tau_{max}=0.678G\theta a$, differs from the above approximation by only 4.7 percent.

By means of a finer network, we expect to improve the result. For example, selecting $h=a/6$, six nodal equations are obtained. It can be shown that the maximum stress in this case, $0.661G\theta a$, is within 2.5 percent of the exact solution. On the basis of results for $h=a/4$ and $h=a/6$, a still better approximation can be found by applying extrapolation techniques.*

7.5 The Relaxation Method

The relaxation method is a procedure by which systems of simultaneous linear equations may be solved by means of successive iteration. This method is among the simplest for solving simultaneous equations by means of a hand calculation. With the widespread availability of computers, it is

---

*For tables of extrapolation coefficients and related formulas see Salvadori and Baron, op. cit., p. 96.

not as appealing as in the past. However, many very efficient techniques are based upon the relaxation process, and we shall describe the method briefly, providing, in Sec. 7.8, one application.

To illustrate the method, consider Poisson's equation $\nabla^2\phi+2G\theta=0$. The equivalent finite difference equation at the central point 0 (Fig. 7.4) is

$$(\phi_1+\phi_2+\phi_3+\phi_4-4\phi_0)+2G\theta h^2=0 \qquad (7.20)$$

The first step toward a solution is the assumption of ϕ at each point. If the values selected are correct, the right hand side of Eq. (7.20) will be zero, indicating that the equation is satisfied. If they are incorrect, the equation will not be satisfied, and there will be a residual R_0 at nodal point 0, indicating the extent of the error:

$$(\phi_1+\phi_2+\phi_3+\phi_4-4\phi_0)+2G\theta h^2=R_0 \qquad (7.21)$$

The initial values of ϕ may be assumed on the basis of information known from experiment or from previous experience with similar problems, or they may be the result of guesswork. Using the initial ϕ's, the residuals at all points are calculated and recorded at the points. If all the ϕ's are correct, all the residuals will be zero. The relaxation procedure has as its goal, then, the reduction to zero, or to an acceptably small value, of the residuals at all points.

Referring to Eq. (7.21), we observe that changes in the residuals follow a definite pattern in relation to alterations in ϕ. Changing ϕ_0 by $+1$ results in a reduction of R_0 by 4. If, on the other hand, ϕ_1, ϕ_2, ϕ_3, or ϕ_4 is altered by $+1$, R_0 is increased by only 1 in each case. These considerations form the basis of the unit relaxation operator (Fig. 7.6), indicating the amount by which the residual at different points is influenced by variations of $+1$ in ϕ.

In the previous paragraphs we discussed the reduction of residuals by manipulating single nodal points while maintaining the others fixed. It is sometimes advantageous in terms of time and effort to deal simultaneously with a group of nodal points. In this case the relaxation process is accomplished by *ordered iterations*.

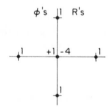

Figure 7.6

7.6 Curved Boundaries

It has already been mentioned that one of the important strengths of numerical analysis is its adaptability to irregular geometries. We now turn, therefore, from the straight and parallel boundaries of previous problems to situations involving curved or irregular boundaries. Examination of one segment of such a boundary (Fig. 7.7) reveals that the standard five-point operator, in which all arms are of equal length, is not appropriate because of the unequal lengths of arms bc, bd, be, and bf. When at least one of the arms is of nonstandard length, the pattern is referred to as an *irregular star*. One method for constructing irregular star operators is discussed below.

Assume that in the vicinity of point b, $w(x, y)$ can be approximated by the second-degree polynomial

$$w(x, y) = w + a_1 x + a_2 y + a_3 x^2 + a_4 y^2 + a_5 xy \qquad (7.22)$$

Referring to Fig. 7.7, the above expression leads to approximations of the function w at points c, d, e, and f:

$$\begin{aligned}
w_c &= w_b + a_1 h_1 + a_3 h_1^2 \\
w_d &= w_b + a_2 h_2 + a_4 h_2^2 \\
w_e &= w_b - a_1 h + a_3 h^2 \\
w_f &= w_b - a_2 h + a_4 h^2
\end{aligned} \qquad \text{(a)}$$

At nodal point $b(x = y = 0)$, Eq. (7.22) yields

$$\left(\frac{\partial^2 w}{\partial x^2} \right)_b = 2a_3, \qquad \left(\frac{\partial^2 w}{\partial y^2} \right)_b = 2a_4 \qquad \text{(b)}$$

Combining Eqs. (a) and (b), we have

$$\left(\frac{\partial^2 w}{\partial x^2} \right)_b = 2 \frac{h(w_c - w_b) + h_1(w_e - w_b)}{hh_1(h + h_1)}$$

$$\left(\frac{\partial^2 w}{\partial y^2} \right)_b = 2 \frac{h(w_d - w_b) + h_2(w_f - w_b)}{hh_2(h + h_2)}$$

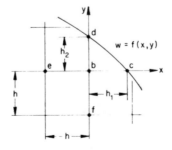

Figure 7.7

Introducing the above into the Laplace operator, we obtain

$$h^2\left(\frac{\partial^2 w}{\partial x^2} + \frac{\partial^2 w}{\partial y^2}\right)_b = \frac{2w_c}{\alpha_1(1+\alpha_1)} + \frac{2w_d}{\alpha_2(1+\alpha_2)}$$

$$+ \frac{2w_e}{1+\alpha_1} + \frac{2w_f}{1+\alpha_2} - \left(\frac{2}{\alpha_1} + \frac{2}{\alpha_2}\right)w_b \quad (7.23)$$

In this expression $\alpha_1 = h_1/h$, $\alpha_2 = h_2/h$. It is clear that for irregular stars, $0 \leqslant \alpha_i \leqslant 1$ $(i=1,2)$.

The foregoing result may readily be reduced for one-dimensional problems with irregularly spaced nodal points. For example, in the case of a beam, Eq. (7.23), with reference to Fig. 7.7, simplifies to

$$h^2\left(\frac{\partial^2 w}{\partial x^2}\right)_b = \frac{2w_c}{\alpha_1(1+\alpha_1)} + \frac{2w_e}{1+\alpha_1} - \frac{2w_b}{\alpha_1} \quad (7.24)$$

where x represents the longitudinal direction. This expression, setting $\alpha = \alpha_1$, may be written

$$h^2\frac{d^2 w}{dx^2} = \frac{2}{\alpha(\alpha+1)}\left[\alpha w_{n-1} - (1+\alpha)w_n + w_{n+1}\right] \quad (7.25)$$

Example 7.3. Find the shearing stresses at the points A and B of the torsional member of elliptical section shown in Fig. 7.8. Let $a = 15$ mm, $b = 10$ mm, and $h = 5$ mm.

SOLUTION. Because of symmetry, only a quarter of the section need be considered. From the equation of the ellipse with the given values of a, b, and h, it is found that $h_1 = 4.4$ mm, $h_2 = 2.45$ mm, $h_3 = 3$ mm. At points b, e, f, g, the standard finite difference equation (7.18) applies, while at c and d, we use a modified equation found from Eq. (7.17) with reference to Eq. (7.23). One can therefore write six

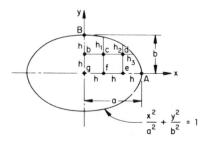

Figure 7.8

equations presented in the following matrix form:

$$\begin{bmatrix} 2 & 0 & 0 & 0 & 2 & -4 \\ 0 & 2 & 0 & 1 & -4 & 1 \\ 0 & 0 & 2 & -4 & 1 & 0 \\ -4 & 2 & 0 & 0 & 0 & 1 \\ 1 & -4.27 & 1 & 0 & 1.06 & 0 \\ 0 & 1.25 & -7.41 & 1.34 & 0 & 0 \end{bmatrix} \begin{Bmatrix} \phi_b \\ \phi_c \\ \phi_d \\ \phi_e \\ \phi_f \\ \phi_g \end{Bmatrix} = \begin{Bmatrix} -2h^2G\theta \\ -2h^2G\theta \\ -2h^2G\theta \\ -2h^2G\theta \\ -2h^2G\theta \\ -2h^2G\theta \end{Bmatrix}$$

The above equations are solved to yield

$$\phi_b = 2.075G\theta h^2, \qquad \phi_c = 1.767G\theta h^2, \qquad \phi_d = 0.843G\theta h^2$$

$$\phi_e = 1.536G\theta h^2, \qquad \phi_f = 2.459G\theta h^2, \qquad \phi_g = 2.767G\theta h^2$$

The solution then proceeds as in Example 7.2. The following forward differences at point B are first evaluated:

$$\Delta\phi_B = 2.075G\theta h^2, \qquad \Delta^3\phi_B = -0.001G\theta h^2$$

$$\Delta^2\phi_B = -1.383G\theta h^2, \qquad \Delta^4\phi_B = 0.002G\theta h^2$$

Similarly, for point A, we obtain

$$\Delta\phi_A = 1.536G\theta h^2, \qquad \Delta^4\phi_A = 0.001G\theta h^2$$

$$\Delta^2\phi_A = -0.613G\theta h^2, \qquad \Delta^5\phi_A = 0.001G\theta h^2$$

$$\Delta^3\phi_A = -0.002G\theta h^2, \qquad \Delta^6\phi_A = -0.002G\theta h^2$$

Thus,

$$\tau_B = \left(\frac{\partial\phi}{\partial y}\right)_B = \left(2.075 + \frac{1.383}{2} - \frac{0.001}{3} + \frac{0.002}{4}\right)G\theta h$$

$$= 2.766G\theta h = 1.383G\theta b = 0.01383G\theta$$

$$\tau_A = \left(\frac{\partial\phi}{\partial x}\right)_A = \left(1.536 + \frac{0.613}{2} - \frac{0.002}{3} - \frac{0.001}{4} + \frac{0.001}{5} + \frac{0.002}{6}\right)G\theta h$$

$$= 1.842G\theta h = 0.614G\theta a = 0.00921G\theta$$

Note that according to the exact theory, the maximum stress occurs at $y = b$ and is equal to $1.384G\theta b$ (see Example 6.1), indicating excellent agreement with τ_B above.

7.7 Boundary Conditions

Our concern has thus far been limited to problems in which the boundaries have been assumed free of constraint. Many practical situations involve boundary conditions related to the deformation, force, or moment at one or more points. Application of numerical methods under these circumstances may become more complex. The following simple examples illustrate the method of solution.

Example 7.4. Use a finite difference approach to determine the deflection of the beam shown in Fig. 7.9a.

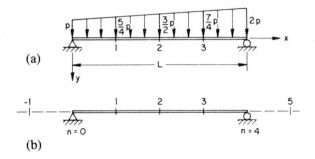

Figure 7.9

SOLUTION. The differential equation for beam deflection, $EI\,d^4v/dx^4=p$, is by means of Eq. (7.11) replaced by the following finite difference equation:

$$v_{n+2}-4v_{n+1}+6v_n-4v_{n-1}+v_{n-2}=h^4\left(\frac{p}{EI}\right)_n \qquad (7.26)$$

For simplicity, take $h=L/4$ (Fig. 7.9b). The boundary conditions $v(0)=v(L)=0$ and $v''(0)=v''(L)=0$ are replaced by finite difference conditions by setting $v(0)=v_0$, $v(L)=v_4$ and applying Eq. (7.9):

$$v_0=0, \qquad v_1=-v_{-1}, \qquad v_4=0, \qquad v_3=-v_5 \qquad (a)$$

When Eq. (7.26) is used at points 1, 2, and 3 the following expressions are obtained:

$$v_3-4v_2+6v_1-4v_0+v_{-1}=\frac{L^4}{256}\frac{5p}{4EI}$$

$$v_4-4v_3+6v_2-4v_1+v_0=\frac{L^4}{256}\frac{3p}{2EI} \qquad (b)$$

$$v_5-4v_4+6v_3-4v_2+v_1=\frac{L^4}{256}\frac{7p}{4EI}$$

Simultaneous solution of Eqs. (a) and (b) yields

$$v_1=0.0139\frac{pL^4}{EI}$$

$$v_2=0.0198\frac{pL^4}{EI}$$

$$v_3=0.0144\frac{pL^4}{EI}$$

Note that by successive integration of $EId^4v/dx^4=p(x)$, the result $v_2=0.0185pL^4/EI$ is obtained. Thus, even a coarse segmentation leads to a satisfactory solution in this case.

Example 7.5. Determine the redundant reaction R for the beam depicted in Fig. 7.10a.

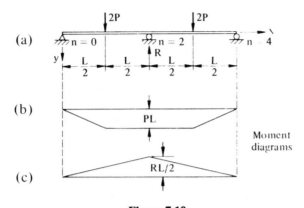

Figure 7.10

SOLUTION. The bending diagrams associated with the applied loads $2P$ and the redundant reaction R are given in Fig. 7.10b and c, respectively. The finite difference form of the differential equation of the beam, $EI(d^2v/dx^2)=M(x)$, is

$$v_{n+1}-2v_n+v_{n-1}=h^2\left[\frac{M(x)}{EI}\right]_n \tag{7.27}$$

For $h=L/4$, Eq. (7.27) results in the following expressions at points 1, 2, and 3:

$$v_0-2v_1+v_2=\frac{L^2}{16EI}\left(-PL+\frac{RL}{4}\right)$$

$$v_1-2v_2+v_3=\frac{L^2}{16EI}\left(-PL+\frac{RL}{2}\right) \tag{c}$$

$$v_2-2v_3+v_4=\frac{L^2}{16EI}\left(-PL+\frac{RL}{4}\right)$$

The number of unknowns in the above set of equations is reduced from five to three through application of the conditions of symmetry, $v_1=v_3$, and the support conditions, $v_0=v_2=v_4=0$. Solution of Eq. (c) now yields $R=8P/3$.

7.8   The Moment Distribution Method

Problems involving statically indeterminate structures or members can readily be solved by application of a relaxation process known as the *moment distribution method*. This type of analysis, particularly suited to the digital computer, has as its object the determination of the bending moments at all the joints of a structure. Once these are known, the conditions of equilibrium are applied to calculate the unknown forces.

Before proceeding to a description of the method, it will prove useful to define the following terms related to a single span beam.

Fixed end moments. Fixed end moments M^f are those existing at the ends of a beam having both ends clamped, subjected to an externally

Table 7.3

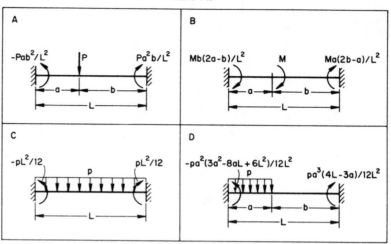

imposed load or deflection. Expressions for such moments for a number of commonly encountered cases, derived from the mechanics of materials, are given in Table 7.3. Note that clockwise acting moments are taken as positive.

The Carry-over Factor. A moment M_A applied at A will cause that end to experience rotation through angle θ_A (Fig. 7.11). In addition, a moment proportional to M_A will be produced at B. The moments at A and B are related by a constant of proportionality termed the *carry-over factor* C_{AB}:

$$C_{AB} = \frac{M_B}{M_A} \tag{7.28}$$

It is clear that C_{AB} represents the extent to which the moment at A is "carried over" to B.

The Stiffness Factor. The angle of rotation at A, θ_A, is dependent upon the magnitude of M_A as well as the stiffness of the beam. The moment and angle are related by the *stiffness factor* K. For end A of beam AB, we have

$$K_{AB} = \frac{M_A}{\theta_A} \tag{7.29}$$

Figure 7.11

The stiffness of the beam is dependent upon its shape and the end condition at B. Application of the mechanics of materials approach permits relatively simple computation of the carry-over and stiffness factors for any member. For example, for the beam shown in Fig. 7.11 it is found that $M_A = 2M_B$ and $\theta_A = M_A L/4EI$. Then, from Eqs. (7.28) and (7.29), $C_{AB} = 0.5$ and $K_{AB} = 4EI/L$. If the end B were simply supported, we would have $C_{AB} = 0$ and $K_{AB} = 3EI/L$.

The Distribution Factor. The ratio of the rotational stiffness of a structural element to the sum of the stiffnesses of all elements attached at a joint is termed the *distribution factor*:

$$D = \frac{K}{\Sigma K} \tag{7.30}$$

The sum of the distribution factors of all elements at a joint is clearly unity.

Application of the Method

The beam shown in Fig. 7.12a will serve to illustrate the moment distribution method. The first step consists of restraining the beam against the rotation at B. After this, the restraint is relaxed in order to effect a balance of the internal moments with the applied load.

The effect of the applied moment M is to produce a rotation θ. In order to prevent joint B from experiencing this rotation, external fixing moments are applied at the ends of the two beam segments, as shown in Fig. 7.12b. These moments M_{AB}, M_{BA}, M_{BC}, and M_{CB} are exerted on the joints by the beam segments. In terms of the stiffnesses K_{BA} and K_{BC} and angle θ, applying Eq. (7.29), the moments are given by

$$M_{BA} = K_{BA}\theta, \qquad M_{BC} = K_{BC}\theta$$

or

$$\frac{M_{BA}}{M_{BC}} = \frac{K_{BA}}{K_{BC}} \tag{a}$$

Equilibrium requires that

$$M_{BA} + M_{BC} = M \tag{b}$$

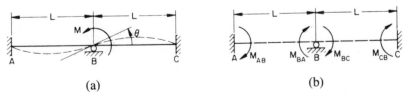

(a)   (b)

Figure 7.12

Combining Eqs. (a) and (b), we have

$$M_{BA} = D_{BA}M, \qquad M_{BC} = D_{BC}M \qquad \text{(c)}$$

where the distribution factors, in accordance with Eq. (7.30), are

$$D_{BA} = \frac{K_{BA}}{K_{BA} + K_{BC}}, \qquad D_{BC} = \frac{K_{BC}}{K_{BA} + K_{BC}} \qquad \text{(d)}$$

Moments M_{AB} and M_{CB}, from Eq. (7.28), are as follows:

$$M_{AB} = C_{BA}M_{BA}, \qquad M_{CB} = C_{BC}M_{BC} \qquad \text{(e)}$$

For the uniform beam under consideration, $C_{AB} = C_{CB} = 0.5$, $K_{BA} = K_{BC} = 4EI/L$, and therefore Eqs. (d) yield $D_{BA} = D_{BC} = 0.5$.

What has been observed in execution of the method is the distribution of the applied moment M to the right and left of point B in accordance with the distribution factors D_{BA} and D_{BC}. In addition, the moments are carried over beam segments AB and BC to the extent governed by the fractions C_{BA} and C_{BC} [Eqs. (e)].

The procedure discussed above may be generalized to apply to many-jointed structures:

1. Consider all joints externally fixed. The fixed end moments (f.e.m.) are determined as in Table 7.3.
2. Next relax (remove) the restraint at each joint, one at a time, until the internal moments at each joint are in equilibrium.
3. In the relaxation of each joint, introduce a carry-over moment at that end of each member which is adjacent to the joint.
4. Repeat steps 2 and 3. The process is convergent if the carry-over moments are observed to decrease with each cycle.
5. By adding the moment changes occuring in n cycles to the original fixed end moment, obtain the final moment at each joint.

Example 7.6. Determine the moments at the supports of the beam shown in Fig. 7.13a.

SOLUTION.   Fixing joint B, the initial fixed end moments at B are, from Table 7.3,

$$M_{AB}^f = -\tfrac{1}{12}pL_1^2 = -\frac{15(3)^2}{12} = -11.25 \text{ kN·m}, \qquad M_{BA}^f = 11.25 \text{ kN·m}$$

$$M_{BC}^f = -\frac{Pab^2}{L_2^2} = \frac{45(3)(1.5)^2}{(4.5)^2} = -15 \text{ kN·m}, \qquad M_{CB}^f = \frac{Pa^2b}{L_2^2} = 30 \text{ kN·m}$$

The effective f.e.m. at joint B is $11.25 - 15 = -3.75$ kN·m, the minus sign indicating a counterclockwise moment about the joint. The restraint at B is now relaxed, i.e., the joint rotates until a change in moment of $+3.75$ kN·m occurs and joint B is in equilibrium. To obtain the final values of the moments at the fixed ends, carry-over moments are added to the original f.e.m.'s.

Proceeding with the solution, the K's and D's are computed in accordance with Eqs. (7.29) and (7.30). When joint B is relaxed, the change of moment in beam

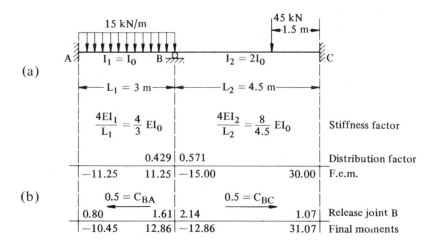

Figure 7.13

segment AB is $(0.429)(+3.75) = 1.61$ kN·m, and the change in BC is $(0.571)(+3.75)$ $= 2.14$ kN·m. The results of all the computations involved in the solution are shown in Fig. 7.13b. It is observed that the solution is complete when each joint is in equilibrium.

7.9   The Finite Element Method—Preliminaries

The powerful finite element method had its beginnings in the 1950s, and with the widespread use of the digital computer has since gained consider- able favor relative to other numerical approaches. The finite element method may be viewed as an approximate Ritz method combined with a variational principle applied to continuum mechanics. It permits the pre- diction of stress and strain in an engineering structure, with unprecedented ease and precision. The general procedures of the finite element and conventional structural matrix methods are similar. In the latter approach, the structure is idealized as an assembly of structural members connected to one another at joints or nodes at which the resultants of the applied forces are assumed to be concentrated. Employing the mechanics of materials or the elasticity approach, the stiffness properties of each element can be ascertained. Equilibrium and compatibility considered at each node lead to a set of algebraic equations, in which the unknowns may be nodal displacements, internal nodal forces, or both, depending upon the specific method used. In the displacement or so-called direct stiffness method, the set of algebraic equations involves the nodal displacements. In the force method, the equations are expressed in terms of unknown internal nodal forces. A mixed method is also used, in which the equations contain both nodal displacements and internal forces.

In contrast with the foregoing approaches, in the finite element method, the solid continuum is discretized by a finite number of elements, connected not only at their nodes, but along the hypothetical interelement boundaries as well. In addition then to nodal compatibility and equilibrium, as in conventional structural analysis, it is clear that compatibility must also be satisfied *along the boundaries* between elements. Because of this, the element stiffness can be only approximately determined in the finite element method. Alternative approaches to those of structural analysis are thus employed.

An essential difference exists between the finite element and *finite difference* analyses. In the latter, the differential equations describing the state of the continuum governing quantities such as stress or deflection are approximated by a set of finite difference equations written for a limited number of nodes. On the other hand, the finite element approach deals with an assembly of elements which replaces the continuous structure, and it is the replaced structure which is then the subject of analysis.

There are a number of finite element methods, as in the structural analyses previously cited. We here treat only the commonly employed finite element displacement approach.*

Now we shall define a number of basic quantities relevant to an individual finite element of an isotropic elastic body. In the interest of simple presentation, in this section the relationships are written only for the two-dimensional case. In the next section is presented the general formulation of the finite element method applicable to any structure. Solutions of plane stress and plane strain problems are illustrated in detail in Sec. 7.11. The analyses of axisymmetric structures and thin plates employing the finite element are given in Chapters 8 and 13, respectively.

To begin with, the relatively thin, continuous body shown in Fig. 7.14a is replaced by an assembly of finite elements (triangles, for example) indicated by the dashed lines (Fig. 7.14b). These elements are connected not only at their corners or nodes, but along the interelement boundaries as well. The basic unknowns are the nodal displacements.

Displacement Matrix

The nodal displacements are related to the internal displacements throughout the entire element by means of a so-called *displacement function*. Consider the typical element e in Fig. 7.14b, shown isolated in Fig. 7.15a. Designating the nodes i, j, m, the element nodal displacement matrix

---

*The literature related to this method is extensive. See, for example, O. C. Zienkiewicz, *The Finite Element Method in Engineering Science*, New York, McGraw-Hill, 1971; J. J. Conner and G. Will, *Computer-Aided Teaching of the Finite Element Displacement Method*, Report 69-23, Massachusetts Institute of Technology, 1969.

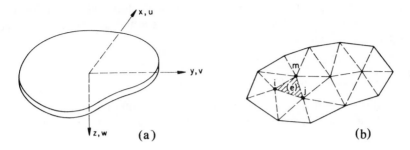

Figure 7.14

is

$$\{\delta\}_e = \{u_i, v_i, u_j, v_j, u_m, v_m\} \qquad (7.31a)$$

or, for convenience, expressed in terms of submatrices δ_u and δ_v,

$$\{\delta\}_e = \left\{ \frac{\delta_u}{\delta_v} \right\} = \{u_i, u_j, u_m, v_i, v_j, v_m\} \qquad (7.31b)$$

where the braces indicate a *column* matrix. The displacement function defining the displacement at any point within the element, $\{f\}_e$, is given by

$$\{f\}_e = \{u(x, y), v(x, y)\} \qquad (7.32)$$

which may also be expressed as

$$\{f\}_e = [N]\{\delta\}_e \qquad (7.33)$$

where the matrix $[N]$ is a function of position, to be obtained later for a specific element. It is, of course, desirable that a displacement function $\{f\}_e$ be selected such that the true displacement field will be as closely

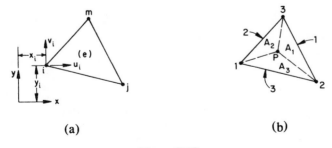

(a)   (b)

Figure 7.15

represented as possible. The approximation should result in a finite element solution which converges to the exact solution as the element size is progressively decreased.

Strain, Stress, and Elasticity Matrices

The strain, and hence the stress, are defined uniquely in terms of displacement functions (see Chapter 2). The strain matrix is of the form

$$\{\varepsilon\}_e = \{\varepsilon_x, \varepsilon_y, \gamma_{xy}\}$$

$$= \left\{\frac{\partial u}{\partial x}, \quad \frac{\partial v}{\partial y}, \quad \frac{\partial u}{\partial y} + \frac{\partial v}{\partial x}\right\} \tag{7.34a}$$

or

$$\{\varepsilon\}_e = [B]\{\delta\}_e \tag{7.34b}$$

where $[B]$ is also yet to be defined.

Similarly, the state of stress throughout the element is, from Hooke's law,

$$\{\sigma\}_e = \frac{E}{1-\nu^2} \begin{bmatrix} 1 & \nu & 0 \\ \nu & 1 & 0 \\ 0 & 0 & (1-\nu)/2 \end{bmatrix} \begin{Bmatrix} \varepsilon_x \\ \varepsilon_y \\ \gamma_{xy} \end{Bmatrix} \tag{a}$$

In general,

$$\{\sigma\}_e = [D]\{\varepsilon\}_e \tag{7.35}$$

where $[D]$, an elasticity matrix, contains material properties. If the element is subjected to thermal or *initial strain*, the stress matrix becomes

$$\{\sigma\}_e = [D](\{\varepsilon\} - \{\varepsilon_0\})_e \tag{7.36}$$

The thermal strain matrix, for the case of plane stress, is given by $\{\varepsilon_0\} = \{\alpha T, \alpha T, 0\}$ (Sec. 3.8). Comparing Eqs. (a) and (7.35), it is clear that

$$[D] = \frac{E}{1-\nu^2} \begin{bmatrix} 1 & \nu & 0 \\ \nu & 1 & 0 \\ 0 & 0 & (1-\nu)/2 \end{bmatrix} = \begin{bmatrix} D_{11} & D_{12} & D_{13} \\ & D_{22} & D_{23} \\ \text{Symm.} & & D_{33} \end{bmatrix} \tag{7.37}$$

where $D_{11} = E/(1-\nu^2)$, $D_{12} = \nu E/(1-\nu^2)$,.... Equation (7.37) is valid for the case of plane stress. A matrix $[D]$ for plane strain is found in a similar fashion.

7.10   Formulation of the Finite Element Method

A convenient method for executing the finite element procedure relies upon the minimization of the total potential energy of the system, expressed in terms of displacement functions. Consider again in this regard an elastic body (Fig. 7.14). The principle of potential energy, from Eq. (10.15), is expressed for the *entire body* as follows:

$$\Delta II = \sum_{1}^{n} \int_{V} (\sigma_x \Delta \varepsilon_x + \cdots + \sigma_z \Delta \varepsilon_z)\, dV$$

$$- \sum_{1}^{n} \int_{V} (F_x \Delta u + F_y \Delta v + F_z \Delta w)\, dV$$

$$- \sum_{1}^{n} \int_{s} (T_x \Delta u + T_y \Delta v + T_z \Delta w)\, ds = 0 \qquad (a)$$

Note that variation notation δ has been replaced by Δ to avoid confusing it with nodal displacement. In the above,

$n =$ number of elements comprising the body

$V =$ volume of a discrete element

$s =$ portion of the boundary surface area over which forces are prescribed

$F =$ Body forces per unit volume

$T =$ prescribed boundary force or surface traction per unit area.

Through the use of Eq. (7.32), Eq. (a) may be expressed in the following matrix form:

$$\sum_{1}^{n} \int_{V} (\{\Delta \varepsilon\}_e^T \{\sigma\}_e - \{\Delta f\}_e^T \{F\}_e)\, dV - \sum_{1}^{n} \int_{s} \{\Delta f\}_e^T \{T\}_e\, ds = 0 \qquad (b)$$

where superscript T denotes the transpose of a matrix. Now, using Eqs. (7.33), (7.34), and (7.36), the above becomes

$$\sum_{1}^{n} \{\Delta \delta\}_e^T ([k]_e \{\delta\}_e - \{Q\}_e) = 0 \qquad (c)$$

The element *stiffness* matrix $[k]_e$ and element *nodal force* matrix $\{Q\}_e$ (due to body force, initial strain, and surface traction) are

$$[k]_e = \int_{V} [B]^T [D][B]\, dV \qquad (7.38)$$

$$\{Q\}_e = \int_{V} [N]^T \{F\}\, dV + \int_{V} [B]^T [D]\{\varepsilon_0\}\, dV + \int_{s} [N]^T \{T\}\, ds \qquad (7.39)$$

It is clear that the variations in $\{\delta\}_e$ are independent and arbitrary, and from Eq. (c) we may therefore write

$$[k]_e\{\delta\}_e = \{Q\}_e \qquad (7.40)$$

We next derive the governing equations appropriate to the entire continuous body. The assembled form of Eq. (c) is

$$(\Delta\delta)^T([K]\{\delta\} - \{Q\}) = 0 \qquad (d)$$

The above expression must be satisfied for arbitrary variations of *all* nodal displacements $\{\Delta\delta\}$. This leads to the following equations of equilibrium for nodal forces for the entire structure, the so-called *system* equations:

$$[K]\{\delta\} = \{Q\} \qquad (7.41)$$

where

$$[K] = \sum_1^n [k]_e, \qquad \{Q\} = \sum_1^n \{Q\}_e \qquad (7.42)$$

It is noted that structural matrix $[K]$ and the total or equivalent nodal force matrix $\{Q\}$ are found by proper superposition of all element stiffness and nodal force matrices, respectively.

We can now summarize the *general procedure* for solving a problem by application of the finite element method:

1. Calculate $[k]_e$ from Eq. (7.38), in terms of the given element properties. Generate $[K] = \Sigma[k]_e$.
2. Calculate $\{Q\}_e$ from Eq. (7.39), in terms of the applied loading. Generate $\{Q\} = \Sigma\{Q\}_e$.
3. Calculate the nodal displacements from Eq. (7.41), by satisfying the boundary conditions: $\{\delta\} = [K]^{-1}\{Q\}$.
4. Calculate the element strain using Eq. (7.34): $\{\varepsilon\}_e = [B]\{\delta\}_e$.
5. Calculate the element stress using Eq. (7.36): $\{\sigma\}_e = [D](\{\varepsilon\} - \{\varepsilon_0\})_e$.

When the stress found is uniform throughout each element, this result is usually interpreted two ways: the stress obtained for an element is assigned to its centroid; if the material properties of the elements connected at a node are the same, the average of the stresses in the elements is assigned to the common node.

The foregoing outline will be better understood when applied to a triangular element in the next section.

7.11   The Triangular Finite Element

Because of the relative ease with which the region within an arbitrary boundary can be approximated, the triangle is used extensively in finite element assemblies. Before deriving the properties of the triangular element, we describe *area* or so-called *triangular* coordinates, quite useful for the simplification of the displacement functions.

Consider the triangular finite element 1 2 3 (where $i=1$, $j=2$, $m=3$) shown in Fig. 7.15b, in which the *counterclockwise* numbering convention of nodes and sides is indicated. A point P located within the element, by connection with the corners of the element, forms three subareas denoted A_1, A_2, A_3. The ratios of these areas to the total area A of the triangle locate P and represent the area coordinates:

$$L_1 = \frac{A_1}{A}, \qquad L_2 = \frac{A_2}{A}, \qquad L_3 = \frac{A_3}{A} \tag{7.43}$$

It follows from the above that

$$L_1 + L_2 + L_3 = 1 \tag{7.44}$$

and consequently only two of the three coordinates are independent. A useful property of area coordinates is observed through reference to Fig. 7.15b and Eq. (7.43):

$$L_n = 0 \quad \text{on side } n \quad (n = 1,2,3)$$

and

$$L_i = 1, \qquad L_j = L_m = 0 \quad \text{at node } i$$
$$L_j = 1, \qquad L_i = L_m = 0 \quad \text{at node } j$$
$$L_m = 1, \qquad L_i = L_j = 0 \quad \text{at node } m$$

The area A of the triangle may be expressed in terms of the coordinates of two sides, e.g., 2 and 3:

$$A = \tfrac{1}{2}\{[(x_2 - x_1)\mathbf{i} + (y_2 - y_1)\mathbf{j}] \times [(x_3 - x_1)\mathbf{i} + (y_3 - y_1)\mathbf{j}]\} \cdot \mathbf{k}$$

or

$$2A = \det \begin{vmatrix} 1 & x_1 & y_1 \\ 1 & x_2 & y_2 \\ 1 & x_3 & y_3 \end{vmatrix} \tag{7.45a}$$

Here $\mathbf{i}$, $\mathbf{j}$, and $\mathbf{k}$ are the unit vectors in the x, y, and z directions, respectively. Two additional expressions are similarly found. In general, we have

$$2A = a_j b_i - a_i b_j = a_m b_j - a_j b_m = a_i b_m - a_m b_i \tag{7.45b}$$

where

$$a_i = x_m - x_j, \quad b_i = y_j - y_m$$
$$a_j = x_i - x_m, \quad b_j = y_m - y_i \tag{7.46}$$
$$a_m = x_j - x_i, \quad b_m = y_i - y_j$$

It is noted that a_j, a_m, b_j, b_m can be found from definitions of a_i and b_i with the permutation of the subscripts in the order $ijmijm$, etc.

Similar equations are derivable for subareas A_1, A_2, and A_3. The resulting expressions, together with Eq. (7.43), lead to the following relationship between area and Cartesian coordinates:

$$\begin{Bmatrix} L_1 \\ L_2 \\ L_3 \end{Bmatrix} = \frac{1}{2A} \begin{bmatrix} 2c_{23} & b_1 & a_1 \\ 2c_{31} & b_2 & a_2 \\ 2c_{12} & b_3 & a_3 \end{bmatrix} \begin{Bmatrix} 1 \\ x \\ y \end{Bmatrix} \tag{7.47}$$

where

$$\begin{aligned} c_{23} &= x_2 y_3 - x_3 y_2 \\ c_{31} &= x_3 y_1 - x_1 y_3 \\ c_{12} &= x_1 y_2 - x_2 y_1 \end{aligned} \tag{7.48}$$

Note again that given any of the above expressions for c_{ij}, the others may be obtained by permutation of the subscripts.

Now we explore the properties of an ordinary triangular element of a continuous body in a state of plane stress or plane strain (Fig. 7.15). The nodal displacements are

$$\{\delta\}_e = \{u_1, u_2, u_3, v_1, v_2, v_3\} \tag{a}$$

The displacement throughout the element is provided by

$$\{f\}_e = \begin{bmatrix} u_1 & u_2 & u_3 \\ v_1 & v_2 & v_3 \end{bmatrix} \begin{Bmatrix} L_1 \\ L_2 \\ L_3 \end{Bmatrix} \tag{7.49}$$

Matrices $[N]$ and $[B]$ of Eqs. (7.33) and (7.34b) are next evaluated, beginning with

$$\{f\}_e = [N]\{\delta\}_e \tag{b}$$

$$\{\varepsilon\}_e = [B]\{\delta\}_e \tag{c}$$

We observe that Eqs. (7.49) and (b) are equal, provided that

$$[N] = \begin{bmatrix} L_1 & L_2 & L_3 & 0 & 0 & 0 \\ 0 & 0 & 0 & L_1 & L_2 & L_3 \end{bmatrix} \tag{7.50}$$

The strain matrix is obtained by substituting Eqs. (7.49) and (7.47) into Eq. (7.34a):

$$\begin{Bmatrix} \varepsilon_x \\ \varepsilon_y \\ \gamma_{xy} \end{Bmatrix}_e = \frac{1}{2A} \begin{bmatrix} b_1 & b_2 & b_3 & 0 & 0 & 0 \\ 0 & 0 & 0 & a_1 & a_2 & a_3 \\ a_1 & a_2 & a_3 & b_1 & b_2 & b_3 \end{bmatrix} \begin{Bmatrix} u_1 \\ u_2 \\ u_3 \\ v_1 \\ v_2 \\ v_3 \end{Bmatrix} \tag{d}$$

Here A, a_i, and b_i are defined by Eqs. (7.45) and (7.46). The strain (stress) is observed to be constant throughout, and the element of Fig. 7.15 is thus referred to as a *constant strain triangle*. Comparing Eqs. (c) and (d), we have

$$[B] = \frac{1}{2A} \begin{bmatrix} b_1 & b_2 & b_3 & 0 & 0 & 0 \\ 0 & 0 & 0 & a_1 & a_2 & a_3 \\ a_1 & a_2 & a_3 & b_1 & b_2 & b_3 \end{bmatrix} \tag{7.51}$$

The stiffness of the element can now be obtained through the use of Eq. (7.38):

$$[k]_e = [B]^T[D][B]tA \tag{e}$$

Let us now define

$$[D^*] = \frac{t[D]}{4A} = \begin{bmatrix} D_{11}^* & D_{12}^* & D_{13}^* \\ & D_{22}^* & D_{23}^* \\ \text{Symm.} & & D_{33}^* \end{bmatrix} \tag{7.52}$$

where $[D]$ is given by Eq. (7.37) for plane stress. Assembling Eq. (e) together with Eqs. (7.51) and (7.52), and expanding, the stiffness matrix is expressed in the following partitioned form of order 6×6:

$$[k]_e = \begin{bmatrix} k_{uu,ln} & k_{uv,ln} \\ \hline k_{uv,ln}^T & k_{vv,ln} \end{bmatrix} \quad (l,n=1,2,3) \tag{7.53}$$

where the submatrices are

$$k_{uu,ln} = D_{11}^* b_l b_n + D_{33}^* a_l a_n + D_{13}^*(b_l a_n + b_n a_l)$$

$$k_{vv,ln} = D_{33}^* b_l b_n + D_{22}^* a_l a_n + D_{23}^*(b_l a_n + b_n a_l) \tag{7.54}$$

$$k_{uv,ln} = D_{13}^* b_l b_n + D_{23}^* a_l b_n + D_{12}^* b_l a_n + D_{33}^* b_n a_l$$

Finally, we consider the determination of the element nodal force matrices. The nodal force owing to a constant *body force* per unit volume is, from Eqs. (7.39) and (7.50),

$$\{Q\}_e^b = \int_V [N]^T\{F\}\, dV = \int_V \begin{bmatrix} L_1 & 0 \\ L_2 & 0 \\ L_3 & 0 \\ 0 & L_1 \\ 0 & L_2 \\ 0 & L_3 \end{bmatrix} \begin{Bmatrix} F_x \\ F_y \end{Bmatrix} dV \tag{f}$$

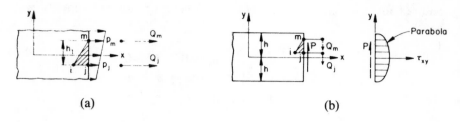

(a)   (b)

Figure 7.16

For an element of constant thickness, this expression is readily integrated to yield*

$$\{Q\}_e = \tfrac{1}{3}At\{F_x, F_x, F_x, F_y, F_y, F_y\} \tag{7.55}$$

The nodal forces associated with the weight of an element are observed to be equally distributed at the nodes.

The element nodal forces attributable to applied external loading may be determined either by evaluating the static resultants or by application of Eq. (7.39). Nodal force expressions for arbitrary nodes j and m are given below for a number of common cases (Problem 7.12).

Linear load, $p(y)$ per unit area, Fig. 7.16a:

$$Q_j = \frac{h_1 t}{6}(2p_j + p_m), \qquad Q_m = \frac{h_1 t}{6}(2p_m + p_j) \tag{7.56a}$$

where t is the thickness of the element.

Uniform load is a special case of the above with $p_j = p_m = p$:

$$Q_j = Q_m = \tfrac{1}{2}ph_1 t \tag{7.56b}$$

End shear load, P, the resultant of a parabolic shear stress distribution defined by Eq. (3.17) (see Fig. 7.16b):

$$
\begin{aligned}
Q_m &= \frac{3P}{4h}\left\{ \tfrac{1}{2}(y_m - y_j) + \frac{1}{3h^2}\left[\frac{y_j}{4}(y_m^2 + y_j y_m + y_j^2) - \tfrac{3}{4}y_m^3\right]\right\} \\
Q_j &= \frac{3P}{4h}\left\{ \tfrac{1}{2}(y_m - y_j) - \frac{1}{3h^2}\left[(y_m - \tfrac{3}{4}y_j)(y_m^2 + y_j y_m + y_j^2) - \tfrac{3}{4}y_m^3\right]\right\}
\end{aligned}
\tag{7.57}
$$

---

*For a general function $f = L_1^\alpha L_2^\beta L_3^\gamma$, defined in area coordinates, the integral of f over any triangular area A is given by

$$\int_A L_1^\alpha L_2^\beta L_3^\gamma \, dA = 2A\frac{\alpha!\beta!\gamma!}{(\alpha+\beta+\gamma)!}$$

where α, β, γ are constants.

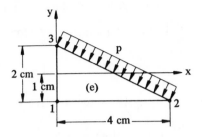

Figure 7.17

Equation (7.53), together with those expressions given above for the nodal forces, characterizes the constant strain element. These are substituted into Eq. (7.42), and subsequently into Eq. (7.41) in order to evaluate the nodal displacements by satisfying the boundary conditions.

The basic procedure employed in the finite element method is illustrated in the following simple problems.

Example 7.7. The element e is shown in Fig. 7.17 represents a segment of a thin elastic plate having side 23 adjacent to its boundary. The plate is subjected to several loads as well as a uniform temperature rise of 50°C. Determine (a) the stiffness matrix and (b) the equivalent (or total) nodal force matrix for the element if a pressure of $p = 14$ MPa acts on side 23. Let $t = 0.3$ cm, $E = 200$ GPa, $\nu = 0.3$, specific weight $\gamma = 77$ kN/m³, and $\alpha = 12 \times 10^{-6}/°C$.

SOLUTION. The origin of the coordinates is located at midlength of side 13, for convenience. However, it may be placed at any point in the x, y plane. Applying Eq. (7.52), we have (in N/cm³):

$$[D^*] = \frac{t[D]}{4A} = \frac{10^6}{8}\begin{bmatrix} 3.33 & 0.99 & 0 \\ 0.99 & 3.3 & 0 \\ 0 & 0 & 1.16 \end{bmatrix} \tag{g}$$

(a) *The stiffness matrix*: The nodal points are located at

$$x_i = x_1 = 0, \qquad y_i = y_1 = -1$$

$$x_j = x_2 = 4, \qquad y_j = y_2 = -1 \tag{h}$$

$$x_m = x_3 = 0, \qquad y_m = y_3 = 1$$

Using Eq. (7.46), and referring to Fig. 7.17, we obtain

$$a_i = a_1 = 0 - 4 = -4, \qquad b_i = b_1 = -1 - 1 = -2$$

$$a_j = a_2 = 0 - 0 = 0, \qquad b_j = b_2 = 1 + 1 = 2 \tag{i}$$

$$a_m = a_3 = 4 - 0 = 4, \qquad b_m = b_3 = -1 + 1 = 0$$

Next the first equation of (7.54), together with Eqs. (g) and (i), yields

$$k_{uu,11} = \frac{10^6}{8}[3.3(4) + 1.16(16)] = 3.97 \times 10^6$$

$$k_{uu,12} = k_{uu,21} = \frac{10^6}{8}[3.3(2)(-2) + 0] = -1.65 \times 10^6$$

$$k_{uu,13} = k_{uu,31} = \frac{10^6}{8}[0 + 1.16(4)(-4)] = -2.32 \times 10^6$$

$$k_{uu,22} = \frac{10^6}{8}[3.3(4) + 0] = 1.65 \times 10^6$$

$$k_{uu,23} = k_{uu,32} = 0$$

$$k_{uu,33} = \frac{10^6}{8}[0 + 1.16(16)] = 2.32 \times 10^6$$

The submatrix k_{uu} is thus

$$k_{uu} = 10^6 \begin{bmatrix} 3.97 & -1.65 & -2.32 \\ -1.65 & 1.65 & 0 \\ -2.32 & 0 & 2.32 \end{bmatrix}$$

Similarly, from the second and third equations of (7.54), we obtain the following matrices:

$$k_{uv} = 10^6 \begin{bmatrix} 2.15 & -1.16 & -0.99 \\ -0.99 & 0 & 0.99 \\ -1.16 & 1.16 & 0 \end{bmatrix}, \qquad k_{vv} = 10^6 \begin{bmatrix} 7.18 & -0.58 & -6.6 \\ -0.58 & 0.58 & 0 \\ -6.6 & 0 & 6.6 \end{bmatrix}$$

Assembling the above, the stiffness matrix of the element (in newtons per centimeter) is

$$[k]_e = 10^6 \left[\begin{array}{ccc:ccc} 3.97 & -1.65 & -2.32 & 2.15 & -1.16 & -0.99 \\ -1.65 & 1.65 & 0 & -0.99 & 0 & 0.99 \\ -2.32 & 0 & 2.32 & -1.16 & 1.16 & 0 \\ \hdashline 2.15 & -0.99 & -1.16 & 7.18 & -0.58 & -6.6 \\ -1.16 & 0 & 1.16 & -0.58 & 0.58 & 0 \\ -0.99 & 0.99 & 0 & -6.6 & 0 & 6.6 \end{array} \right] \qquad (j)$$

(b) We next determine the nodal forces of the element owing to various loadings. The components of body force are $F_x = 0$ and $F_y = 0.077$ N/cm³.

Body force effects: Through the application of Eq. (7.55), it is found that

$$\{Q\}_e^b = \{0, 0, 0, -0.0308, -0.0308, -0.0308\} \text{ N}$$

Surface traction effects: The total load, $st\{p\} = \sqrt{20}(0.3)\{-2 \times 1400/\sqrt{20}, -4 \times 1400/\sqrt{20}\} = \{-840, -1680\}$, is equally divided between nodes 2 and 3. The nodal forces can therefore be expressed as

$$\{Q\}_3^p = \{0, -420, -420, 0, -840, -840\} \text{ N}$$

Thermal strain effects: The initial strain associated with the 50°C temperature rise is $\varepsilon_0 = \alpha T = 0.0006$. From Eq. (7.39),

$$\{Q\}_e^t = [B]^T[D]\{\varepsilon_0\}(At)$$

Substituting into the above the matrix $[B]$, given by Eq. (7.51), and the values of the other constants already determined, the nodal force is calculated as follows:

$$\{Q\}'_e = \frac{1}{8} \begin{bmatrix} -2 & 0 & -4 \\ 2 & 0 & 0 \\ 0 & 0 & 4 \\ 0 & -4 & -2 \\ 0 & 0 & 2 \\ 0 & 4 & 0 \end{bmatrix} \frac{200 \times 10^5}{0.91} \begin{bmatrix} 1 & 0.3 & 0 \\ 0.3 & 1 & 0 \\ 0 & 0 & 0.35 \end{bmatrix} \begin{Bmatrix} 0.0006 \\ 0.0006 \\ 0 \end{Bmatrix} \quad (1.2)$$

or

$$\{Q\}'_e = \{-5142.85, \; 5142.85, \; 0, \; -10{,}285.70 \; 0, \; 10{,}285.70\} \text{ N}$$

The equivalent nodal force matrix: Summation of the nodal matrices due to the several effects yields the total element nodal force matrix:

$$\{Q\}_e = \begin{Bmatrix} Q_{x1} \\ Q_{x2} \\ Q_{x3} \\ Q_{y1} \\ Q_{y2} \\ Q_{y3} \end{Bmatrix} = \begin{Bmatrix} -5142.85 \\ 4742.85 \\ -400 \\ -10{,}285.73 \\ -840.03 \\ 9445.67 \end{Bmatrix} \text{ N}$$

If, in addition, there are any actual nodal forces, these must also, of course, be added to the values obtained above.

Example 7.8. A 0.3-cm-thick cantilever beam is subjected to a parabolically varying end shearing stress resulting in a total load of 5000 N (Fig. 7.18a). Divide the beam into two constant strain triangles and calculate the deflections, Let $E = 200$ GPa and $\nu = 0.3$.

SOLUTION. The discretized beam is shown in Fig. 7.18b.

The stiffness matrix: It is noted inasmuch as the dimensions and material properties of element a are the same as that given in the previous example, $[k]_a$ is defined by Eq. (j). For element b, assignment of $i = 2$, $j = 4$, and $m = 3$ [Eq. (7.46)]

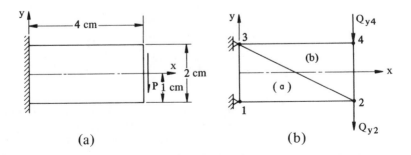

(a)   (b)

Figure 7.18

leads to

$$a_1 = a_2 = 0 - 4 = -4, \qquad b_i = b_2 = 1 - 1 = 0$$
$$a_j = a_4 = 4 - 0 = 4, \qquad b_j = b_4 = 1 + 1 = 2$$
$$a_m = a_3 = 4 - 4 = 0, \qquad b_m = b_3 = -1 - 1 = -2$$

Substitution of the above and Eq. (g) into Eqs. (7.54) yields

$$k_{uu,22} = \frac{10^6}{8}[3.3(0) + 1.16(16)] = 2.32 \times 10^6$$

$$k_{uu,44} = \frac{10^6}{8}[3.3(4) + 1.16(16)] = 3.97 \times 10^6$$

$$k_{uu,23} = k_{uu,32} = 0$$

$$k_{uu,24} = k_{uu,42} = \frac{10^6}{8}[3.3(0) + 1.16(-4)(4)] = -2.32 \times 10^6$$

$$k_{uu,43} = k_{uu,34} = \frac{10^6}{8}[3.3(2)(-2) + 1.16(0)] = -1.65 \times 10^6$$

$$k_{uu,33} = \frac{10^6}{8}[3.3(4) + 1.16(0)] = 1.65 \times 10^6$$

Thus,

$$k_{uu} = 10^6 \begin{bmatrix} 2.32 & 0 & -2.32 \\ 0 & 1.65 & -1.65 \\ -2.32 & -1.65 & 3.97 \end{bmatrix}$$

Similarly, we obtain

$$k_{uv} = 10^6 \begin{bmatrix} 0 & 1.16 & -1.16 \\ 0.99 & 0 & -0.99 \\ -0.99 & -1.16 & 2.15 \end{bmatrix}$$

$$k_{vv} = 10^6 \begin{bmatrix} 6.6 & 0 & -6.6 \\ 0 & 0.58 & -0.58 \\ -6.6 & -0.58 & 7.18 \end{bmatrix}$$

The stiffness matrix of element b is therefore

$$[k]_b = 10^6 \begin{bmatrix} 2.32 & 0 & -2.32 & 0 & 1.16 & -1.16 \\ 0 & 1.65 & -1.65 & 0.99 & 0 & -0.99 \\ -2.32 & -1.65 & 3.97 & -0.99 & -1.16 & 2.15 \\ 0 & 0.99 & -0.99 & 6.6 & 0 & -6.6 \\ 1.16 & 0 & -1.16 & 0 & 0.58 & -0.58 \\ -1.16 & -0.99 & 2.15 & -6.6 & -0.58 & 7.18 \end{bmatrix} \qquad \text{(k)}$$

The displacements u_4, v_4 and u_1, v_1 are not involved in elements a and b, respectively. Therefore, prior to the addition of $[k]_a$ and $[k]_b$ to form the system stiffness matrix, rows and columns of zeros must be added to each of the element matrices to account for the absence of these displacements. In doing so, Eqs. (j) and

(k) become

$$[k]_a = 10^6 \begin{bmatrix} 3.97 & -1.65 & -2.32 & 0 & | & 2.15 & -1.16 & -0.99 & 0 \\ -1.65 & 1.65 & 0 & 0 & | & -0.99 & 0 & 0.99 & 0 \\ -2.32 & 0 & 2.32 & 0 & | & -1.16 & 1.16 & 0 & 0 \\ 0 & 0 & 0 & 0 & | & 0 & 0 & 0 & 0 \\ -\!-\!- & & & & & & & \\ 2.15 & -0.99 & -1.16 & 0 & | & 7.18 & -0.58 & -6.6 & 0 \\ -1.16 & 0 & 1.16 & 0 & | & -0.58 & 0.58 & 0 & 0 \\ -0.99 & 0.99 & 0 & 0 & | & -6.6 & 0 & 6.6 & 0 \\ 0 & 0 & 0 & 0 & | & 0 & 0 & 0 & 0 \end{bmatrix} \quad \text{(l)}$$

and

$$[k]_b = 10^6 \begin{bmatrix} 0 & 0 & 0 & 0 & | & 0 & 0 & 0 & 0 \\ 0 & 2.32 & 0 & -2.32 & | & 0 & 0 & 1.16 & -1.16 \\ 0 & 0 & 1.65 & -1.65 & | & 0 & 0.99 & 0 & -0.99 \\ 0 & -2.32 & -1.65 & 3.97 & | & 0 & -0.99 & -1.16 & 2.15 \\ -\!-\!- & & & & & & & \\ 0 & 0 & 0 & 0 & | & 0 & 0 & 0 & 0 \\ 0 & 0 & 0.99 & -0.99 & | & 0 & 6.6 & 0 & -6.6 \\ 0 & 1.16 & 0 & -1.16 & | & 0 & 0 & 0.58 & -0.58 \\ 0 & -1.16 & -0.99 & 2.15 & | & 0 & -6.6 & -0.58 & 7.18 \end{bmatrix} \quad \text{(m)}$$

Then addition of Eqs. (l) and (m) yields the system matrix (in newtons per centimeter):

$$[K] = 10^6 \begin{bmatrix} 3.97 & -1.65 & -2.32 & 0 & | & 2.15 & -1.16 & -0.99 & 0 \\ -1.65 & 3.97 & 0 & -2.32 & | & -0.99 & 0 & 2.15 & -1.16 \\ -2.32 & 0 & 3.97 & -1.65 & | & -1.16 & 2.15 & 0 & -0.99 \\ 0 & -2.32 & -1.65 & 3.97 & | & 0 & -0.99 & -1.16 & 2.15 \\ -\!-\!- & & & & & & & \\ 2.15 & -0.99 & -1.16 & 0 & | & 7.18 & -0.58 & -6.6 & 0 \\ -1.16 & 0 & 2.15 & -0.99 & | & -0.58 & 7.18 & 0 & -6.6 \\ -0.99 & 2.15 & 0 & -1.16 & | & -6.6 & 0 & 7.18 & -0.58 \\ 0 & -1.16 & -0.99 & 2.15 & | & 0 & -6.6 & -0.58 & 7.18 \end{bmatrix} \quad \text{(n)}$$

Nodal Forces: Referring to Fig. 7.18b and applying Eq. (7.57), we obtain

$$Q_{y4} = \frac{3(-5000)}{4(1)} \left\{ \tfrac{1}{2}(1+1) + \tfrac{1}{3}\left[-\tfrac{1}{4}(1-1+1) - \tfrac{3}{4} \right] \right\} = -2500 \text{ N}$$

$$Q_{y2} = \frac{3(-5000)}{4(1)} \left\{ \tfrac{1}{2}(1+1) - \tfrac{1}{3}\left[1 + \tfrac{3}{4}(1-1+1) - \tfrac{3}{4} \right] \right\} = -2500 \text{ N}$$

As no other external force exists, the system nodal force matrix is

$$\{Q\} = \{0,\ 0,\ 0,\ 0,\ 0,\ -2500,\ 0,\ -2500\}$$

Nodal Displacements: The boundary condition are

$$u_1 = u_3 = v_1 = v_3 = 0 \quad \text{(o)}$$

The force-displacement relationship of the system is therefore

$$\{0,\ 0,\ 0,\ 0,\ 0,\ -2500,\ 0,\ -2500\} = [K]\{0,\ u_2,\ 0,\ u_4,\ 0,\ v_2,\ 0,\ v_4\} \quad \text{(p)}$$

Equation (p) is readily reduced to the form

$$\begin{Bmatrix} 0 \\ 0 \\ -2500 \\ -2500 \end{Bmatrix} = 10^6 \begin{bmatrix} 3.97 & -2.32 & 0 & -1.16 \\ -2.32 & 3.97 & -0.99 & 2.15 \\ 0 & -0.99 & 7.18 & -6.6 \\ -1.16 & 2.15 & -6.6 & 7.18 \end{bmatrix} \begin{Bmatrix} u_2 \\ u_4 \\ v_2 \\ v_4 \end{Bmatrix} \quad \text{(q)}$$

From the above we obtain

$$\begin{Bmatrix} u_2 \\ u_4 \\ v_2 \\ v_4 \end{Bmatrix} = 10^{-6} \begin{bmatrix} 0.429 & 0.180 & 0.252 & 0.247 \\ 0.180 & 0.483 & -0.256 & -0.351 \\ 0.252 & -0.256 & 1.366 & 1.373 \\ 0.247 & -0.351 & 1.373 & 1.546 \end{bmatrix} \begin{Bmatrix} 0 \\ 0 \\ -2500 \\ -2500 \end{Bmatrix}$$

$$= \begin{Bmatrix} -0.0012 \\ 0.0015 \\ -0.0068 \\ -0.0073 \end{Bmatrix} cm \qquad (r)$$

The strains $\{\varepsilon\}_a$ may now be found upon introduction of Eqs. (i) and (r) into (d) as

$$\begin{Bmatrix} \varepsilon_x \\ \varepsilon_y \\ \gamma_{xy} \end{Bmatrix} = \frac{1}{8} \begin{bmatrix} -2 & 2 & 0 & 0 & 0 & 0 \\ 0 & 0 & 0 & -4 & 0 & 4 \\ -4 & 0 & 4 & -2 & 2 & 0 \end{bmatrix} \begin{Bmatrix} 0 \\ -0.0012 \\ 0 \\ 0 \\ -0.0068 \\ 0 \end{Bmatrix} = \begin{Bmatrix} -3 \\ 0 \\ -17 \end{Bmatrix} 10^{-4}$$

Finally, the stress is determined by multiplying $[D]$ by $\{\varepsilon\}_a$:

$$\begin{Bmatrix} \sigma_x \\ \sigma_y \\ \tau_{xy} \end{Bmatrix}_a = \frac{200 \times 10^9}{0.91} \begin{bmatrix} 1 & 0.3 & 0 \\ 0.3 & 1 & 0 \\ 0 & 0 & 0.35 \end{bmatrix} \begin{Bmatrix} -3 \\ 0 \\ -17 \end{Bmatrix} 10^{-4} = \begin{Bmatrix} -66.67 \\ -20.00 \\ -132.22 \end{Bmatrix} MPa$$

Element b is treated similarly.

It should be noted that the model employed in the foregoing solution is quite crude. The effect of element size on solution accuracy is illustrated in the following example.

Example 7.9. By means of (a) exact and (b) finite element approaches, investigate the stresses and displacements in a thin plate or thin beam subjected to end moments applied about the centroidal axis (Fig. 7.19a). Let $L = 76.2$ mm, $h = 50.8$ mm, thickness $t = 25.4$ mm, $p = 6895$ kPa, $E = 207$ GPa, and $\nu = 0.15$. Neglect the ~~ight of the member.

SOLUTION

(a) *Exact solution*: Replacing the end moments with the statically equivalent load per unit area $p = Mh/I$ (Fig. 7.19b), the stress distribution from Eq. (5.5) is

$$\sigma_x = -\frac{y}{h} p, \qquad \sigma_y = \tau_{xy} = 0 \qquad (s)$$

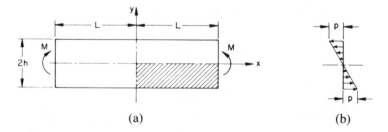

(a)   (b)

Figure 7.19

From Hooke's law and Eq. (2.3), we have

$$\frac{\partial u}{\partial x} = -\frac{yp}{Eh}, \qquad \frac{\partial v}{\partial y} = \frac{\nu yp}{Eh}, \qquad \frac{\partial u}{\partial y} + \frac{\partial v}{\partial x} = 0$$

By now following a procedure similar to that of Sec. 5.4, satisfying the conditions $u(0,0) = v(0,0) = 0$ and $u(L,0) = 0$, we obtain

$$u = -\frac{p}{Eh} xy, \qquad v = -\frac{p}{2Eh}(x^2 + \nu y^2) \tag{t}$$

Substituting the data into Eqs. (s) and (t), the results are

$$\sigma_x = -\frac{1}{0.0508} yp, \qquad \sigma_{x,\max} = 6895 \text{ kPa} \tag{u}$$

$$u(0.0762, -0.0254) = 25.4 \times 10^{-6} \text{ m}, \qquad v(0.0762, 0) = 1.905 \times 10^{-6} \text{ m}$$

(b) *Finite element solution*: Considerations of symmetry and antisymmetry indicate that only any one quarter (shown as shaded portion in the figure) of the beam need be analyzed.

Displacement boundary conditions: Fig. 7.20a shows the quarter plate discretized to contain 12 triangular elements. The origin of coordinates is located at node 3. As no axial deformation occurs along the x and y axes, nodes 1, 2, 6, 9, and 12 are restrained against u deformation; node 3 is restrained against both u and v deformation. The boundary conditions are thus

$$u_1 = u_2 = u_3 = u_6 = u_9 = u_{12} = 0, \qquad v_3 = 0$$

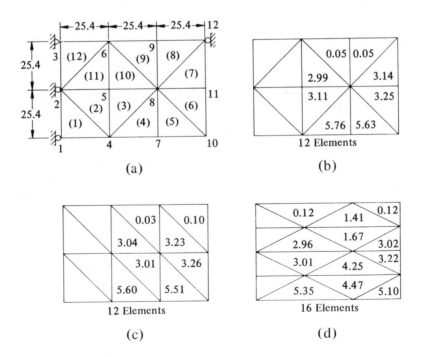

(a)

(b)

(c)

(d)

Figure 7.20

Nodal forces: For the loading system of Fig. 7.19b, Eq. (7.56a) applies. Upon substitution of numerical values,

$$Q_{x10} = \frac{0.0254 \times 0.0254}{6}(2 \times 6895 + 3447.5) = 1853.5 \text{ N}$$

$$Q_{x11} = \frac{0.0254 \times 0.0254}{6}(2 \times 3447.5 + 6895) + \frac{0.0254 \times 0.0254}{6}(2 \times 3447.5 + 0) = 2224 \text{ N}$$

$$Q_{x12} = \frac{0.0254 \times 0.0254}{6}(0 + 3447.5) = 370.5 \text{ N}$$

The remaining Q's are zero.

Results: The nodal displacements are determined following a procedure similar to that of Example 7.8. The stresses are then evaluated and representative values (in megapascals) given in Fig. 7.20b. Note that the stress obtained for an element is assigned to the centroid. It is observed that there is considerable difference between the exact solution, Eq. (u), and that resulting from the coarse mesh arrangement employed. To demonstrate the influence of element size and orientation, calculations have also been carried out for the grid configurations shown in Figs. 7.20c and d. For purposes of comparison, the deflections corresponding to Figs. 7.20b through d are presented in Table 7.4.

The effect of element orientation is shown in Figs. 7.20b and c, for an equal number of elements and node locations. Figure 7.20d reveals that elements characterized by large differences between their side lengths, *weak elements*, lead to unfavorable results even though the number of nodes is larger than those of Figs. 7.20b and c. The employment of equilateral or nearly equilateral so-called *well-formed elements* of finer mesh leads to solutions approaching the exact values.

It is noted that owing to the approximate nature of the finite element method, nonzero values are found for σ_y and τ_{xy}. These are not listed in the figures. As the mesh becomes finer, these stresses do essentially vanish.

It should be mentioned that when the three midside points as well as the three corners of a triangular element are taken as the nodes, a second-order polynomial can be selected for $\{f\}_e$ such that it ensures the compatibility of interelement displacement and a linear variation of strain throughout the element. The element is thus referred to as a *linear strain triangle*. To arrive at the characteristics of this element, the same procedure is followed as for the constant strain triangle. Two other common finite element types are the quadrilateral, formed by four constant strain triangles, and the

Table 7.4

Case	Number of nodes	Deflection (m)	
		$v_{12}(10^6)$	$u_{10}(10^6)$
Fig. 7.20b	12	1.547	2.133
Fig. 7.20c	12	1.745	2.062
Fig. 7.20d	15	1.572	1.976
Exact solution	—	1.905	2.540

rectangular element, a special case of the quadrilateral. The factors such as size, arrangement, and shape of the finite element influence directly the accuracy of the finite element solution. It is clear that one cannot reduce element size to extremely small values, as this would tend to increase to significant magnitudes the computer error incurred. An "exact" solution is thus unattainable, and one therefore seeks instead an *acceptable* solution. The goal is then the establishment of a finite element which ensures convergence to the exact solution in the absence of round-off error. The literature contains many comparisons between the various basic elements. The efficiency of a finite element solution can, in certain situations, be enhanced through the use of a "mix" of elements. For example, a denser mesh within a region of severely changing or localized stress may save much time and effort.

7.12 Use of Digital Computers

The finite difference method as well as the finite element method leads to systems of linear algebraic equations. The digital computer is often employed to provide rapid solutions of these simultaneous equations, usually by means of matrix methods. If the number of equations is not great, the computer solution may be expected to be quite accurate. Errors associated with round-off and truncation are negligible for matrices as large as 15×15, provided that the elements contain eight significant figures. This may also be true for well-conditioned matrices of order higher than 15, that is, those with dominant diagonal terms. In cases involving higher order matrices, iteration procedures, as well as other techniques of numerical analysis, are used.

Compared with analytical techniques, the finite element method offers a distinct advantage in the treatment of anisotropic situations, three-dimensional and eigenvalue problems, and two-dimensional problems involving irregular geometries and loads. It is evident, however, that the finite element approach, even in the simplest of cases, requires considerable matrix algebra. For any significant problem, the electronic digital computer must be used. A variety of techniques are applicable to the efficient execution of a finite element solution. It is worthwhile to mention that there have been developed many large-scale general purpose system programs for finite element analysis. Such systems permit structural data to be processed in a general and flexible manner for the purpose of fulfilling the specific requirements of the user. System programs involve almost all known finite element models and can be stored on the disk of a large computer.

The digital computer not only offers extreme speed, but tends to encourage the analyst to retain the complexities of a problem. The latter is in contrast with the tendency to make overly simplifying assumptions in the interest of a tractable hand calculation. The many advantages of digital computation more than outweigh the disadvantage sometimes cited, that digital computation may cause one to lose one's physical grasp of a problem.

Chapter 7—Problems

Secs. 7.1 to 7.6

7.1. Divide the beam of Fig. P7.1 into four segments of equal length, each having a constant section. Determine the deflection at the free end. Dimensions are given in meters.

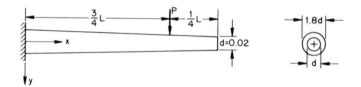

Figure P7.1

7.2. Redo Example 7.1(a), taking $\Delta x = L/5$, and compare the results with those previously found.

7.3. Referring to Fig. 7.4, demonstrate that the biharmonic equation

$$\nabla^4 w = \frac{\partial^4 w}{\partial x^4} + 2\frac{\partial^4 w}{\partial x^2 \partial y^2} + \frac{\partial^4 w}{\partial y^4} = 0$$

takes the following finite difference form:

$$h^4 \nabla^4 w = 20 w_0 - 8(w_1 + w_2 + w_3 + w_4)$$

$$+ 2(w_5 + w_6 + w_7 + w_8) + w_9 + w_{10} + w_{11} + w_{12} \qquad (P7.3)$$

7.4. Consider a torsional bar having rectangular cross section of width $4a$ and depth $2a$. Divide the cross section into equal nets with $h = a/2$. Assume that the origin of coordinates is located at the centroid. Find the shear stresses at points $x = \pm 2a$ and $y = \pm a$. Use the direct finite difference approach. Note that the exact value of stress at $y = \pm a$ is, from Table 6.1, $\tau_{max} = 1.860 G\theta a$.

7.5. For the torsional member of cross section shown in Fig. P7.5, find the shear stresses at point B. Take $h = 5$ mm and $h_1 = h_2 = 3.5$ mm.

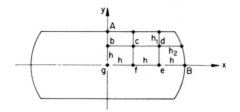

Figure P7.5

7.6. Redo Problem 7.5 to find the shear stress at point A. Let $h=4.25$ mm; then $h_1=h$ and $h_2=2.25$ mm.

Secs. 7.7 to 7.8

7.7. A force P is applied at the free end of a stepped cantilever beam of length L (Fig. P7.7). Determine the deflection of the free end using the finite difference method, taking $n=3$. Compare the result with the exact solution $v(L)=3PL^3/16EI_1$.

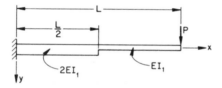

Figure P7.7

7.8. Redo Problem 7.7 with the beam subjected to a uniform load p per unit length and $P=0$. The exact solution is $v(L)=3pL^4/32EI_1$.

7.9. Obtain the joint moments in the frame of Fig. P7.9. Assume all members to have constant flexural rigidity EI. Note that end F is free to move horizontally.

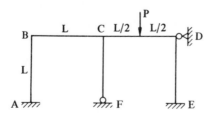

Figure P7.9

7.10. Calculate the joint moments for the frame shown in Fig. P7.10. Determine solution accurate to within $0.01 \text{ kN} \cdot \text{m}$.

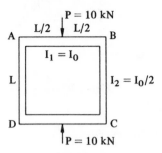

Figure P7.10

7.11. Determine the moments in each member composing the structure shown in Fig. P7.11a. Note that the analyses of the frame (Fig. P7.11a) and the continuous beam (Fig. P7.11b) proceed in identical fashion. Obtain a solution accurate to within $0.1 \text{ kN} \cdot \text{m}$.

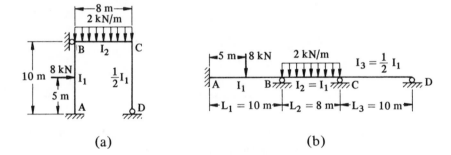

(a)   (b)

Figure P7.11

Secs. 7.9 to 7.12

7.12. Verify Eqs. (7.56) and (7.57) by determining the static resultant of the applied loading. [Hint: For Eq. (7.57), apply the principle of virtual work:

$$Q_j \delta v_j + Q_m \delta v_m - \int_{y_j}^{y_m} \tau_{xy} t \, \delta v \, dy = 0$$

with

$$v = v_j + \frac{v_m - v_j}{y_m - y_j}(y - y_j)$$

to obtain

$$Q_j + Q_m = \int_{y_j}^{y_m} \tau_{xy} t \, dy, \qquad Q_m = \frac{1}{y_m - y_j} \int_{y_j}^{y_m} \tau_{xy} t (y - y_j) \, dy \Bigg]$$

7.13. Redo Example 7.8 for the beam subjected to a uniform additional load throughout its span, $p = -7$ MPa, and a temperature rise of 50°C. Let $\gamma = 77$ kN/m^3 and $\alpha = 12 \times 10^{-6}/°$C.

7.14. A $\frac{2}{7}$-cm thick cantilever beam is subjected to a parabolically varying end shear stress resulting in a load of P N, and a linearly distributed load p N/cm (Fig. P7.14). Dividing the beam into two triangles as shown, calculate the stresses in the member. The beam is made of a transversely isotropic material, in which a rotational symmetry of properties exists within the xz plane:

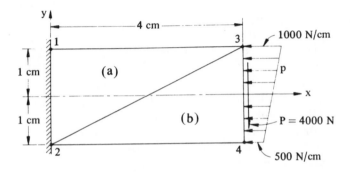

Figure P7.14

$$E_1 = 210 \text{ GPa}, \qquad \nu_2 = 0.1$$
$$E_2 = 70 \text{ GPa}, \qquad G_2 = 28 \text{ GPa}$$

Here E_1 is associated with the behavior in the xz plane, and E_2, G_2, ν_2 with the direction perpendicular to the xz plane. Now the elasticity matrix, Eq. (7.37), becomes

$$[D] = \frac{E_2}{1 - n\nu_2^2} \begin{bmatrix} n & n\nu_2 & 0 \\ n\nu_2 & 1 & 0 \\ 0 & 0 & m(1 - n\nu_2^2) \end{bmatrix}$$

where $n = E_1/E_2$ and $m = G_2/E_2$.

7.15. Determine the maximum and minimum stresses in an infinite thin plate containing a circular hole. Begin with a coarse grid and use an additional discretization to improve the result. The exact solution is given in Sec. 3.6.

Chapter 8

Axisymmetrically Loaded Members

8.1 Introduction

There are a host of practical situations in which the distribution of stress manifests symmetry about an axis. Examples include pressure vessels, compound cylinders, clad reactor elements, chemical reaction vessels, heat exchanger tubes, flywheels, solid or hollow spherical structures, and turbine disks.

Consider an infinite thin plate having a small circular hole subjected to uniform pressure, as shown in Fig. 8.1. Note that axial loading is absent and therefore $\sigma_z = 0$. The stresses are clearly symmetrical about the z axis, and the deformations likewise display θ independence. The symmetry argument also dictates that the shearing stresses $\tau_{r\theta}$ must be zero. Assuming z independence for this thin plate, the polar equations of equilibrium (3.18), reduce to

$$\frac{d\sigma_r}{dr} + \frac{\sigma_r - \sigma_\theta}{r} + F_r = 0 \qquad (8.1)$$

Here σ_θ and σ_r denote the tangential (circumferential) and radial stresses acting normal to the sides of the element, and F_r represents the radial body force per unit volume, e.g., the inertia force associated with rotation. In the absence of body forces, Eq. (8.1) reduces to

$$\frac{d\sigma_r}{dr} + \frac{\sigma_r - \sigma_\theta}{r} = 0 \qquad (8.2)$$

Consider now the radial and tangential displacements, u and v, respectively. There can be *no tangential displacement* in the symmetrical field, i.e., $v = 0$. A point represented by the element *abcd* in the figure will thus move radially as a consequence of loading, but not tangentially. On the basis of displacements indicated, the strains given by Eqs. (3.20) become

$$\varepsilon_r = \frac{du}{dr}, \qquad \varepsilon_\theta = \frac{u}{r}, \qquad \gamma_{r\theta} = 0 \qquad (8.3)$$

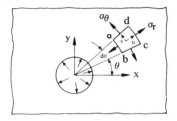

Figure 8.1

Substituting $u = r\varepsilon_\theta$ into the first expression above, a simple compatibility equation is obtained:

$$\frac{du}{dr} - \varepsilon_r = \frac{d}{dr}(r\varepsilon_\theta) - \varepsilon_r = 0$$

or

$$r\frac{d\varepsilon_\theta}{dr} + \varepsilon_\theta - \varepsilon_r = 0 \qquad (8.4)$$

The equation of equilibrium [Eq. (8.1) or (8.2)], the strain-displacement or compatibility relations [Eqs. (8.3) or (8.4)], and Hooke's law are sufficient to obtain a unique solution to any axisymmetrical problem with specified boundary conditions.

8.2 Thick-Walled Cylinders

The circular cylinder, of special importance in engineering, is usually divided into *thin-walled* and *thick-walled* classifications. A thin-walled cylinder is defined as one in which the tangential stress may, within certain prescribed limits, be regarded as constant with thickness. The following familiar expression applies to the case of a thin-walled cylinder subject to internal pressure:

$$\sigma_\theta = \frac{pr}{t}$$

Here p is the internal pressure, r the mean radius (see Sec. 13.11), and t the thickness. If the wall thickness exceeds the inner radius by more than approximately 10%, the cylinder is generally classified as thick-walled, and the variation of stress with radius can no longer be disregarded (see Problem 8.1).

In the case of a thick-walled cylinder subject to uniform internal or external pressure, the deformation is symmetrical about the z axis. Therefore the equilibrium and strain-displacement equations, Eqs. (8.2) and (8.3), apply to any point on a ring of unit length cut from the cylinder (Fig. 8.2). Assuming that the ends of cylinder are *unconstrained*, $\sigma_z = 0$, as shall be subsequently demonstrated. Thus according to Hooke's law (3.21), the strains are given by

$$\frac{du}{dr} = \frac{1}{E}(\sigma_r - \nu\sigma_\theta)$$

$$\frac{u}{r} = \frac{1}{E}(\sigma_\theta - \nu\sigma_r)$$

(8.5)

from which σ_r and σ_θ are as follows:

$$\sigma_r = \frac{E}{1-\nu^2}(\varepsilon_r + \nu\varepsilon_\theta) = \frac{E}{1-\nu^2}\left(\frac{du}{dr} + \nu\frac{u}{r}\right)$$

$$\sigma_\theta = \frac{E}{1-\nu^2}(\varepsilon_\theta + \nu\varepsilon_r) = \frac{E}{1-\nu^2}\left(\frac{u}{r} + \nu\frac{du}{dr}\right)$$

(8.6)

Substituting the above into Eq. (8.2) results in the following *equidimensional equation* in radial displacement:

$$\frac{d^2u}{dr^2} + \frac{1}{r}\frac{du}{dr} - \frac{u}{r^2} = 0$$

(8.7)

having a solution

$$u = c_1 r + \frac{c_2}{r}$$

(a)

The radial and tangential stresses may now be written in terms of the constants of integration c_1 and c_2 by combining Eqs. (a) and (8.6):

$$\sigma_r = \frac{E}{1-\nu^2}\left[c_1(1+\nu) - c_2\left(\frac{1-\nu}{r^2}\right)\right]$$

(b)

$$\sigma_\theta = \frac{E}{1-\nu^2}\left[c_1(1+\nu) + c_2\left(\frac{1-\nu}{r^2}\right)\right]$$

(c)

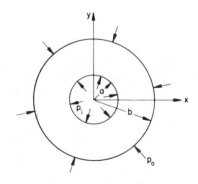

Figure 8.2

The constants are determined from consideration of the conditions pertaining to the inner and outer surfaces.

It is observed that the sum of the radial and tangential stresses is constant, regardless of radial position: $\sigma_r + \sigma_\theta = 2Ec_1/(1-\nu)$. Hence, the longitudinal strain is constant:

$$\varepsilon_z = -\frac{\nu}{E}(\sigma_r + \sigma_\theta) = \text{constant}$$

We conclude therefore that *plane sections remain plane* subsequent to loading. Then $\sigma_z = E\varepsilon_z = \text{constant} = c$. But, if the ends of the cylinder are free,

$$\int_a^b \sigma_z \cdot 2\pi r \, dr = \pi c(b^2 - a^2) = 0$$

or $c = \sigma_z = 0$, as already assumed above.

For a cylinder subjected to internal and external pressures p_i and p_o, respectively, the boundary conditions are

$$(\sigma_r)_{r=a} = -p_i, \qquad (\sigma_r)_{r=b} = -p_o \tag{d}$$

where the negative sign connotes compressive stress. The constants are evaluated by substitution of Eqs. (d) into (b):

$$c_1 = \frac{1-\nu}{E}\frac{a^2 p_i - b^2 p_o}{b^2 - a^2}, \qquad c_2 = \frac{1+\nu}{E}\frac{a^2 b^2 (p_i - p_o)}{b^2 - a^2} \tag{e}$$

leading finally to

$$\sigma_r = \frac{a^2 p_i - b^2 p_o}{b^2 - a^2} - \frac{(p_i - p_o)a^2 b^2}{(b^2 - a^2)r^2}$$

$$\sigma_\theta = \frac{a^2 p_i - b^2 p_o}{b^2 - a^2} + \frac{(p_i - p_o)a^2 b^2}{(b^2 - a^2)r^2} \tag{8.8}$$

$$u = \frac{1-\nu}{E}\frac{(a^2 p_i - b^2 p_o)r}{b^2 - a^2} + \frac{1+\nu}{E}\frac{(p_i - p_o)a^2 b^2}{(b^2 - a^2)r}$$

These expressions were first derived by G. Lamé, for whom they are named. The maximum numerical value of σ_r is found at $r=a$ to be p_i, provided that p_i exceeds p_o. If $p_o > p_i$, the maximum σ_r occurs at $r=b$ and equals p_o. On the other hand, the maximum σ_θ occurs at either the inner or outer edge according to the pressure ratio, as discussed in Sec. 8.3.

Recall that the maximum shearing stress at any point equals one-half the algebraic difference between the maximum and minimum principal stresses. At any point in the cylinder, we may therefore state that

$$\tau_{\max} = \tfrac{1}{2}(\sigma_\theta - \sigma_r) = \frac{(p_i - p_o)a^2 b^2}{(b^2 - a^2)r^2} \tag{8.9}$$

The largest value of $\tau_{\max}$ is found at $r=a$, the inner surface. The effect of reducing p_o is clearly to increase τ_{max}. Consequently, the greatest τ_{max} corresponds to $r=a$ and $p_o=0$:

$$\tau_{max} = \frac{p_i b^2}{b^2 - a^2} \tag{f}$$

Because σ_r and σ_θ are principal stresses, τ_{max} occurs on planes making an angle of $45°$ with the plane on which σ_r and σ_θ act. This is quickly confirmed by a Mohr's circle construction.

Special Cases

Internal pressure only. If only internal pressure acts, Eqs. (8.8) reduce to

$$\sigma_r = \frac{a^2 p_i}{b^2 - a^2} \left(1 - \frac{b^2}{r^2} \right) \tag{8.10}$$

$$\sigma_\theta = \frac{a^2 p_i}{b^2 - a^2} \left(1 + \frac{b^2}{r^2} \right) \tag{8.11}$$

$$u = \frac{a^2 p_i r}{E(b^2 - a^2)} \left[(1 - \nu) + (1 + \nu)\frac{b^2}{r^2} \right] \tag{8.12}$$

Since $b^2/r^2 \geq 1$, σ_r is negative (compressive) for all r except $r=b$, in which case $\sigma_r=0$. The maximum radial stress occurs at $r=a$. As for σ_θ, it is positive (tensile) for all radii, and also has a maximum at $r=a$.

To illustrate the variation of stress and radial displacement for the case of zero external pressure, dimensionless stress and displacement are plotted against dimensionless radius in Fig. 8.3a for $b/a=4$.

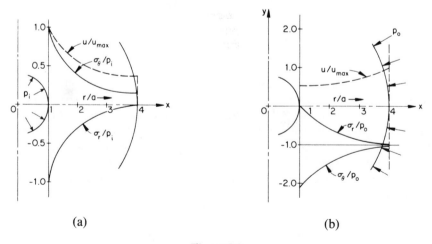

(a)   (b)

Figure 8.3

External pressure only. In this case, $p_i = 0$, and the equations (8.8) become

$$\sigma_r = -\frac{p_o b^2}{b^2 - a^2}\left(1 - \frac{a^2}{r^2}\right) \tag{8.13}$$

$$\sigma_\theta = -\frac{p_o b^2}{b^2 - a^2}\left(1 + \frac{a^2}{r^2}\right) \tag{8.14}$$

$$u = -\frac{b^2 p_o r}{E(b^2 - a^2)}\left[(1 - \nu) + (1 + \nu)\frac{a^2}{r^2}\right] \tag{8.15}$$

The maximum radial stress occurs at $r = b$ and is compressive for all r. The maximum σ_θ is found at $r = a$ and is likewise compressive.

For a cylinder with $b/a = 4$ and subjected to external pressure only, the stress and displacement variations over the wall thickness are shown in Fig. 8.3b.

Example 8.1. A thick-walled cylinder with 0.3-m and 0.4-m internal and external diameters is fabricated of a material whose elastic limit is 250 MPa. Let $\nu = 0.3$. Determine (a) for $p_o = 0$, the maximum internal pressure to which the cylinder may be subjected without exceeding the elastic limit, (b) for $p_i = 0$, the maximum external pressure to which the cylinder can be subjected without exceeding the elastic limit, and (c) the radial displacement of a point on the inner surface for case (a).

SOLUTION

(a) From Eq. (8.11), with $r = a$,

$$\sigma_{\theta, \max} = p_i \frac{b^2 + a^2}{b^2 - a^2} \tag{8.16}$$

or

$$p_i = \sigma_{\theta, \max}\frac{b^2 - a^2}{b^2 + a^2} = (250 \times 10^6)\frac{0.2^2 - 0.15^2}{0.2^2 + 0.15^2} = 70 \text{ MPa}$$

(b) From Eq. (8.14), with $r = a$,

$$\sigma_{\theta, \max} = -2p_o\frac{b^2}{b^2 - a^2} \tag{8.17}$$

Then

$$p_o = -\sigma_{\theta, \max}\frac{b^2 - a^2}{2b^2} = -(-250 \times 10^6)\frac{0.2^2 - 0.15^2}{(2)0.2^2} = 54.7 \text{ MPa}$$

(c) Using Eq. (8.12), we obtain

$$(u)_{r=a} = \frac{0.15^3 \times 70 \times 10^6}{E(0.2^2 - 0.15^2)}\left[0.7 + 1.3\frac{0.2^2}{0.15^2}\right] = 4.065 \times 10^7/E \text{ m}$$

In the case of a closed-ended cylinder subjected to internal and external pressures, longitudinal or z-directed stresses exist in addition to the radial and tangential stresses. For a transverse section some distance from the ends, this stress may be assumed uniformly distributed over the wall thickness. The magnitude of σ_z is then determined by equating the net force acting on an end attributable to pressure loading, to the internal z directed force in the cylinder wall:

$$p_i \pi a^2 - p_o \pi b^2 = (\pi b^2 - \pi a^2)\sigma_z$$

The resulting expression for longitudinal stress, applicable only away from the ends, is

$$\sigma_z = \frac{p_i a^2 - p_o b^2}{b^2 - a^2} \tag{8.18}$$

Clearly, here it is again assumed that the ends of cylinder are not constrained: $\varepsilon_z \neq 0$ (see Problem 8.8).

8.3 Maximum Tangential Stress

An examination of Figs. 8.3a and b shows that if either internal pressure or external pressure acts alone, the maximum tangential stress occurs at the innermost fibers, $r = a$. This conclusion is not always valid, however, if both internal and external pressures act simultaneously. There are situations, explored below, in which the maximum tangential stress occurs at $r = b$.*

Consider a thick-walled cylinder, as in Fig. 8.2, subject to p_i and p_o. Denote the ratio b/a by R, p_o/p_i by P, and the ratio of tangential stress at the inner and outer surfaces by S. The tangential stress, given by Eqs. (8.8), is written

$$\sigma_\theta = p_i \frac{1 - PR^2}{R^2 - 1} + p_i b^2 \frac{1 - P}{(R^2 - 1)r^2} \tag{8.19}$$

Hence

$$S = \frac{\sigma_{\theta i}}{\sigma_{\theta o}} = \frac{1 + \dfrac{R^2(1 - P)}{1 - PR^2}}{1 + \dfrac{1 - P}{1 - PR^2}} \tag{8.20}$$

The variation of the tangential stress σ_θ over the wall thickness is shown in Fig. 8.4 for several values of S and P. Note that for pressure ratios P indicated by dashed lines, the maximum magnitude of the circumferential stress occurs at the outer surface of the cylinder.

---

*T. Ranov and F. R. Park, On the numerical value of the tangential stress on thick-walled cylinders, *J. Appl. Mech.* (March 1953).

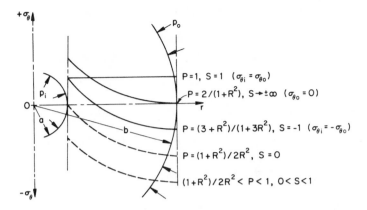

Figure 8.4

8.4 Application of Failure Theories

Unless one is content to grossly overdesign, it is necessary to predict, as best possible, the most probable failure mechanism. Thus, while examination of Fig. 8.4 indicates that failure is likely to originate at the innermost or outermost fibers of the cylinder, it cannot predict at what pressure or stresses failure will occur. To do this, consideration must be given the stresses determined from Lamé's equations, the material strength, and an appropriate theory of failure consistent with the nature of the material.

Example 8.2. A steel cylinder is subjected to an internal pressure four times greater than the external pressure. The tensile elastic strength of the steel is $\sigma_{yp} = 340$ MPa, and the shearing elastic strength $\tau_{yp} = \sigma_{yp}/2 = 170$ MPa. Calculate the allowable internal pressure according to the various theories of failure. The dimensions are $a = 0.1$ m, $b = 0.15$ m. Let $\nu = 0.3$.

SOLUTION.   The maximum stresses occur at the innermost fibers. From Eqs. (8.8), for $r = a$ and $p_i = 4p_o$, we have

$$\sigma_\theta = \frac{p_i(a^2 + b^2) - 2p_o b^2}{b^2 - a^2} = 1.7p_i, \qquad \sigma_r = -p_i \qquad (a)$$

The value of internal pressure at which yielding begins is predicted according to the various theories of failure, as follows:

(a) *Maximum principal stress theory*:
$$1.7p_i = 340 \times 10^6, \qquad p_i = 200 \text{ MPa}$$

(b) *Maximum shearing stress theory*:
$$\frac{\sigma_\theta - \sigma_r}{2} = 1.35p_i = 170 \times 10^6, \qquad p_i = 125.9 \text{ MPa}$$

(c) *Energy of distortion theory* [Eq. (4.9a)]:

$$\sigma_{yp} = (\sigma_\theta^2 + \sigma_r^2 - \sigma_\theta \sigma_r)^{1/2} = p_i \left[(1.7)^2 + (-1)^2 - (-1.7) \right]^{1/2}$$

$$340 \times 10^6 = 2.364 p_i, \qquad p_i = 143.8 \text{ MPa}$$

(d) *Maximum principal strain theory*:

$$\frac{1}{E}(\sigma_\theta - \nu \sigma_r) = \frac{340 \times 10^6}{E} = \frac{2}{E} p_i, \qquad p_i = 170 \text{ MPa}$$

(e) *Octahedral shearing stress theory*: By use of Eqs. (4.10) and (1.25), we have

$$\frac{\sqrt{2}}{3} \sigma_{yp} = \frac{1}{3} \left[(\sigma_\theta - \sigma_r)^2 + (\sigma_r)^2 + (-\sigma_\theta)^2 \right]^{1/2}$$

$$\frac{\sqrt{2}}{3}(340 \times 10^6) = \frac{p_i}{3} \left[(2.7)^2 + (-1)^2 + (-1.7)^2 \right]^{1/2}, \qquad p_i = 143.8 \text{ MPa}$$

The results found in (c) and (e) are indentical as expected (Sec. 2.8). As the cylinder is made of a ductile material, the onset of inelastic action is governed by the maximum shearing stress. The allowable value of internal pressure is therefore limited to 125.9 MPa, modified by an appropriate factor of safety.

8.5   Compound Cylinders

If properly designed, a system of multiple cylinders resists relatively large pressures more efficiently, that is, requires less material, than a single cylinder. To assure the integrity of the compound cylinder, one of several methods of prestressing is employed. For example, the inner radius of the outer member or jacket may be made smaller than the outer radius of the inner cylinder. The cylinders are assembled after the outer cylinder is heated, contact being effected upon cooling. The magnitude of the resulting *contact pressure p* between members may be calculated by use of the equations of Sec. 8.2. Examples of compound cylinders, carrying very high pressures, are seen in compressors, extrusion presses, and the like.

Referring to Fig. 8.5a, assume the external radius of the inner cylinder to be larger, in its unstressed state, than the internal radius of the jacket, by an amount δ. This quantity is the *shrinking allowance*. Subsequent to assembly, the *contact pressure*, acting equally on both members, causes the sum of the increase in inner radius of the jacket and decrease in outer radius of the inner member to exactly equal δ. By using Eqs. (8.12) and (8.15), we obtain

$$\delta = \frac{bp}{E} \left(\frac{b^2 + c^2}{c^2 - b^2} + \nu \right) + \frac{bp}{E} \left(\frac{a^2 + b^2}{b^2 - a^2} - \nu \right) \tag{8.21}$$

From the above expression, the contact pressure is then found as a function of the shrinking allowance:

$$p = \frac{E\delta}{b} \frac{(b^2 - a^2)(c^2 - b^2)}{2b^2(c^2 - a^2)} \tag{8.22}$$

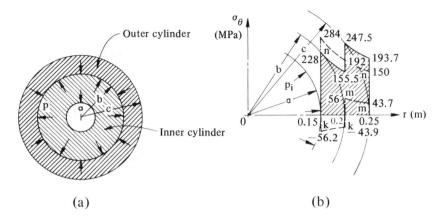

Figure 8.5

The stresses in the jacket are then determined from Eqs. (8.10) and (8.11) by treating the contact pressure as p_i. Similarly, by regarding the contact pressure as p_o, the stresses in the inner cylinder are calculated from Eqs. (8.13) and (8.14).

Example 8.3. A compound cylinder with $a = 150$ mm, $b = 200$ mm, $c = 250$ mm, $E = 200$ GPa, and $\delta = 0.1$ mm is subjected to an internal pressure of 140 MPa. Determine the distribution of tangential stress throughout the composite wall.

SOLUTION. In the absence of applied internal pressure, the contact pressure is, from Eq. (8.22),

$$p = \frac{200 \times 10^9 \times 0.0001}{8} \frac{(0.2^2 - 0.15^2)(0.25^2 - 0.2^2)}{2(0.2^2)(0.25^2 - 0.15^2)} = 12.3 \text{ MPa}$$

The tangential stresses in the outer cylinder associated with this pressure are found by using Eq. (8.11):

$$(\sigma_\theta)_{r=0.2} = p \frac{b^2 + c^2}{c^2 - b^2} = 12.3 \times 10^6 \frac{0.2^2 + 0.25^2}{0.25^2 - 0.2^2} = 56.0 \text{ MPa}$$

$$(\sigma_\theta)_{r=0.25} = \frac{2pb^2}{c^2 - b^2} = \frac{2(12.3 \times 10^6)(0.2^2)}{0.25^2 - 0.2^2} = 43.7 \text{ MPa}$$

The stresses in the inner cylinder are, from Eq. (8.14),

$$(\sigma_\theta)_{r=0.15} = -\frac{2pb^2}{b^2 - a^2} = -\frac{2(12.3 \times 10^6)(0.2^2)}{0.2^2 - 0.15^2} = -56.2 \text{ MPa}$$

$$(\sigma_\theta)_{r=0.2} = -p \frac{b^2 + a^2}{b^2 - a^2} = -12.3 \times 10^6 \frac{0.2^2 + 0.15^2}{0.2^2 - 0.15^2} = -43.9 \text{ MPa}$$

These stresses are plotted in Fig. 8.5b, indicated by the dashed lines kk and mm. The stresses owing to internal pressure alone, through the use of Eq. (8.11) with $b^2 = c^2$, are found to be $(\sigma_\theta)_{r=0.15} = 284$ MPa, $(\sigma_\theta)_{r=0.2} = 192$ MPa, $(\sigma_\theta)_{r=0.25} = 150$

MPa, and are shown as the dashed line *nn*. The stress resultant is obtained by superposition of the two distributions, represented by the solid line. The use of a compound prestressed cylinder has thus reduced the maximum stress from 284 MPa to 247.5 MPa. Based upon the maximum principal stress theory of elastic failure, significant weight savings can apparently be effected through such configurations as above. It is interesting to note that additional jackets prove not as effective, in that regard, as the first one. Multilayered shrink fit cylinders, each of small wall thickness, are, however, considerably stronger than a single jacket of the same total thickness. These assemblies can, in fact, be designed so that prestressing owing to shrinking combines with stresses due to loading to produce a nearly uniform distribution of stress throughout.* The closer this uniform stress is to the allowable stress for the given material, the more efficiently is the material utilized. A single cylinder cannot be uniformly stressed and consequently must be stressed considerably below its allowable value, contributing to inefficient use of material.

8.6 Rotating Disks of Constant Thickness

The equation of equilibrium, Eq. (8.1), can be used to treat the case of a rotating disk, provided that the centrifugal "inertia force" is included as a body force. Again, stresses induced by rotation are distributed symmetrically about the axis of rotation and assumed independent of disk thickness. Thus, application of Eq. (8.1), with the body force per unit volume F_r equated to the centrifugal force $\rho \omega^2 r$, yields

$$\frac{d\sigma_r}{dr} + \frac{\sigma_r - \sigma_\theta}{r} + \rho\omega^2 r = 0 \tag{8.23}$$

where ρ is the mass *density* and ω the constant *angular speed* of the disk in *radians per second*. Note that the gravitational body force ρg has been neglected. Substituting Eq. (8.6) into Eq. (8.23), we have

$$\frac{d^2u}{dr^2} + \frac{1}{r}\frac{du}{dr} - \frac{u}{r^2} = -(1-\nu^2)\rho\omega^2 r/E \tag{a}$$

requiring a homogeneous and particular solution. The former is given by Eq. (a) of Sec. 8.2. It is easily demonstrated that the particular solution is

$$u_p = -(1-\nu^2)\frac{\rho\omega^2 r^3}{8E}$$

The complete solution is therefore

$$u = -\frac{\rho\omega^2 r^3(1-\nu^2)}{8E} + c_1 r + \frac{c^2}{r} \tag{8.24a}$$

which, upon substitution into Eq. (8.6), provides the following expressions

---

*Refer, for example, to S. J. Becker and L. Mollick, The theory of the ideal design of a compound vessel, *Trans. ASME, J. Engineering for Industry*, p. 136, (May 1960).

for radial and tangential stress:

$$\sigma_r = \frac{E}{1-\nu^2}\left[\frac{-(3+\nu)(1-\nu^2)\rho\omega^2 r^2}{8E} + (1+\nu)c_1 - (1-\nu)\frac{c_2}{r^2}\right] \quad (8.24b)$$

$$\sigma_\theta = \frac{E}{1-\nu^2}\left[\frac{-(1+3\nu)(1-\nu^2)\rho\omega^2 r^2}{8E} + (1+\nu)c_1 + (1-\nu)\frac{c_2}{r^2}\right] \quad (8.24c)$$

The constants of integration may now be evaluated on the basis of the boundary conditions.

Annular Disk

In the case of an annular disk with zero pressure at the inner and outer boundaries, the distribution of stress is due entirely to rotational effects. The boundary conditions are

$$(\sigma_r)_{r=a} = 0, \qquad (\sigma_r)_{r=b} = 0 \quad \text{(b)}$$

These conditions, combined with Eq. (8.24b), yield two equations in the two unknown constants,

$$0 = -\rho\omega^2\frac{a^2}{E}\frac{(1-\nu^2)(3+\nu)}{8} + (1+\nu)c_1 - (1-\nu)\frac{c_2}{a^2}$$

$$0 = -\rho\omega^2\frac{b^2}{E}\frac{(1-\nu^2)(3+\nu)}{8} + (1+\nu)c_1 - (1-\nu)\frac{c_2}{b^2}$$
$$\text{(c)}$$

from which

$$c_1 = \rho\omega^2\frac{(a^2+b^2)}{E}\frac{(1-\nu)(3+\nu)}{8}$$

$$c_2 = \rho\omega^2\left(\frac{a^2b^2}{E}\right)\frac{(1+\nu)(3+\nu)}{8}$$
$$\text{(d)}$$

The stresses and displacement are therefore

$$\sigma_r = \frac{3+\nu}{8}\left(a^2+b^2-r^2-\frac{a^2b^2}{r^2}\right)\rho\omega^2$$

$$\sigma_\theta = \frac{3+\nu}{8}\left(a^2+b^2-\frac{1+3\nu}{3+\nu}r^2+\frac{a^2b^2}{r^2}\right)\rho\omega^2 \quad (8.25)$$

$$u = \frac{(3+\nu)(1-\nu)}{8E}\left(a^2+b^2-\frac{1+\nu}{3+\nu}r^2+\frac{1+\nu}{1-\nu}\frac{a^2b^2}{r^2}\right)\rho\omega^2 r$$

Applying the condition $d\sigma_r/dr=0$ to the first of the above equations, it is readily verified that the *maximum* radial stress occurs at $r=\sqrt{ab}$. Figure 8.6a is a dimensionless representation of stress and displacement as a function of radius for an annular disk described by $b/a=4$.

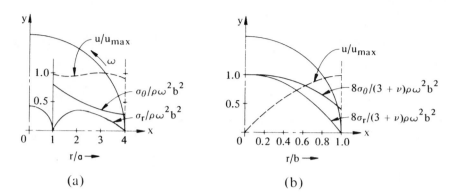

Figure 8.6

Solid Disk

In this case, $a = 0$, and the boundary conditions are

$$(\sigma_r)_{r=b} = 0, \qquad (u)_{r=0} = 0 \qquad \text{(e)}$$

In order to satisfy the condition on the displacement it is clear from Eq. (8.24a) that c_2 must be zero. The remaining constant is now evaluated from the first expression of Eq. (d):

$$c_1 = \rho \omega^2 \frac{b^2}{E} \frac{(1-\nu)(3+\nu)}{8}$$

Combining these constants with Eqs. (8.24), the following results are obtained:

$$\sigma_r = \frac{3+\nu}{8}(b^2 - r^2)\rho\omega^2$$

$$\sigma_\theta = \frac{3+\nu}{8}\left(b^2 - \frac{1+3\nu}{3+\nu}r^2\right)\rho\omega^2 \qquad \text{(8.26)}$$

$$u = \frac{1-\nu}{8E}\left[(3+\nu)b^2 - (1+\nu)r^2\right]\rho\omega^2 r$$

The stress and displacement of a solid rotating disk are displayed in a dimensionless representation, Fig. 8.6b, as functions of radial location.

The constant thickness disks discussed in this section are generally employed when stresses or speeds are low, and, as is clearly shown in Fig. 8.6b, do not make optimum use of material. Other types of rotating disks, offering many advantages over flat disks, are discussed in the sections to follow.

Example 8.4. A flat 0.5-m outer diameter, 0.1-m inner diameter, and 0.08-m thick steel disk is shrunk onto a steel shaft (Fig. 8.7). If the assembly is to run at speeds up to 6900 rpm, determine (a) the shrinking allowance, (b) the maximum stress

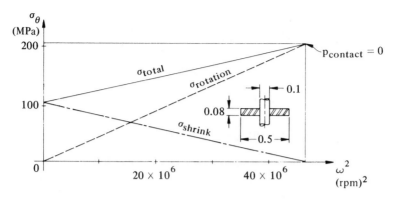

Figure 8.7

when not rotating, and (c) the maximum stress when rotating. The material properties are $\rho = 7.8$ kN·s^2/m^4, $E = 200$ GPa, and $\nu = 0.3$.

SOLUTION

(a) The radial displacements of the disk (u_d) and shaft (u_s) are, from Eqs. (8.25) and (8.26),

$$u_d = 0.05 \times \frac{3.3 \times 0.7}{8E}\left(0.0025 + 0.0625 - \frac{1.3}{3.3} \times 0.0025 + \frac{1.3}{0.7} \times 0.0625\right)\rho\omega^2$$

$$= 0.0026\rho\frac{\omega^2}{E}$$

$$u_s = 0.05 \times \frac{0.7}{8E}(3.3 \times 0.0025 - 1.3 \times 0.0025)\rho\omega^2$$

$$= 2.1875 \times 10^{-5}\rho\frac{\omega^2}{E}$$

We observe that u_s may be neglected as it is less than 1% of u_d of the disk at the common radius. The exact allowance is

$$\delta = u_d - u_s = \frac{(0.0026 - 2.1875 \times 10^{-5}) \times 7.8 \times 10^3(6900 \times 2\pi/60)^2}{200 \times 10^9}$$

$$= 5.25 \times 10^{-5} \text{ m}$$

(b) Applying Eq. (8.22), we have

$$p = \frac{E\delta}{2b}\frac{c^2 - b^2}{c^2} = \frac{200 \times 10^9 \times 5.25 \times 10^{-5}(0.0625 - 0.0025)}{2 \times 0.05 \times 0.0625} = 100.8 \text{ MPa}$$

Therefore, from Eq. (8.16),

$$\sigma_{\theta,\max} = p\frac{c^2 + b^2}{c^2 - b^2} = 100.8 \times 10^6\frac{0.0625 + 0.0025}{0.0625 - 0.0025} = 109.2 \text{ MPa}$$

(c) From Eq. (8.25), for $r = 0.05$,

$$\sigma_{\theta,\,\mathrm{max}} = \frac{3.3}{8}\left(0.0025 + 0.0625 - \frac{1.9}{3.3} \times 0.0025 + 0.0625\right)\rho\omega^2$$

$$= 0.052\rho\omega^2 = 211.78 \text{ MPa}$$

A plot of the variation of stress in the rotating disk is shown in Fig. 8.7.

8.7 Rotating Disks of Variable Thickness

In the previous section, the maximum stress in a flat rotating disk was observed to occur at the innermost fibers. This explains the general shape of many disks: thick near the hub, tapering down in thickness toward the periphery, as in a steam turbine. This not only has the effect of reducing weight, but results in lower rotational inertia as well.

The approach employed in the analysis of flat disks can be extended to variable thickness disks. Let the profile of a radial section be represented by the *general hyperbola* (Fig. 8.8),

$$t = t_1 r^{-s} \tag{8.27}$$

where t_1 represents a constant and s a positive number. The shape of the curve depends upon the value selected for s, e.g., for $s = 1$, the profile is that of an *equilateral hyperbola*. The constant t_1 is simply the thickness at radius equal to unity. If the thickness at $r = a$ is t_i and that at $r = b$ is t_o, as shown in the figure, the hyperbolic curve is fitted by forming the ratio

$$\frac{t_i}{t_o} = \frac{t_1 a^{-s}}{t_1 b^{-s}}$$

and solving for s. Clearly, Eq. (8.27) does not apply to solid disks, as all values of s except zero yield infinite thickness at $r = 0$.

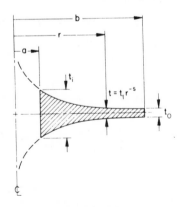

Figure 8.8

In a turbine application, the actual configuration may have a thickened outer rim to which blades are affixed, and a hub for attachment to a shaft. The hyperbolic relationship cannot, of course, describe such a situation exactly, but sometimes serves as an adequate approximation. If greater accuracy is required, the hub and outer ring may be approximated as flat disks, with the elements of the assembly related by the appropriate boundary conditions.

The differential equation of equilibrium (Eq. 8.23) must now include $t(r)$ and takes the form

$$\frac{d}{dr}(tr\sigma_r) - t\sigma_\theta + t\rho\omega^2 r^2 = 0 \tag{8.28}$$

Equation (8.28) is satisfied by a stress function of the form

$$\phi = tr\sigma_r, \qquad \frac{d\phi}{dr} + t\rho\omega^2 r^2 = t\sigma_\theta \tag{a}$$

Then the compatibility equation (8.4), using Eqs. (a) and Hooke's law, becomes

$$r^2 \frac{d^2\phi}{dr^2} + \left(1 - \frac{r}{t}\frac{dt}{dr}\right) r \frac{d\phi}{dr} + \left(\nu \frac{r}{t}\frac{dt}{dr} - 1\right)\phi = -(3+\nu)\rho\omega^2 tr^3 \tag{b}$$

Introducing Eq. (8.27), we have

$$r^2 \frac{d^2\phi}{dr^2} + (1+s)r \frac{d\phi}{dr} - (1+\nu s)\phi = -(3+\nu)\rho\omega^2 t_1 r^{3-s} \tag{c}$$

This is an *equidimensional equation* which the transformation $r = e^\alpha$ reduces to a linear differential equation with constant coefficients:

$$\frac{d^2\phi}{d\alpha^2} + s\frac{d\phi}{d\alpha} - (1+\nu s)\phi = -(3+\nu)t_1\rho\omega^2 e^{(3-s)\alpha} \tag{d}$$

The auxiliary equation corresponding to Eq. (d) is given by

$$m^2 + sm - (1+\nu s) = 0$$

and has the roots

$$m_{1,2} = -\frac{s}{2} \pm \left[\left(\frac{s}{2}\right)^2 + (1+\nu s)\right]^{1/2} \tag{8.29}$$

The general solution of Eq. (d) is then

$$\phi = c_1 r^{m_1} + c_2 r^{m_2} - \frac{3+\nu}{8-(3+\nu)s} t_1\rho\omega^2 r^{3-s} \tag{e}$$

The stress components for a disk of variable thickness are therefore, from Eqs. (a),

$$\sigma_r = \frac{c_1}{t_1} r^{m_1+s-1} + \frac{c_2}{t_1} r^{m_2+s-1} - \frac{3+\nu}{8-(3+\nu)s} \rho\omega^2 r^2$$

$$\sigma_\theta = \frac{c_1}{t_1} m_1 r^{m_1+s-1} + \frac{c_2}{t_1} m_2 r^{m_2+s-1} - \frac{1+3\nu}{8-(3+\nu)s} \rho\omega^2 r^2 \tag{8.30}$$

Note that for a flat disk, $t=$ constant; consequently, $s=0$ in Eq. (8.27) and $m=1$ in Eq. (8.29). Thus Eqs. (8.30) reduce to Eqs. (8.26), as expected. The constants c_1 and c_2 are determined from the boundary conditions

$$(\sigma_r)_{r=a}=(\sigma_r)_{r=b}=0 \tag{f}$$

The evaluation of the constants is illustrated in the example below.

Example 8.5. The cross section of the disk in the assembly given in Example 8.4 is hyperbolic with $t_i=0.075$ m, $t_o=0.015$ m; $a=0.05$ m, $b=0.25$ m, and $\delta=0.05$ mm. The rotational speed is 6900 rpm. Determine (a) the maximum stress owing to rotation and (b) the maximum radial displacement at the bore of the disk.

SOLUTION

(a) The value of the positive number s is obtained by the use of Eq. (8.27):

$$\frac{t_i}{t_o}=\frac{t_1 a^{-s}}{t_1 b^{-s}}=\left(\frac{b}{a}\right)^s$$

Substituting $t_i/t_o=5$ and $b/a=5$, we obtain $s=1$. The profile will thus be given by $t=t_i/r$. From Eq. (8.29) we have

$$m_{1,2}=-\tfrac{1}{2}\pm\left[\left(\tfrac{1}{2}\right)^2+(1+0.3\times1)\right]^{1/2}, \qquad m_1=0.745, \quad m_2=-1.745$$

Hence the radial stresses, using Eqs. (8.30) and (f) for $r=0.05$ and 0.25, are

$$(\sigma_r)_{r=0.05}=0=\frac{c_1}{t_1}0.05^{0.745}+\frac{c_2}{t_1}0.05^{-1.745}-0.00176\rho\omega^2$$

$$(\sigma_r)_{r=0.25}=0=\frac{c_1}{t_1}0.25^{0.745}+\frac{c^2}{t_1}0.25^{-1.745}-0.0439\rho\omega^2$$

from which

$$\frac{c_1}{t_1}=0.12529\rho\omega^2, \qquad \frac{c_2}{t_1}=-6.272\times10^{-5}\rho\omega^2$$

The stress components in the disk, substituting the above values into Eqs. (8.30), are therefore

$$\sigma_r=(0.12529r^{0.745}-6.272\times10^{-5}r^{-1.745}-0.70r^2)\rho\omega^2$$

$$\sigma_\theta=(0.09334r^{0.745}+1.095\times10^{-4}r^{-1.745}-0.40r^2)\rho\omega^2 \tag{g}$$

The maximum stress occurs at the bore of the disk, and from Eqs. (g) is equal to $(\sigma_\theta)_{r=0.05}=0.0294\rho\omega^2$. Note that it was $0.052\rho\omega^2$ in Example 8.4. For the same speed we conclude that the maximum stress is reduced considerably by tapering the disk.

(b) The radial displacement is obtained from the second equation of (8.5), which together with Eq. (g) gives $u_{r=0.05}=(r\sigma_\theta/E)_{r=0.05}=0.00147\rho\omega^2/E$. Again, this is quite advantageous relative to the value of $0.0026\rho\omega^2/E$ found in Example 8.4.

8.8   Rotating Disks of Uniform Stress

If every element of a rotating disk is stressed to a prescribed allowable value, presumed constant throughout, the disk material will clearly be used in the most efficient manner. For a given material, such a design is of minimum mass, offering the distinct advantage of reduced inertia loading, as well as lower weight. What is sought then, is a thickness variation $t(r)$ such that $\sigma_r = \sigma_\theta = \sigma = $ constant everywhere in the body. Under such a condition of stress, the strains, according to Hooke's law, are $\varepsilon_r = \varepsilon_\theta = \varepsilon = $ constant, and the compatibility equation (8.4) is satisfied.

The equation of equilibrium (8.28) may, under the conditions outlined, be written

$$\sigma \frac{d}{dr}(rt) - t\sigma + \rho\omega^2 t r^2 = 0$$

or

$$\frac{dt}{dr} + \frac{\rho\omega^2}{\sigma} rt = 0 \tag{8.31}$$

which is easily integrated to yield

$$t = c_1 e^{-(\rho\omega^2/2\sigma)r^2} \tag{8.32}$$

The above variation assures that $\sigma_\theta = \sigma_r = \sigma = $ constant throughout the disk. To obtain the value of the constant in Eq. (8.32), the boundary condition $t = t_1$ at $r = 0$ is applied, resulting in $c_1 = t_1$ (Fig. 8.9).

Example 8.6. A steel disk of the same outer radius, $b = 0.25$ m, and rotational speed, 6900 rpm, as the disk of Example 8.4, is to be designed for uniform stress. The thicknesses are $t_1 = 0.075$ m at the center and $t_2 = 0.015$ m at the periphery. Determine the stress and disk profile.

SOLUTION.   From Eq. (8.32),

$$\frac{t_2}{t_1} = \frac{c_1 e^{-(\rho\omega^2/2\sigma)b^2}}{c_1} = e^{-\rho(\omega^2/2\sigma)b^2} = \frac{1}{5}$$

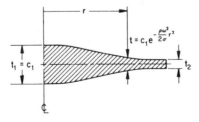

Figure 8.9

or

$$\ln\left(\frac{1}{5}\right) = -\rho\frac{\omega^2}{2\sigma}b^2, \qquad \sigma = \rho\frac{\omega^2 b^2}{3.218} = 0.0194\rho\omega^2 = 79.08 \text{ MPa}$$

Thus

$$t = c_1 e^{-\rho(\omega^2/2\sigma)r^2} = 0.075 e^{-25.752r^2}$$

Recall that the maximum stress in the hyperbolic disk of Example 8.5 was $0.0294\rho\omega^2$. The uniformly stressed disk is thus about 34% stronger than a hyperbolic disk with a small hole at its center.

In actual practice, fabrication and design constraints make it impractical to produce a section of exactly constant stress in a solid disk. On the other hand, in an annular disk if the boundary condition is applied that the radial stress is zero at the inner radius, constancy of stress dictates that σ_θ and σ_r be zero everywhere. This is clearly not a useful result for the situation as described. For these reasons, the hyperbolic variation in thickness is often used.

8.9 A Numerical Approach to Rotating Disk Analysis

In this section is presented a practical numerical method for computation of radial and tangential stress in a disk of nonuniform thickness, which does not rely upon a knowledge of the analytical function $t(r)$ nor of the solution of the resulting differential equation,* as in Sec. 8.7.

To make Eq. (8.24) of Sec. 8.6 appear less cumbersome, we use the form

$$\sigma_r = -\left(\frac{3+\nu}{8}\right)Nr^2 + B_1 - \frac{B_2}{r^2} \tag{a}$$

$$\sigma_\theta = -\left(\frac{1+3\nu}{8}\right)Nr^2 + B_1 + \frac{B_2}{r^2} \tag{b}$$

where $N = \rho\omega^2$, and B_1, B_2 are constants of integration. They are constant only within a given section of an element and determined from boundary conditions on the inner edge of each element. If now the disk of smoothly varying thickness is divided into a number of rectangular ring elements, then, denoting the *outer* and *inner* edges of each element by subscripts o and i, respectively, Eqs. (a) and (b) lead to

$$\sigma_{rm} = -\left(\frac{3+\nu}{8}\right)Nr_m^2 + B_1 - \frac{B_2}{r_m^2} \tag{c}$$

$$\sigma_{\theta m} = -\left(\frac{1+3\nu}{8}\right)Nr_m^2 + B_1 + \frac{B_2}{r_m^2} \tag{d}$$

---

*T. W. Gawain and E. C. Curry, Stresses in rotating disks of nonuniform thickness, *Product Engineering* (July 1951), pp. 152–155.

where $m = o, i$. From these four equations, upon subtraction of $\sigma_{\theta i}$ from σ_{ri}, the B_1 is determined and the result is solved to yield B_2. The stress variation occurring in an element of the section is determined by the operations $\Delta\sigma_r = \sigma_{ro} - \sigma_{ri}$ and $\Delta\sigma_\theta = \sigma_{\theta o} - \sigma_{\theta i}$, collecting the terms suitably, as follows:

$$\Delta\sigma_r = \frac{1}{2}(\sigma_{\theta i} - \sigma_{ri})\left(1 - \frac{r_i^2}{r_o^2}\right) - \left[\left(\frac{3+\nu}{8}\right) + \left(\frac{1-\nu}{8}\right)\frac{r_i^2}{r_o^2}\right]N\overset{\cdot}{r_o^2}\left(1 - \frac{r_i^2}{r_o^2}\right) \quad \text{(e)}$$

$$\Delta\sigma_\theta = -\frac{1}{2}(\sigma_{\theta i} - \sigma_{ri})\left(1 - \frac{r_i^2}{r_o^2}\right) - \left[\frac{1+3\nu}{8} - \left(\frac{1-\nu}{8}\right)\frac{r_i^2}{r_o^2}\right]Nr_o^2\left(1 - \frac{r_i^2}{r_o^2}\right) \quad \text{(f)}$$

These may be converted to their working forms through the introduction of the following abbreviations:

$$R = \frac{r_o - r_i}{r_o} = \frac{\Delta r}{r_o}$$

$$\zeta = \left(1 - \frac{R}{2}\right)R$$

$$L_r = \left(1 - \frac{1-\nu}{2}\zeta\right)Nr_o^2 \qquad (8.33)$$

$$L_\theta = \left(\nu + \frac{1-\nu}{2}\zeta\right)Nr_o^2$$

$$\lambda = \sigma_{\theta i} - \sigma_{ri}$$

Substituting the above into Eqs. (e) and (f), we obtain

$$\Delta\sigma_r = \left(1 - \frac{R}{2}\right)R\left\{(\sigma_{\theta i} - \sigma_{ri}) - \left[1 - \left(\frac{1-\nu}{2}\right)\left(1 - \frac{R}{2}\right)R\right]Nr_o^2\right\}$$

$$= (\lambda - L_r)\zeta \qquad (8.34)$$

$$\Delta\sigma_\theta = \left(1 - \frac{R}{2}\right)R\left\{-(\sigma_{\theta i} - \sigma_{ri}) - \left[\nu + \left(\frac{1-\nu}{2}\right)\left(1 - \frac{R}{2}\right)\right]Nr_o^2\right\}$$

$$= -(\lambda + L_\theta)\zeta \qquad (8.35)$$

The stresses at the outer edge are therefore

$$\sigma_{ro} = \sigma_{ri} + \Delta\sigma_r$$
$$\sigma_{\theta o} = \sigma_{\theta i} + \Delta\sigma_\theta \qquad (8.36)$$

At the interface between contiguous sections, a radial force balance must exist, leading to the following expression, in which the subscripts n and $n+1$ refer to consecutive element numbers, and h_n and h_{n+1} are the uniform thicknesses of consecutive elements n and $n+1$:

$$(\sigma_{ro}h)_n = (\sigma_{ri}h)_{n+1} \qquad \text{(g)}$$

Another condition that must be satisfied is the equality of tangential deflections at the interface of contiguous sections. From Eq. (8.5),

$$(\sigma_{\theta o} - \nu \sigma_{ro})_n = (\sigma_{\theta i} - \nu \sigma_{ri})_{n+1} \tag{h}$$

Through the use of Eqs. (g) and (h), the variations of radial and tangential stress across the interfaces are

$$\Delta \bar{\sigma}_r = \delta(\sigma_{ro})_n = (\sigma_{ri})_{n+1} - (\sigma_{ro})_n \tag{i}$$

$$\Delta \bar{\sigma}_\theta = \nu \Delta \bar{\sigma}_r = (\sigma_{\theta i})_{n+1} - (\sigma_{\theta o})_n = \nu \delta(\sigma_{ro})_n \tag{j}$$

where

$$\delta = (h_n / h_{n+1}) - 1$$

Hence, the knowledge of the stresses at the outer edge of section n leads to the stresses at the inner edge of section $n+1$:

$$
\begin{aligned}
(\sigma_{ri})_{n+1} &= (\sigma_{ro})_n + \Delta \bar{\sigma}_r \\
(\sigma_{\theta i})_{n+1} &= (\sigma_{\theta o})_n + \Delta \bar{\sigma}_\theta
\end{aligned}
\tag{8.37}
$$

Referring again to Eqs. (a) and (b), consider the *particular* solution for a given element to be

$$\sigma_r' = -\frac{3+\nu}{8} N r^2 + B_1' - \frac{B_2'}{r^2} \tag{k}$$

$$\sigma_\theta' = -\frac{1+3\nu}{8} N r^2 + B_1' + \frac{B_2'}{r^2} \tag{l}$$

The *homogeneous* solution for $N=0$ is

$$\sigma_r'' = B_1'' - \frac{B_2''}{r^2} \tag{m}$$

$$\sigma_\theta'' = B_1'' + \frac{B_2''}{r^2} \tag{n}$$

We shall employ the subscripts a and b to denote the *inner* edge (or center) and the *outer* edge of the disk, respectively. The thickness of the *actual* disk is labeled h^*. The *complete* solution requires addition of Eqs. (m) and (n) in the following manner:

$$
\begin{aligned}
\sigma_r &= \sigma_r' + k \sigma_r'' \\
\sigma_\theta &= \sigma_\theta' + k \sigma_\theta''
\end{aligned}
\tag{8.38}
$$

To evaluate k, the actual thickness h_b^* and the prescribed radial stress σ_{rb}^* are used. Equating the true radial stress at $r=b$ to the stress at $r=b$ for the constant thickness section,

$$\sigma_{rb} h_b = (\sigma_{rb}' + k \sigma_{rb}'') h_b = h_b^* \sigma_{rb}^*$$

The value of the adjustment constant is thus

$$k = \frac{(h_b^* / h_b) \sigma_{rb}^* - \sigma_{rb}'}{\sigma_{rb}''} \tag{8.39}$$

The *calculating procedure*, based upon the foregoing analysis, is as follows:

a. Beginning with initially assumed values of radial and tangential stress at the *inner* edge of an annular disk $(r=a)$, or at the center of a solid disk, calculate a particular solution, denoting these stresses by a single prime. The initially taken values are therefore σ'_{ra} and $\sigma'_{\theta a}$ (Table 8.1). Then Eqs. (8.34), (8.35), (8.36), and (8.37) are used to determine $\sigma'_{ri}, \sigma'_{ro}, \sigma'_{\theta i}, \sigma'_{\theta o}$ for successive interfaces.

b. Beginning with assumed values of radial and tangential stress as in Table 8.1, calculate a complementary solution $\sigma''_{ri}, \sigma''_{ro}, \sigma''_{\theta i}, \sigma''_{\theta o}$ from Eqs. (8.34), (8.35), (8.36), and (8.37) with $L_r = L_\theta = 0$.

c. Determine the final solution by superposition in accordance with Eqs. (8.38). These require calculation of k using Eq. (8.39).

d. If a check of the foregoing computations is desired, the final solution obtained in (c) for inner edge stresses may be used instead of arbitrary initial values in (a). This should provide the same results as (c).

e. Because the results of (c) apply to a disk with stepped rather than smoothly varying sections, an adjustment may be made to the values $(\sigma_r)_n, (\sigma_\theta)_n$:

$$\sigma_r^* = \left(\frac{h}{h^*}\right)\sigma_r \tag{8.40}$$

$$\sigma_\theta^* = \nu\sigma_r^* + (\sigma_\theta - \nu\sigma_r) \tag{8.41}$$

where the starred stresses are the adjusted final values and h^* is the true disk thickness at a given radius. Care should be taken to use the value of h corresponding to σ_r and σ_θ.

The technique outlined is highly amenable to computerization, and is illustrated in the following example.

Table 8.1

Given condition at inner edge or center	Assumed values at inner edge or center (A and B are any arbitrary values, $B \neq 0$)	
	Particular solution	Homogeneous solution
True radial stress, σ_{ra}^*	$\sigma'_{ra} = (h_a/h_a^*)\sigma_{ra}^*$ $\sigma'_{\theta a} = A$	$\sigma''_{ra} = 0$ $\sigma''_{\theta a} = B$
Clamped	$\sigma'_{ra} = A$ $\sigma'_{\theta a} = \nu\sigma'_{ra}$	$\sigma''_{ra} = B$ $\sigma''_{\theta a} = \nu\sigma''_{ra}$
Solid disk	$\sigma'_{ra} = A$ $\sigma'_{\theta a} = \sigma'_{ra}$	$\sigma''_{ra} = B$ $\sigma''_{\theta a} = \sigma''_{ra}$

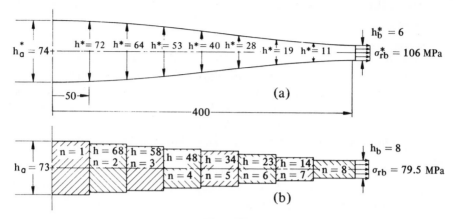

Figure 8.10

Example 8.7. Determine the stresses in the steel disk shown in Fig. 8.10a. The rotational speed is 6000 rpm, and a load of 106 MPa is applied to the periphery. Use $\rho = 7.8 \text{ kN} \cdot \text{s}^2/\text{m}^4$ and $\nu = 0.3$. All dimensions shown in the figure are in millimeters.

SOLUTION. Calculations proceed as outlined. The results are tabulated as follows:

1. We have, $N = \rho \omega^2 = 3.076 \text{ MN/m}^4$. The dimensions (see Fig. 8.10b) and the values of various coefficients are listed in Table 8.2(a).
2. The particular solution, for an assumed stress $\sigma'_{ra} = \sigma'_{\theta a} = 85$ MPa at $r = 0$, is given in Table 8.2(b).
3. The homogeneous solution, for an assumed stress $\sigma''_{ra} = \sigma''_{\theta a} = 7$ MPa at $r = 0$ is given in Table 8.2(c).
4. The constant k, for a given radial stress $\sigma_{rb} = 106$ MPa at $r = b$, calculated from Eq. (8.39), is 2.50. The stresses obtained by superposition of particular and homogeneous solutions from Eqs. (8.38) are shown in Table 8.2(d).
5. The final true stresses in the smooth disk, found by use of Eqs. (8.40) and (8.41), are shown in Table 8.2(e).

The results of this example are observed to differ not very appreciably from the exact analytical values.

Table 8.2(a) Properties of Elements of the Disk Shown in Fig. 8.10

Symbol unit	Element number n							
	1	2	3	4	5	6	7	8
r_o m	0.05	0.1	0.15	0.20	0.25	0.30	0.35	0.40
R	1.0	0.50	0.333	0.250	0.200	0.167	0.143	0.125
ζ	0.500	0.375	0.277	0.219	0.180	0.153	0.133	0.117
L, MPa	6.34	26.7	62.5	113.	180.	262.	359.	472.
L_θ MPa	8.65	13.3	27.4	46.2	69.7	97.8	130.6	167.8
h m	0.073	0.068	0.058	0.048	0.034	0.023	0.014	0.008
δ	0.074	0.172	0.208	0.412	0.478	0.642	0.75	—
h_o^* m	0.072	0.064	0.053	0.040	0.028	0.019	0.011	0.006

Table 8.2(b) Particular Solution (Where $\sigma'_{ra} = \sigma'_{\theta a} = 85$ MPa)

Symbol	Element number n							
	1	2	3	4	5	6	7	8
σ'_{ri}	85.00	87.80	89.92	86.06	85.63	77.49	61.28	25.55
σ'_{ro}	81.83	76.72	71.24	60.69	52.43	37.32	14.60	−37.27
$\sigma'_{\theta i}$	85.00	84.96	85.00	83.22	81.20	76.97	69.32	54.25
$\sigma'_{\theta o}$	83.17	81.04	78.77	73.72	69.45	62.09	50.96	31.26

Table 8.2(c) Homogeneous Solution (Where $\sigma''_{ra} = \sigma''_{\theta a} = 7$ MPa)

Symbol	Element number n							
	1	2	3	4	5	6	7	8
σ''_{ri}	7.00	7.51	8.64	10.11	13.75	19.33	30.05	49.58
σ''_{ro}	7.00	7.37	8.37	9.75	13.08	18.30	28.33	46.73
$\sigma''_{\theta i}$	7.00	7.15	7.67	8.46	10.02	12.57	17.13	25.23
$\sigma''_{\theta o}$	7.00	7.29	7.94	8.82	10.69	13.60	18.85	28.08

Table 8.2(d) Superposition of Trial Solution

Symbol	Element number n							
	1	2	3	4	5	6	7	8
σ_{ri}	102.50	106.58	111.52	111.34	120.00	125.82	136.41	149.50
σ_{ro}	99.33	95.15	92.17	85.07	85.13	83.07	85.43	79.56
$\sigma_{\theta i}$	102.50	102.84	104.18	104.40	106.25	108.40	112.15	117.33
$\sigma_{\theta o}$	100.67	99.27	98.62	95.77	96.18	96.13	98.09	101.46

Table 8.2(e) Final True Stresses

Symbol	Element number n							
	1	2	3	4	5	6	7	8
σ^*_{ri}	101.11	100.66	101.07	100.84	102.00	103.35	100.51	108.73
σ^*_{ro}	100.71	101.10	100.87	102.08	103.37	100.56	108.73	106.08
$\sigma^*_{\theta i}$	102.08	101.06	101.19	101.25	100.85	100.66	101.38	105.10
$\sigma^*_{\theta o}$	101.08	101.06	101.23	100.87	100.81	101.34	105.08	109.42

8.10   Thermal Stresses in Thin Disks

In this section, our concern is with the stresses associated with a radial temperature field $T(r)$ which is independent of the axial dimension. The practical applications are numerous and include annular fins and turbine disks. Because the temperature field is symmetrical with respect to a central axis, it is valid to assume that the stresses and displacements are distributed in the same way as those of Sec. 8.1, and therefore the equations of that section apply here as well.

In this case of plane stress, the applicable equations of stress and strain are obtained from Eq. (3.21) with reference to Eq. (3.37):

$$\sigma_r = \frac{E}{1-\nu^2}\left[\varepsilon_r + \nu\varepsilon_\theta - (1+\nu)\alpha T\right]$$

$$\sigma_\theta = \frac{E}{1-\nu^2}\left[\varepsilon_\theta + \nu\varepsilon_r - (1+\nu)\alpha T\right]$$

(8.42)

The equation of equilibrium, Eq. (8.2), is now

$$r\frac{d}{dr}(\varepsilon_r + \nu\varepsilon_\theta) + (1-\nu)(\varepsilon_r - \varepsilon_\theta) = (1+\nu)\alpha r\frac{dT}{dr}$$

(a)

Introduction of Eq. (8.3) into the above expression yields the following differential equation in radial displacement:

$$\frac{d^2u}{dr^2} + \frac{1}{r}\frac{du}{dr} - \frac{u}{r^2} = (1+\nu)\alpha\frac{dT}{dr}$$

(b)

This is rewritten

$$\frac{d}{dr}\left[\frac{1}{r}\frac{d(ru)}{dr}\right] = (1+\nu)\alpha\frac{dT}{dr}$$

(c)

to render it easily integrable. The solution is

$$u = \frac{(1+\nu)\alpha}{r}\int_a^r Tr\,dr + c_1 r + \frac{c_2}{r}$$

(d)

where a, the inner radius of an annular disk, is taken as zero for a solid disk, and ν and α have been treated as constants.

Annular Disk

The radial and tangential stresses in the annular disk may be found by substituting Eq. (d) into Eq. (8.6):

$$\sigma_r = -\frac{\alpha E}{r^2}\int_a^r Tr\,dr + \frac{E}{1-\nu^2}\left[c_1(1+\nu) - \frac{c_2(1-\nu)}{r^2}\right]$$

(e)

$$\sigma_\theta = \frac{\alpha E}{r^2}\int_a^r Tr\,dr - \alpha ET + \frac{E}{1-\nu^2}\left[c_1(1+\nu) + \frac{c_2(1-\nu)}{r^2}\right]$$

(f)

The constants c_1 and c_2 are determined on the basis of the boundary conditions $(\sigma_r)_{r=a} = (\sigma_r)_{r=b} = 0$. Equation (e) thus gives

$$c_1 = \frac{(1-\nu)\alpha}{b^2-a^2}\int_a^b Tr\,dr, \qquad c_2 = \frac{(1+\nu)a^2\alpha}{b^2-a^2}\int_a^b Tr\,dr$$

The stresses are therefore

$$\sigma_r = \alpha E \left[-\frac{1}{r^2} \int_a^r Tr\,dr + \frac{r^2-a^2}{r^2(b^2-a^2)} \int_a^b Tr\,dr \right]$$

$$\sigma_\theta = \alpha E \left[-T + \frac{1}{r^2} \int_a^r Tr\,dr + \frac{r^2+a^2}{r^2(b^2-a^2)} \int_a^b Tr\,dr \right] \tag{8.43}$$

Solid Disk

In the case of a solid disk, the displacement must vanish at $r=0$ in order to preserve the continuity of material. The value of c_2 in Eq. (d) must therefore be zero. To evaluate c_1, the boundary condition $(\sigma_r)_{r=b}=0$ is employed, and Eq. (e) now gives

$$c_1 = \frac{(1-\nu)\alpha}{b^2} \int_0^b Tr\,dr$$

Substituting c_1 and c_2 into Eqs. (e) and (f), the stresses in a solid disk are found to be

$$\sigma_r = \alpha E \left[\frac{1}{b^2} \int_0^b Tr\,dr - \frac{1}{r^2} \int_0^r Tr\,dr \right]$$

$$\sigma_\theta = \alpha E \left[-T + \frac{1}{b^2} \int_0^b Tr\,dr + \frac{1}{r^2} \int_0^r Tr\,dr \right] \tag{8.44}$$

Given a temperature distribution $T(r)$, the stresses in a solid or annular disk can thus be determined from Eqs. (8.43) or (8.44). Note that $T(r)$ need not be limited to those functions which can be analytically integrated. A numerical integration can easily be carried out for σ_r, σ_θ to provide results of acceptable accuracy.

8.11   Thermal Stress in Long Circular Cylinders

Consider a long cylinder with *ends assumed restrained* so that $w=0$. This is another example of plane strain, for which $\varepsilon_z=0$. The stress–strain relations are, from Hooke's law:

$$\varepsilon_r = \frac{1}{E} \left[\sigma_r - \nu(\sigma_\theta + \sigma_z) \right] + \alpha T$$

$$\varepsilon_\theta = \frac{1}{E} \left[\sigma_\theta - \nu(\sigma_r + \sigma_z) \right] + \alpha T \tag{8.45}$$

$$\varepsilon_z = \frac{1}{E} \left[\sigma_z - \nu(\sigma_r + \sigma_\theta) \right] + \alpha T$$

For $\varepsilon_z=0$, the final expression above yields

$$\sigma_z = \nu(\sigma_r + \sigma_\theta) - \alpha E T \tag{a}$$

Substitution of Eq. (a) into the first two of Eqs. (8.45) leads to the following forms in which z stress does not appear:

$$\varepsilon_r = \frac{1+\nu}{E}\left[(1-\nu)\sigma_r - \nu\sigma_\theta + \alpha ET\right]$$

$$\varepsilon_\theta = \frac{1+\nu}{E}\left[(1-\nu)\sigma_\theta - \nu\sigma_r + \alpha ET\right] \tag{b}$$

Inasmuch as Eqs. (8.2) and (8.3) are valid for the case under discussion, the solutions for u, σ_r and σ_θ proceed as in Sec. 8.10, resulting in

$$u = \frac{(1+\nu)\alpha}{(1-\nu)r}\int_a^r Tr\,dr + c_1 r + \frac{c_2}{r} \tag{c}$$

$$\sigma_r = \frac{E}{1+\nu}\left[-\frac{(1+\nu)\alpha}{(1-\nu)r^2}\int_a^r Tr\,dr + \frac{c_1}{1-2\nu} - \frac{c_2}{r^2}\right] \tag{d}$$

$$\sigma_\theta = \sigma_r + r\frac{d\sigma_r}{dr} \tag{e}$$

Finally, from Eq. (a) we obtain

$$\sigma_z = -\frac{\alpha ET}{1-\nu} + \frac{2\nu E c_1}{(1+\nu)(1-2\nu)} \tag{f}$$

Solid Cylinder

In order for the radial displacement of a solid cylinder to vanish at $r=0$, the constant c_2 in Eq. (c) must clearly be zero. Applying the boundary condition $(\sigma_r)_{r=b}=0$, Eq. (d) may be solved for the remaining constant of integration,

$$c_1 = \frac{(1+\nu)(1-2\nu)\alpha}{(1-\nu)b^2}\int_0^b Tr\,dr \tag{g}$$

and the stress distributions determined from Eqs. (d), (e), and (f):

$$\sigma_r = \frac{\alpha E}{1-\nu}\left[\frac{1}{b^2}\int_0^b Tr\,dr - \frac{1}{r^2}\int_0^r Tr\,dr\right]$$

$$\sigma_\theta = \frac{\alpha E}{1-\nu}\left[-T + \frac{1}{b^2}\int_0^b Tr\,dr + \frac{1}{r^2}\int_0^r Tr\,dr\right] \tag{8.46}$$

$$\sigma_z = \frac{\alpha E}{1-\nu}\left[\frac{2\nu}{b^2}\int_0^b Tr\,dr - T\right] \tag{8.47a}$$

To derive an expression for the radial displacement, $c_2=0$ and Eq. (g) are introduced into Eq. (c).

The longitudinal stress given by Eq. (8.47a) is valid only for the case of a fixed-ended cylinder. In the event the *ends are free*, a uniform axial stress

$\sigma_z = s_0$ may be superimposed to cause the force resultant at each end to vanish:

$$s_0 \pi b^2 - \int_0^b \sigma_z (2\pi r)\, dr = 0$$

This expression together with Eq. (f) yields

$$s_0 = \frac{2\alpha E}{b^2(1-\nu)} \int_0^b Tr\, dr - \frac{2\nu E c_1}{(1+\nu)(1-2\nu)} \tag{h}$$

The longitudinal stress for a free-ended cylinder is now obtained by adding s_0 to the stress given by Eq. (f):

$$\sigma_z = \frac{\alpha E}{1-\nu}\left(\frac{2}{b^2}\int_0^b Tr\, dr - T\right) \tag{8.47b}$$

Stress components σ_r and σ_θ remain as before. The axial displacement is obtained by adding to the right-hand side of Eq. (c), $u_0 = -\nu s_0 r / E$, a displacement due to uniform axial stress s_0.

Cylinder with Central Circular Hole

When the inner and outer surfaces of a hollow cylinder are free of applied load, the boundary conditions $(\sigma_r)_{r=a} = (\sigma_r)_{r=b} = 0$ apply. Introducing these into Eq. (d), the constants of integration are

$$c_1 = \frac{(1+\nu)(1-2\nu)\alpha}{(1-\nu)(b^2-a^2)}\int_a^b Tr\, dr - \nu\varepsilon_z$$

$$c_2 = \frac{(1+\nu)a^2\alpha}{(1-\nu)(b^2-a^2)}\int_a^b Tr\, dr \tag{i}$$

Equations (d), (e) and (f) thus provide

$$\sigma_r = \frac{E}{1-\nu}\left[-\frac{\alpha}{r^2}\int_a^r Tr\, dr + \frac{(r^2-a^2)\alpha}{r^2(b^2-a^2)}\int_a^b Tr\, dr\right]$$

$$\sigma_\theta = \frac{E}{1-\nu}\left[-T+\frac{\alpha}{r^2}\int_a^r Tr\, dr + \frac{(r^2+a^2)\alpha}{r^2(b^2-a^2)}\int_a^b Tr\, dr\right] \tag{8.48}$$

$$\sigma_z = \frac{\alpha E}{1-\nu}\left[\frac{2\nu}{b^2-a^2}\int_a^b Tr\, dr - T\right] \tag{8.49a}$$

If the *ends are free*, proceeding as in the case of a solid cylinder, the longitudinal stress is described by

$$\sigma_z = \frac{\alpha E}{1-\nu}\left[\frac{2}{b^2-a^2}\int_a^b Tr\, dr - T\right] \tag{8.49b}$$

Implementation of the foregoing analyses is dependent upon a knowledge of the radial distribution of temperature $T(r)$.

Example 8.8. Determine the stress distribution in a hollow free-ended cylinder, subject to constant temperatures at the inner and outer surfaces.

SOLUTION. The radial steady-state heat flow through an arbitrary internal cylindrical surface is given by Fourier's law of conduction:

$$Q = -2\pi rK\frac{dT}{dr} \qquad (j)$$

Here Q is the heat flow per unit axial length, and K the thermal conductivity. Assuming Q and K to be constant,

$$r\frac{dT}{dr} = -\frac{Q}{2\pi K} = \text{constant} = c$$

The above is easily integrated upon separation:

$$T = c_1\frac{dr}{r} + c_2 = c_1 \ln r + c_2 \qquad (k)$$

Applying the temperature boundary conditions $(T)_{r=a} = T_1, (T)_{r=b} = T_2$, Eq. (k) may be written

$$T = \frac{T_2 \ln(r/a) + T_1 \ln(b/r)}{\ln(b/a)} = \frac{T_1 - T_2}{\ln(b/a)} \ln\frac{b}{r} \qquad (8.50)$$

When the above is substituted into Eqs. (8.48) and (8.49a), the following stresses are obtained:

$$\sigma_r = \frac{\alpha E(T_1 - T_2)}{2(1-\nu)\ln(b/a)}\left[-\ln\frac{b}{r} - \frac{a^2(r^2-b^2)}{r^2(b^2-a^2)}\ln\frac{b}{a} \right]$$

$$\sigma_\theta = \frac{\alpha E(T_1 - T_2)}{2(1-\nu)\ln(b/a)}\left[1-\ln\frac{b}{r} - \frac{a^2(r^2+b^2)}{r^2(b^2-a^2)}\ln\frac{b}{a} \right] \qquad (8.51)$$

$$\sigma_z = \frac{\alpha E(T_1 - T_2)}{2(1-\nu)\ln(b/a)}\left[1-2\ln\frac{b}{r} - \frac{2a^2}{b^2-a^2}\ln\frac{b}{a} \right]$$

We note that for $\nu = 0$, Eqs. (8.51) provide a solution for a thin hollow disk. In the event the heat flow is outward, i.e., $T_1 > T_2$, examination of Eqs. (8.51) indicates that the stresses are compressive (negative) on the inner surface and tensile (positive) on the outer surface.

In practice, a pressure loading is often superimposed upon the thermal stresses, as in the case of chemical reaction vessels. An approach is following similar to that already discussed, with boundary conditions modified to reflect the pressure, e.g., $(\sigma_r)_{r=a} = -p_i, (\sigma_r)_{r=b} = 0$. In this instance, the internal pressure results in a circumferential tensile stress, Fig. 8.3a, causing a partial cancellation of compressive stress owing to temperature.

Note that when a rotating disk is subjected to inertial loading combined with an axisymmetrical distribution of temperature $T(r)$, the stresses and displacements may be determined by superposition of the two cases.

8.12   The Finite Element Solution

In the previous sections, the cases of axisymmetry discussed were ones in which along the axis of revolution (z), there was uniformity of structural geometry and loading. In this section, the finite element approach of Chapter 7 is applied for computation of displacement, strain, and stress in a general axisymmetric structural system, formed as a solid of revolution having material properties, support conditions, and loading, all of which are symmetrical about the z axis but which may vary along this axis. A simple illustration of this situation is a sphere, uniformly loaded by gravity forces.

"Elements" of the body of revolution (rings, or more generally, tori) are used to discretize the axisymmetric structure. We shall here employ an element of triangular cross section, as shown in Fig. 8.11. It should be noted that a node is now in fact a circle, e.g., node i is the circle with r_i as radius. Thus, the elemental volume dV appearing in the expressions of Sec. 7.10 is the volume of the ring element ($2\pi r\, dr\, dz$). The element clearly lies in three-dimensional space. Any randomly selected vertical cross section of the element, however, is a *plane* triangle.

As already discussed in Sec. 8.1, no tangential displacement can exist in the symmetrical system, i.e., $v=0$. Inasmuch as only the radial displacement u and the axial displacement w in a plane are involved (rz plane), the expressions for displacement established for plane stress and plane strain may readily be extended to the axisymmetric analysis. The theoretical development will therefore follow essentially the procedure given in Chapter 7, with the exception that in the present case, cylindrical coordinates will be employed (r, θ, z).

Strain, Stress, and Elasticity Matrices

The strain matrix, from Eqs. (2.3), (3.20a), and (3.20b), may be defined as follows:

$$\{\varepsilon\}_e = \{\varepsilon_r, \varepsilon_z, \varepsilon_\theta, \gamma_{rz}\} = \left\{ \frac{\partial u}{\partial r}, \ \frac{\partial w}{\partial z}, \ \frac{u}{r}, \ \frac{\partial u}{\partial z} + \frac{\partial w}{\partial r} \right\} \qquad (8.52)$$

The initial strain owing to a temperature change is expressed in the form

$$\{\varepsilon_0\}_e = \{\alpha T, \alpha T, \alpha T, 0\}$$

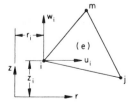

Figure 8.11

It is observed from Eq. (8.52) that the tangential strain ε_θ becomes infinite for a zero value of r. Thus, if the structural geometry is continuous at the z axis, as in the case of a solid sphere, r is generally assigned a small value (e.g., 0.0001 m) for the node located at this axis.

It can be demonstrated that the state of stress throughout the element $\{\sigma\}_e$ is expressible as follows:

$$\{\sigma_r, \sigma_z, \sigma_\theta, \tau_{rz}\}_e = \frac{E}{(1+\nu)\,(1-2\nu)}$$

$$\times \begin{bmatrix} \nu & 1-\nu & \nu & 0 \\ 1-\nu & \nu & \nu & 0 \\ \nu & \nu & 1-\nu & 0 \\ 0 & 0 & 0 & (1-2\nu)/2 \end{bmatrix} \begin{Bmatrix} \varepsilon_r - \alpha T \\ \varepsilon_z - \alpha T \\ \varepsilon_\theta - \alpha T \\ \gamma_{rz} \end{Bmatrix} \qquad (8.53)$$

A comparison of Eqs. (8.53) and (7.35) yields the elasticity matrix

$$[D] = \frac{E}{(1+\nu)(1-2\nu)} \begin{bmatrix} \nu & 1-\nu & \nu & 0 \\ & \nu & \nu & 0 \\ & & 1-\nu & 0 \\ \text{Symm.} & & & (1-2\nu)/2 \end{bmatrix} \qquad (8.54)$$

Displacement Function

The nodal displacements of the element are written in terms of submatrices δ_u and δ_v:

$$\{\delta\}_e = \{u_i, u_j, u_m, w_i, w_j, w_m\} \qquad \text{(a)}$$

The displacement function $\{f\}_e$ which describes the behavior of the element is given by

$$\{f\}_e = \begin{Bmatrix} u \\ w \end{Bmatrix} = \begin{Bmatrix} \alpha_1 + \alpha_2 r + \alpha_3 z \\ \alpha_4 + \alpha_5 r + \alpha_6 z \end{Bmatrix} \qquad (8.55a)$$

or

$$\{f\}_e = \begin{bmatrix} 1 & r & z & 0 & 0 & 0 \\ 0 & 0 & 0 & 1 & r & z \end{bmatrix} \{\alpha_1, \alpha_2, \alpha_3, \alpha_4, \alpha_5, \alpha_6\} \qquad (8.55b)$$

Here the α's are the constants, which can be evaluated as follows. First, we express the nodal displacements $\{\delta\}_e$:

$$u_i = \alpha_1 + \alpha_2 r_i + \alpha_3 z_i$$
$$u_j = \alpha_1 + \alpha_2 r_j + \alpha_3 z_j$$
$$u_m = \alpha_1 + \alpha_2 r_m + \alpha_3 z_m$$
$$w_i = \alpha_4 + \alpha_5 r_i + \alpha_6 z_i$$
$$w_j = \alpha_4 + \alpha_5 r_j + \alpha_6 z_j$$
$$w_m = \alpha_4 + \alpha_5 r_m + \alpha_6 z_m$$

Then, by the inversion of these linear equations,

$$
\begin{Bmatrix} \alpha_1 \\ \alpha_2 \\ \alpha_3 \end{Bmatrix} = \frac{1}{2A} \begin{bmatrix} a_i & a_j & a_m \\ b_i & b_j & b_m \\ c_i & c_j & c_m \end{bmatrix} \begin{Bmatrix} u_i \\ u_j \\ u_m \end{Bmatrix}
$$

$$
\begin{Bmatrix} \alpha_4 \\ \alpha_5 \\ \alpha_6 \end{Bmatrix} = \frac{1}{2A} \begin{bmatrix} a_i & a_j & a_m \\ b_i & b_j & b_m \\ c_i & c_j & c_m \end{bmatrix} \begin{Bmatrix} w_i \\ w_j \\ w_m \end{Bmatrix}
$$
(8.56)

where A is defined by Eq. (7.45) and

$$
\begin{aligned}
a_i &= r_j z_m - z_j r_m, & b_i &= z_j - z_m, & c_i &= r_m - r_j \\
a_j &= r_m z_i - z_m r_i, & b_j &= z_m - z_i, & c_j &= r_i - r_m \\
a_m &= r_i z_j - z_i r_j, & b_m &= z_i - z_j, & c_m &= r_j - r_i
\end{aligned}
$$
(8.57)

Finally, upon substitution of Eqs. (8.56) into (8.55), the displacement function is represented in the following convenient form:

$$
\{f\}_e = \begin{bmatrix} N_i & N_j & N_m & 0 & 0 & 0 \\ 0 & 0 & 0 & N_i & N_j & N_m \end{bmatrix} \{u_i, u_j, u_m, w_i, w_j, w_m\}
$$
(8.58)

or

$$
\{f\}_e = [N]\{\delta\}_e
$$

with

$$
\begin{aligned}
N_i &= \frac{1}{2A}(a_i + b_i r + c_i z) \\
N_j &= \frac{1}{2A}(a_j + b_j r + c_j z) \\
N_m &= \frac{1}{2A}(a_m + b_m r + c_m z)
\end{aligned}
$$
(8.59)

The element strain matrix is found by introducing Eq. (8.58) together with (8.59) into (8.52):

$$
\{\varepsilon_r, \varepsilon_z, \varepsilon_\theta, \gamma_{rz}\}_e = [B]\{u_i, u_j, u_m, w_i, w_j, w_m\}
$$
(8.60)

where

$$
[B] = \frac{1}{2A} \begin{bmatrix} b_i & b_j & b_m & 0 & 0 & 0 \\ 0 & 0 & 0 & c_i & c_j & c_m \\ d_i & d_j & d_m & 0 & 0 & 0 \\ c_i & c_j & c_m & b_i & b_j & b_m \end{bmatrix}
$$
(8.61)

with

$$
d_n = \frac{a_n}{r} + b_n + \frac{c_n z}{r} \qquad (n = i, j, m)
$$

It is observed that the matrix $[B]$ includes the coordinates r and z. Thus, *the strains are not constant* as is the case with plane stress and plane strain.

The Stiffness Matrix

The element stiffness matrix, from Eq. (7.38), is given by

$$[k]_e = \int_V [B]^T [D][B] \, dV \qquad (8.62a)$$

and must be integrated along the circumferential or ring boundary. The above may thus be rewritten

$$[k]_e = 2\pi \int r[B]^T [D][B] \, dr \, dz \qquad (8.62b)$$

where the matrices $[D]$ and $[B]$ are defined by Eqs. (8.54) and (8.61), respectively. It is observed that integration is not as easily performed as in case of plane stress problems, because $[B]$ is also a function of r and z. Although tedious, the integration can be carried out explicitly. Alternatively, approximate numerical approaches may be used. In a simple approximate procedure, $[B]$ is evaluated for a centroidal point of the element. To accomplish this, we substitute fixed centroidal coordinates of the element,

$$\bar{r} = \tfrac{1}{3}(r_i + r_j + r_m), \qquad \bar{z} = \tfrac{1}{3}(z_i + z_j + z_m) \qquad (8.63)$$

into Eq. (8.61) in place of r and z, to obtain $[\bar{B}]$. Then by letting $dV = 2\pi \bar{r} A$, from Eqs. (8.62), the element stiffness is found to be

$$[k]_e = 2\pi \bar{r} A [\bar{B}]^T [D][\bar{B}] \qquad (8.64)$$

This simple procedure leads to results of acceptable accuracy.

External Nodal Forces

In the axisymmetric case, "concentrated" or "nodal" forces are actually loads axisymmetrically located around the body. Let q_r and q_z represent the radial and axial components of force per unit length, respectively, of the circumferential boundary of a node or a radius r. The total nodal force in the radial direction is

$$Q_r = 2\pi r q_r$$

Similarly, the total nodal force in the axial direction is

$$Q_z = 2\pi r q_z$$

Other external load components can be treated analogously. When the approximation leading to Eq. (8.64) is used we can, from Eq. (7.39), readily obtain expressions for nodal forces owing to the initial strains, body forces, and any surface tractions (see Problem 8.23).

In summary, the solution of an axisymmetric problem can be obtained, having generated the total stiffness matrix $[K]$, from Eq. (8.62) or (8.64), and the load matrices $\{Q\}$. Then Eq. (7.41) provides the numerical values of nodal displacements $\{\delta\} = [K]^{-1}\{Q\}$. The expression (8.60) together with Eq. (8.61) yields values of the element strains. Finally, Eq. (8.53), upon substitution of Eq. (8.60), is used to determine the element stresses.

Chapter 8—Problems

Secs. 8.1 to 8.5

8.1. A cylinder of internal radius a and external radius $b = 1.10a$ is subjected to (a) internal pressure p_i only and (b) external pressure p_o only. Determine for each case the ratio of maximum to minimum tangential stress.

8.2. A cylinder of inner radius a and outer radius na, where n is an integer, has been designed to resist a specific internal pressure, but reboring becomes necessary. (a) find the new inner radius r_x required so that the maximum tangential stress does not exceed the previous value by more than $\Delta\sigma_\theta$ while the internal pressure is the same as before. (b) If $a = 25$ mm, $n = 2$, and after reboring the tangential stress is increased by 10%, determine the new diameter.

8.3. A steel tank having an internal diameter of 1.2 m is subjected to an internal pressure of 7 MPa. The tensile and compressive elastic strengths of the material are 280 MPa. Assuming a factor of safety of 2, determine the wall thickness.

8.4. Two thick-walled, closed-ended cylinders of the same dimensions are subjected to internal and external pressure, respectively. The outer diameter of each is twice the inner diameter. What is the ratio of the pressures for the following cases? (a) The maximum tangential stress has the same absolute value in each cylinder. (b) The maximum tangential strain has the same absolute value in each cylinder. Take $\nu = \frac{1}{3}$.

8.5. Determine the radial displacement of a point on the inner surface of the tank described in Problem 8.3. Assume that outer diameter $2b = 1.2616$ m, $E = 200$ GPa, and $\nu = 0.3$.

8.6. A steel cylinder is subjected to an internal pressure only. (a) Obtain the ratio of the wall thickness to the inner diameter, if the internal pressure is three-quarters of the maximum allowable tangential stress. (b) Determine the increase in inner diameter of such a cylinder, 0.15 m in internal diameter, for an internal pressure of 6.3 MPa. Take $E = 210$ GPa and $\nu = \frac{1}{3}$.

8.7. Verify the results shown in Fig. 8.4 using Eqs. (8.19) and (8.20).

8.8. A thick-walled cylinder is subjected to internal pressure p_i and external pressure p_o. Find (a) the longitudinal stress σ_z if the longitudinal strain is zero and (b) the longitudinal strain if σ_z is zero.

8.9. A cylinder, subjected to internal pressure only, is constructed of aluminum having a tensile strength σ_{yp}. The internal radius of the cylinder is a, and the outer radius is $2a$. Based upon the maximum stress, maximum strain, and maximum strain energy theories of failure, predict the limiting values of internal pressure.

8.10. A flywheel of 0.5-m outer diameter and 0.1-m inner diameter is pressed onto a solid shaft. The maximum tangential stress induced in the flywheel is 35 MPa. The length of the flywheel parallel to the shaft axis is 0.05 m. Assuming a coefficient of static friction of 0.2 at the common surface, find the maximum torque that may be transmitted by the flywheel without slippage.

8.11. A solid steel shaft of 0.1-m diameter is pressed onto a steel cylinder, inducing a contact pressure p_1 and a maximum tangential stress $2p_1$ in the cylinder. If an axial tensile load of $P_L = 45$ kN is applied to the shaft, what change in contact pressure occurs? Let $\nu = \frac{1}{3}$.

8.12. When a steel sleeve of external diameter $3b$ is shrunk onto a solid shaft of diameter $2b$, the internal diameter of the sleeve is increased by an amount δ_0. What reduction occurs in the diameter of the shaft?

8.13. A cylinder of inner diameter b is shrunk onto a solid shaft. Find (a) the difference in diameters when the contact pressure is p and the maximum tangential stress is $2p$ in the cylinder and (b) the axial compressive load that should be applied to the shaft to increase the contact pressure from p to p_1.

8.14. A brass solid cylinder is a firm fit within a steel tube of inner diameter $2b$ and outer diameter $4b$ at a temperature $T_1°$C. If now the temperature of both elements is increased to $T_2°$C find the maximum tangential stresses in the cylinder and in the tube. Take $\alpha_s = 11.7 \times 10^{-6}/°$C, $\alpha_b = 19.5 \times 10^{-6}/°$C, and neglect longitudinal friction forces at the interface.

8.15. A gear of inner and outer radii 0.1 m and 0.15 m, respectively, is shrunk onto a hollow shaft of inner radius 0.05 m. The maximum tangential stress induced in the gear wheel is 0.21 MPa. The length of the gear wheel parallel to shaft axis is 0.1 m. Assuming a coefficient of static friction of 0.2 at the common surface, what maximum torque may be transmitted by the gear without slip?

Secs. 8.6 to 8.12

8.16. Show that for an annular rotating disk, the ratio of the maximum tangential stress to the maximum radial stress is given by

$$\frac{\sigma_{\theta,\max}}{\sigma_{r,\max}} = 2b^2 + \frac{1+\nu}{3+\nu}a^2(b^2 - a^2) \qquad (P8.16)$$

8.17. Show that in a solid disk of diameter $2b$, rotating with a tangential velocity V, the maximum stress is $\sigma_{\max} = \frac{5}{12}\rho V^2$.

8.18. Consider a steel rotating disk of hyperbolic cross section (Fig. 8.8) with $a=0.125$ m, $b=0.625$ m, $t_i=0.125$ m, and $t_o=0.0625$ m. Determine the maximum tangential force that can occur at the outer surface in newtons per meter of circumference if the maximum stress at the bore is not to exceed 140 MPa. Assume that outer and inner edges are free of pressure.

8.19. A steel turbine disk with $b=0.5$ m, $a=0.0625$ m, and $t_o=0.05$ m rotates at 5000 rpm carrying blades weighting a total of 540 N. The center of gravity of each blade lies on a circle of 0.575 m radius. Assuming zero pressure at the bore, determine (a) the maximum stress for a disk of constant thickness and (b) the maximum stress for a disk of hyperbolic cross section. The thickness at the hub and tip are $t_i=0.4$ m and $t_o=0.05$ m, respectively. (c) For a thickness at the axis $t_i=0.02425$ m, determine the thickness at the outer edge, t_o, for a disk under uniform stress, 84 MPa. Take $\rho=7.8$ kN·s^2/m^4 and $g=9.81$ m/s.^{2}

8.20. Redo Example 8.7 for a rotational speed of 12,000 rpm and zero load at the periphery of the disk.

8.21. Show that for a hollow disk, when subjected to a temperature distribution given by $T=(T_a-T_b)\ln(b/r)/\ln(b/a)$, the maximum radial stress occurs at

$$r=ab\left(\frac{2}{b^2-a^2}\ln\frac{b}{a}\right)^{1/2} \qquad \text{(P8.21)}$$

8.22. Verify that the distribution of stress in a solid disk in which the temperature varies linearly with the radial dimension, $T(r)=T_0(b-r)/b$, is given by

$$\sigma_r=\tfrac{1}{3}T_0\left(\frac{r}{b}-1\right)\alpha E \qquad \text{(P8.22)}$$

Here T_0 represents the temperature rise at $r=0$.

8.23. Redo Example 7.7, with the element shown in Fig. 7.17 representing a segment adjacent to the boundary of a sphere subjected to external pressure $p=14$ MPa.

Chapter 9

Beams on Elastic Foundations

9.1 General Theory

In the problems involving beams previously considered, support was provided at a number of discrete locations, and the beam was usually assumed to suffer no deflection at these points of support. We now explore the case of a prismatic beam supported continuously along its length by a foundation, itself assumed to experience elastic deformation. We shall take the reaction forces to be *linearly* proportional to the beam deflection at any point. This assumption not only leads to equations amenable to solution, but represents an idealization closely approximating many real situations. Examples include a railroad track, where the so-called elastic support consists of the cross ties, the ballast, and the subgrade; concrete footings on an earth foundation; long steel pipes resting on earth or on a series of elastic springs; a bridgedeck or floor structure consisting of a network of closely spaced bars. The force q per unit length, resisting the displacement of the beam, is equal to $-kv$. Here v is the beam deflection, positive downward as in Fig. 9.1, and k is a constant, usually referred to as the *modulus of the foundation*, possessing the dimensions of force per unit length of beam per unit of deflection (e.g., newtons per square meter or pascals).

The analysis of a beam whose length is very much greater than its depth and width serves as the basis of the treatment of all beams on elastic foundations. Referring again to Fig. 9.1, which shows a beam of constant section supported by an elastic foundation, the x axis passes through the centroid and the y axis is a principal axis of the cross section. The deflection v, subject to reaction q and applied load per unit length p, for a condition of small slope, must satisfy the beam equation:

$$EI\frac{d^4v}{dx^4} + kv = p \tag{9.1}$$

For those parts of the beam on which no distributed load acts, $p = 0$, and

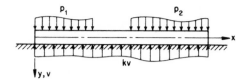

Figure 9.1

Eq. (9.1) takes the form

$$EI\frac{d^4v}{dx^4} + kv = 0 \qquad (9.2)$$

It will suffice to consider the general solution of Eq. (9.2) only, requiring the addition of a particular integral to satisfy Eq. (9.1) as well. Selecting $v = e^{ax}$ as a trial solution, it is found that Eq. (9.2) is satisfied if

$$\left(a^4 + \frac{k}{EI}\right)v = 0$$

requiring that

$$a = \pm\beta(1 \pm i)$$

where

$$\beta = \left(\frac{k}{4EI}\right)^{1/4} \qquad (9.3)$$

The general solution of Eq. (9.2) may now be written

$$v = e^{\beta x}[A\cos\beta x + B\sin\beta x] + e^{-\beta x}[C\cos\beta x + D\sin\beta x] \qquad (9.4)$$

where A, B, C, and D are the constants of integration.

In the developments which follow, the case of a single load acting on an infinitely long beam will be treated first. The solution of problems involving a variety of loading combinations will then rely upon the principle of superposition.

9.2 Infinite Beams

Consider an infinitely long beam resting on a continuous elastic foundation, loaded by a concentrated force P (Fig. 9.2). The variation of the reaction kv is unknown, and the equations of static equilibrium are not sufficient for its determination. The problem is therefore *statically indeterminate* and requires additional formulation, which is available from the equation of the deflection curve of the beam. Owing to beam symmetry, only that portion to the right of the load P need be considered. The two boundary conditions for this segment are deduced from the fact that as

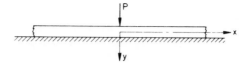

Figure 9.2

$x \to \infty$, the deflection and all derivatives of v with respect to x must vanish. On this basis, it is clear that the constants A and B in Eq. (9.4) must equal zero. What remains is

$$v = e^{-\beta x}(C\cos \beta x + D\sin \beta x) \tag{9.5}$$

The conditions applicable a very small distance to the right of P are

$$v'(0) = 0, \qquad V = -EIv'''(0) = -\frac{P}{2} \tag{a}$$

where the minus sign is consistent with the general convention adopted in Sec. 1.3. Substitution of Eq. (a) into Eq. (9.5) yields

$$C = D = \frac{P}{8\beta^3 EI} = \frac{P\beta}{2k}$$

Introduction of the expressions for the constants into Eq. (9.5) provides the following equation, applicable to an infinite beam subject to a concentrated force P at midlength:

$$v = \frac{P\beta}{2k} e^{-\beta x}(\cos \beta x + \sin \beta x) \tag{9.6a}$$

Table 9.1

βx	$f_1(\beta x)$	$f_2(\beta x)$	$f_3(\beta x)$	$f_4(\beta x)$	βx	$f_1(\beta x)$	$f_2(\beta x)$	$f_3(\beta x)$	$f_4(\beta x)$
0.0	1.000	0.000	1.000	1.000	3.0	−0.042	0.007	−0.056	−0.049
0.2	0.965	0.163	0.640	0.802	3.2	−0.043	−0.002	−0.038	−0.041
0.4	0.878	0.261	0.356	0.617	3.4	−0.041	−0.009	−0.024	−0.032
0.6	0.763	0.310	0.143	0.453	3.6	−0.037	−0.012	−0.012	−0.024
0.8	0.635	0.322	−0.009	0.313	3.8	−0.031	−0.014	−0.004	−0.018
1.0	0.508	0.310	−0.111	0.199	4.0	−0.026	−0.014	0.002	−0.012
1.2	0.390	0.281	−0.172	0.109	4.2	−0.020	−0.013	−0.006	−0.007
1.4	0.285	0.243	−0.201	0.042	4.4	−0.016	−0.012	0.008	−0.004
1.6	0.196	0.202	−0.208	−0.006	4.6	−0.011	−0.010	0.009	−0.001
1.8	0.123	0.161	−0.199	−0.038	4.8	−0.008	−0.008	0.009	0.001
2.0	0.067	0.123	−0.179	−0.056	5.0	−0.005	−0.007	0.008	0.002
2.2	0.024	0.090	−0.155	−0.065	5.5	0.000	−0.003	0.006	0.003
2.4	−0.006	0.061	−0.128	−0.067	6.0	0.002	−0.001	0.003	0.002
2.6	−0.025	0.038	−0.102	−0.064	6.5	0.002	0.000	0.001	0.001
2.8	−0.037	0.020	−0.078	−0.057	7.0	0.001	0.001	0.000	0.001

or

$$v = \frac{P\beta}{2k} e^{-\beta x} \left[\sqrt{2} \sin\left(\beta x + \frac{\pi}{4} \right) \right] \qquad (9.6b)$$

Equation (9.6b) indicates clearly that the characteristic of the deflection is an exponential decay of a sine wave of wavelength

$$\lambda = \frac{2\pi}{\beta} = 2\pi \left(\frac{4EI}{k} \right)^{1/4}$$

In order to simplify the equations for deflection, rotation, moment, and shear, the following notations and relationships are introduced:

$$f_1(\beta x) = e^{-\beta x}(\cos \beta x + \sin \beta x)$$

$$f_2(\beta x) = e^{-\beta x} \sin \beta x = -\frac{1}{2\beta} f_1'$$

$$f_3(\beta x) = e^{-\beta x}(\cos \beta x - \sin \beta x) = \frac{1}{\beta} f_2' = -\frac{1}{2\beta^2} f_1'' \qquad (9.7)$$

$$f_4(\beta x) = e^{-\beta x} \cos \beta x = -\frac{1}{2\beta} f_3' = -\frac{1}{2\beta^2} f_2'' = \frac{1}{4\beta^3} f_1'''$$

$$f_1(\beta_x) = -\frac{1}{\beta} f_4'$$

Table 9.1 lists numerical values of the foregoing functions for various values of the argument βx. Equation (9.6) and its derivatives, together with Eq. (9.7), yield the following expressions for deflection, slope, moment, and shearing force:

$$v = \frac{P\beta}{2k} f_1$$

$$\theta = v' = -\frac{P\beta^2}{k} f_2$$

$$M = EIv'' = -\frac{P}{4\beta} f_3 \qquad (9.8)$$

$$V = -EIv''' = -\frac{P}{2} f_4$$

The above expressions are valid for $x \geq 0$.

Example 9.1. A very long rectangular beam of width 0.1 m and depth 0.15 m (Fig. 9.3) is subject to a uniform loading over 4 m of its length of $p = 175$ kN/m. The beam is supported on an elastic foundation having a modulus $k = 14$ MPa. Derive an expression for the deflection at an arbitrary point A within length L. Calculate

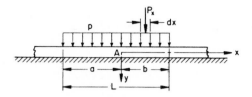

Figure 9.3

the maximum deflection and the maximum force per unit length between beam and foundation. Use $E = 200$ GPa.

SOLUTION. The deflection Δv at point A due to the load $P_x = p\, dx$ is, from Eq. (9.8),

$$\Delta v = \frac{p\, dx}{2k} \beta e^{-\beta x}(\cos \beta x + \sin \beta x)$$

The deflection at point A resulting from the entire distributed load is then

$$v_A = \int_0^a \frac{p\, dx}{2k} \beta e^{-\beta x}(\cos \beta x + \sin \beta x) + \int_0^b \frac{p\, dx}{2k} \beta e^{-\beta x}(\cos \beta x + \sin \beta x)$$

$$= \frac{p}{2k}(2 - e^{-\beta a}\cos \beta a - e^{-\beta b}\cos \beta b)$$

or

$$v_A = \frac{p}{2k}[2 - f_4(\beta a) - f_4(\beta b)] \tag{b}$$

Although the algebraic sign of the distance a in Eq. (b) is negative, in accordance with the placement of the origin in Fig. 9.3, we shall treat it as a positive number because Eq. (9.8) gives the deflection for positive x only. This is justified on the basis that the beam deflection under a concentrated load is the same at equal distances from the load, whether these distances are positive or negative. By the use of Eq. (9.3)

$$\beta = \left(\frac{k}{4EI}\right)^{1/4} = \left(\frac{14 \times 10^6}{4 \times 200 \times 10^9 \times 0.1 \times 0.15^3/12}\right)^{1/4} = 0.888 \text{ m}^{-1}$$

From the above value of β, $\beta L = (0.888)(4) = 3.552 = \beta(a + b)$. We are interested in the *maximum* deflection and therefore locate the origin at point A, the center of the distributed loading. Now a and b represent equal lengths, so that $\beta a = \beta b = 1.776$ and Eq. (b) gives

$$v_{\max} = \frac{175}{2(14{,}000)}[2 - (-0.0345) - (-0.0345)] = 0.0129 \text{ m}$$

The maximum force per unit of length between beam and foundation is then $kv_{\max} = 14 \times 10^6(0.0129) = 180.6$ kN/m.

Example 9.2. A very long beam is supported on an elastic foundation and is subjected to a concentrated moment M_0 (Fig. 9.4). Determine the equations describing the deflection, slope, moment, and shear.

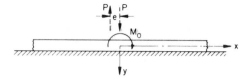

Figure 9.4

SOLUTION.   Observe that the couple $P \cdot e$ is equivalent to M_0 for the case in which e approaches zero (indicated by the dashed lines in the figure). Applying Eq. (9.8), we have therefore

$$v = \frac{P\beta}{2k}\{f_1(\beta x) - f_1[\beta(x+e)]\} = -\frac{M_0 \beta}{2k} \frac{f_1[\beta(x+e)] - f_1(\beta x)}{e}$$

$$= -\frac{M_0 \beta}{2k} \lim_{e \to 0} \frac{f_1[\beta(x+e)] - f_1(\beta x)}{e} = -\frac{M_0 \beta}{2k} \frac{df_1(\beta x)}{dx} = \frac{M_0 \beta^2}{k} f_2(\beta x)$$

Successive differentiation yields

$$v = \frac{M_0}{k}\beta^2 f_2(\beta x)$$

$$\theta = \frac{M_0}{k}\beta^3 f_3(\beta x)$$

$$M = EIv'' = -\frac{M_0}{2}f_4(\beta x) \tag{9.9}$$

$$V = -EIv''' = -\frac{M_0 \beta}{2}f_1(\beta x)$$

which are the deflection, slope, moment, and shear, respectively.

9.3   Semi-infinite Beams

The theory of Sec. 9.2 is now applied to a semi-infinite beam, having one end at the origin and the other extending indefinitely in a positive x direction, as in Fig. 9.5. At $x = 0$, the beam is subjected to a concentrated load P and a moment M_A. The constants C and D of Eq. (9.5) can be ascertained by applying the following conditions at the left end of the beam:

$$EIv'' = M_A, \qquad EIv''' = -V = P$$

The results are

$$C = \frac{P + \beta M_A}{2\beta^3 EI}, \qquad D = -\frac{M_A}{2\beta^2 EI}$$

The deflection is now found by substituting C and D into Eq. (9.5):

$$v = \frac{e^{-\beta x}}{2\beta^3 EI}[P\cos\beta x + \beta M_A(\cos\beta x - \sin\beta x)] \tag{9.10}$$

Figure 9.5

At $x = 0$,

$$\delta = v(0) = \frac{2\beta}{k}(P + \beta M_A) \tag{9.11}$$

Finally, successively differentiating Eq. (9.10) yields expressions for slope, moment, and shear:

$$\begin{aligned}
v &= \frac{2\beta}{k}\left[\, Pf_4(\beta x) + \beta M_A f_3(\beta x)\,\right] \\
\theta &= -\frac{2\beta^2}{k}\left[\, Pf_1(\beta x) + 2\beta M_A f_4(\beta x)\,\right] \\
M &= \frac{P}{k} f_2(\beta x) + M_A f_1(\beta x) \\
V &= -Pf_3(\beta x) + 2\beta M_A f_2(\beta x)
\end{aligned} \tag{9.12}$$

Application of the above equations together with the principle of super-position permits the solution of more complex problems, as is illustrated below.

Example 9.3. Determine the equation of the deflection curve of a semi-infinite beam on an elastic foundation loaded by concentrated force P a distance c from the free end (Fig. 9.6a).

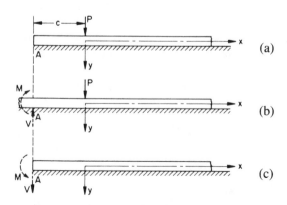

Figure 9.6

SOLUTION. The problem may be restated as the sum of the cases shown in Figs. 9.6b and c. Applying Eqs. (9.8) and the conditions of symmetry, the reactions appropriate to the infinite beam of Fig. 9.6b are

$$M = -\frac{P}{4\beta} f_3(\beta c)$$

$$V = -\frac{P}{2} f_4(\beta c)$$

(a)

Superposition of the deflections of Fig. 9.6b and c (see Eqs. 9.8 and 9.12) results in

$$v = v_{\text{inf}} + v_{\text{semi-inf}} = \frac{P\beta}{2k} f_1(\beta x) + \frac{2\beta}{k} \left\{ -V f_4[\beta(x+c)] + \beta M f_3[\beta(x+c)] \right\}$$

Introducing Eqs. (a) into the above, the following expression for deflection, applicable for positive x, is obtained:

$$v = \frac{P\beta}{2k} f_1(\beta x) + \frac{P\beta}{k} \left\{ f_4(\beta c) f_4[\beta(x+c)] - \tfrac{1}{2} f_3(\beta c) f_3[\beta(x+c)] \right\}$$

The above is clearly applicable for negative x as well, provided that x is replaced by $|x|$.

9.4 Finite Beams. Classification of Beams

The bending of a finite beam on elastic foundation may also be treated by application of the general solution, Eq. (9.4). In this instance there are four constants of integration to be evaluated. To accomplish this, two boundary conditions at each end may be applied, usually resulting in rather lengthy formulations. An alternate approach to the solution of finite beam problems employs equations derived for infinite and semi-infinite beams, together with the principle of superposition.

To permit the establishment of a stiffness criterion for beams, it serves well to consider a finite beam on an elastic foundation, centrally loaded by a concentrated force P (Fig. 9.7), and to compare the deflections occurring at the center and end of the beam. It is noted that the beam deflection is symmetrical with respect to C. The appropriate boundary conditions are, for $x \geqslant 0$: $v'(L/2) = 0$, $EIv'''(L/2) = P/2$, $EIv''(0) = 0$, and $EIv'''(0) = 0$. Substituting these into the proper derivatives of Eq. (9.4) leads to four equations with unknown constants A, B, C, and D. After routine but somewhat lengthy algebraic manipulation, the following expressions are determined:[*]

$$v_C = \frac{P\beta}{2k} \frac{2 + \cos\beta L + \cosh\beta L}{\sin\beta L + \sinh\beta L}$$

(9.13)

$$v_E = \frac{2P\beta}{k} \frac{\cos(\beta L/2)\cosh(\beta L/2)}{\sin\beta L + \sinh\beta L}$$

(9.14)

---

[*]See for details, M. Hetényi, *Beams on Elastic Foundation*, New York: McGraw-Hill, 1960.

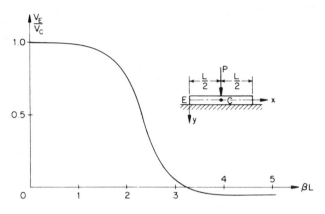

Figure 9.7

From Eqs. (9.13) and (9.14), we have

$$\frac{v_E}{v_C} = \frac{4\cos(\beta L/2)\cosh(\beta L/2)}{2+\cos\beta L+\cosh\beta L} \tag{9.15}$$

Figure 9.7, a plot of Eq. (9.15), enables us to readily discern a rationale for the classification of beams:

  a. *Short beams*, $\beta L < 1$: Inasmuch as the end deflection is essentially equal to that at the center, the deflection of the foundation can be determined to good accuracy by regarding the beam as infinitely rigid.
  b. *Intermediate beams*, $1 < \beta L < 3$: In this region, the influence of the central force at the ends of the beam is substantial, and the beam must be treated as finite in length.
  c. *Long beams*, $\beta L > 3$: It is clear from the figure that the ends are not affected appreciably by the central loading. Therefore, if we are concerned with one end of the beam, the other end or the middle may be regarded as being an infinite distance away, i.e., the beam may be treated as infinite in length.

The foregoing remarks do not relate only to the special case of loading shown in Fig. 9.7, but are quite general. Should greater accuracy be required, the upper limit of group (a) may be placed at $\beta L = 0.6$ and the lower limit of group (c) at $\beta L = 5$.

9.5 Beams Supported by Equally Spaced Elastic Elements

In the event that a long beam is supported by individual elastic elements, as in Fig. 9.8a, the problem is simplified if the separate supports are replaced by an equivalent continuous elastic foundation. To accomplish

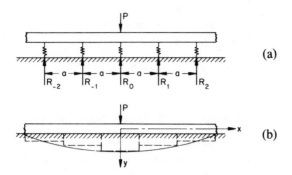

Figure 9.8

this, it is assumed that the distance a between each support and the next is small, and that the concentrated reactions $R_i = Kv_i$ are replaced by equivalent uniform loadings shown by the dashed lines of Fig. 9.8b. For practical calculations, the usual limitation is $a \leqslant \pi/4\beta$. Here K represents a spring constant (e.g., newtons per meter). The average continuous load distribution is shown by the solid line in the figure. The intensity of the latter distribution is ascertained as follows:

$$\frac{R}{a} = \frac{K}{a}v = q$$

or

$$q = kv \tag{a}$$

where the foundation modulus of the equivalent continuous elastic support is

$$k = \frac{K}{a} \tag{9.16}$$

The solution for the case of a beam on individual elastic supports is then obtained through the use of Eq. 9.2, in which the value of k is that given by Eq. (9.16).

Example 9.4. A series of springs, spaced so that $a = 1.5$ m, supports a long thin-walled steel tube having $E = 206.8$ GPa. A weight of 6700 N acts down at midlength of the tube. The average diameter of the tube is 0.1 m, and the moment of inertia of its section is 6×10^{-6} m^4. Take the spring constant of each support to be $K = 10,000$ N/m. Find the maximum moment and the maximum deflection assuming tube weight negligible.

SOLUTION. Applying Eqs. (9.3) and (9.16), we obtain

$$k = \frac{K}{a} = \frac{10,000}{1.5} = 6667 \text{ Pa}$$

$$\beta = \left(\frac{6667}{4 \times 206.8 \times 10^9 \times 6 \times 10^{-6}} \right)^{1/4} = 0.1913 \text{ m}^{-1}$$

From Eq. (9.8),

$$\sigma_{max} = \frac{Mc}{I} = \frac{Pc}{4\beta I} = \frac{6700 \times 0.05}{4 \times 0.1913 \times 6 \times 10^{-6}} = 72.969 \text{ MPa}$$

$$v_{max} = \frac{P\beta}{2k} = \frac{6700 \times 0.1913}{2 \times 6667} = 0.0961 \text{ m}$$

9.6   Simplified Solutions for Relatively Stiff Beams

Examination of the analyses of the previous sections and of Fig. 9.7 leads one to conclude that the distribution of force acting on the beam by the foundation is, in general, a nonlinear function of the beam length coordinate. This distribution approaches linearity as the beam length decreases or as the beam becomes stiffer. Reasonably good results can be expected, therefore, by assuming a linearized elastic foundation pressure for stiff beams. The foundation pressure is then predicated upon beam displacement in the manner of a rigid body,* and the reaction is, as a consequence, statically determinate.

To illustrate the approach, consider once more the beam of Fig. 9.7, this time with a linearized foundation pressure (Fig. 9.9). Because of loading symmetry, the foundation pressure is, in this case, not only linear but constant as well. We shall compare the results thus obtained with those found earlier.

The *exact theory* states that points E and C deflect in accordance with Eqs. (9.13) and (9.14). The relative deflection of these points is simply

$$v = v_C - v_E \qquad (a)$$

For the simplified load configuration shown in Fig. 9.9, the relative beam deflection may be determined by considering the elementary solution for a beam subjected to a uniformly distributed loading and a concentrated force. For this case, we label the relative deflection v_1 as follows:

$$v_1 = \frac{PL^3}{48EI} - \frac{5(P/L)L^4}{384EI} = \frac{PL^3}{128EI} \qquad (b)$$

The ratio of the relative deflections obtained by the exact and approximate analyses now serves to indicate the validity of the approximations. Consider

$$\frac{v}{v_1} = \frac{32}{(\beta L)^3} \frac{\frac{1}{2}\cosh \beta L + \frac{1}{2}\cos \beta L + 1 - 2\cosh\frac{1}{2}\beta L \cos\frac{1}{2}\beta L}{\sinh \beta L + \sin \beta L} \qquad (c)$$

---

*B. W. Shaffer, Some simplified solutions for relatively stiff beams on elastic foundations, *Trans. ASME, Journal of Engineering for Industry*, pp. 1–5 (February 1963).

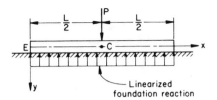

Figure 9.9

where v and v_1 are given by Eqs. (a) and (b). The trigonometric and hyperbolic functions may be expanded as follows:

$$\sin \beta L = \beta L - \frac{(\beta L)^3}{3!} + \frac{(\beta L)^5}{5!} - \frac{(\beta L)^7}{7!} + \cdots$$

$$\cos \beta L = 1 - \frac{(\beta L)^2}{2!} + \frac{(\beta L)^4}{4!} - \frac{(\beta L)^6}{6!} + \cdots$$

$$\sinh \beta L = \beta L + \frac{(\beta L)^3}{3!} + \frac{(\beta L)^5}{5!} + \cdots \tag{d}$$

$$\cosh \beta L = 1 + \frac{(\beta L)^2}{2!} + \frac{(\beta L)^4}{4!} + \cdots$$

Introducing the above into Eq. (c), we obtain

$$\frac{v}{v_1} = 1 - \frac{23}{120} \frac{(\beta L)^4}{4!} + \frac{51}{20} \frac{(\beta L)^3}{8!} + \cdots \tag{e}$$

Substituting various values of βL into Eq. (e) discloses that for $\beta L < 1.0$, v/v_1 differs from unity by no more than 1%, and the linearization is seen to yield good results. It can be shown that for values of $\beta L < 1$, the ratio of the moment (or slope) obtained by the linearized analysis to that obtained from the exact analysis differs from unity by less than 1%.

Analysis of a finite beam, centrally loaded by a concentrated moment, also reveals results similar to those given above. We conclude therefore, that when βL is small (< 1.0), no significant error is introduced by assuming a linear distribution of foundation pressure.

9.7   Solution by Finite Differences

Because of the considerable time and effort required in the analytical solution of practical problems involving beams on elastic foundations, approximate methods employing numerical analysis are frequently applied. A solution utilizing the method of finite differences is illustrated in the example which follows.

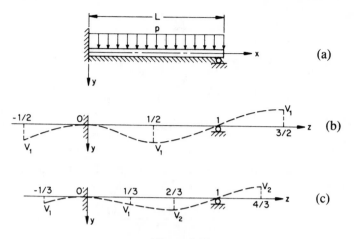

Figure 9.10

Example 9.5. Determine the deflection of the built-in beam on an elastic foundation shown in Fig. 9.10a. The beam is subjected to a uniformly distributed loading p and is simply supported at $x = L$.

SOLUTION. The deflection is governed by Eq. (9.1), for which the applicable boundary conditions are

$$v(0) = v(L) = 0, \qquad v'(0) = 0, \qquad v''(L) = 0 \tag{a}$$

The solution will be obtained by replacing Eqs. (9.1) and (a) by a system of finite difference equations. It is convenient to first transform Eq. (9.1) into dimensionless form through the introduction of the following quantities:[*]

$$z = \frac{x}{L}, \qquad \frac{d}{dx} = \frac{1}{L}\frac{d}{dz} \qquad (\text{at } x = 0, \quad z = 0; \quad x = L, z = 1)$$

The deflection equation is therefore

$$EI\frac{d^4v}{dz^4} + kL^4v = pL^4 \tag{b}$$

We next divide the interval of $z(0, 1)$ into n equal parts of length $h = 1/m$, where m represents an integer. Multiplying Eq. (b) by $h^4 = 1/m^4$, we have

$$h^4\frac{d^4v}{dz^4} + \frac{kL^4}{m^4EI}v = \frac{pL^4}{m^4EI} \tag{c}$$

Employing Eq. (7.12), Eq. (c) assumes the following finite difference form:

$$v_{n-2} - 4v_{n-1} + 6v_n - 4v_{n+1} + v_{n+2} + \frac{kL^4}{m^4EI}v_n = \frac{pL^4}{m^4EI} \tag{d}$$

---

[*]M. G. Salvadori and M. L. Baron, *Numerical Methods in Engineering*, Englewood Cliffs, NJ: Prentice-Hall, 1961, pp. 148–151.

Upon setting $C = kL^4/EI$, the above becomes

$$v_{n-2} - 4v_{n-1} + \left[\frac{C}{m^4} + 6\right]v_n - 4v_{n+1} + v_{n+2} = \frac{pL^4}{m^4 EI} \tag{e}$$

The boundary conditions, Eqs. (a), are transformed into difference conditions by employing Eq. (7.6):

$$v_0 = 0, \qquad v_{-1} = v_1, \qquad v_n = 0, \qquad v_{n-1} = -v_{n+1} \tag{f}$$

Equations (e) and (f) represent the set required for a solution, with the degree of accuracy increased as the magnitude of m is increased. Any desired accuracy can thus be attained.

For purposes of illustration, let $k = 2.1$ MPa, $E = 200$ GPa, $I = 3.5 \times 10^{-4}$ m^4, $L = 3.8$ m, and $p = 540$ kN/m. Determine the deflections for $m = 2$, $m = 3$, and $m = 4$. Equation (e) thus becomes

$$v_{n-2} - 4v_{n-1} + 6\left(\frac{m^4 + 1}{m^4}\right)v_n - 4v_{n+1} + v_{n+2} = \frac{1.6}{m^4} \tag{g}$$

For $m = 2$, the deflection curve, satisfying Eq. (f), is sketched in Fig. 9.10b. At $z = \frac{1}{2}$ we have $v_n = v_1$. Equation (g) then yields

$$v_1 - 4(0) + 6\frac{2^4 + 1}{2^4}v_1 - 4(0) - v_1 = \frac{1.6}{2^4}$$

from which $v_1 = 16$ mm.

For $m = 3$, the deflection curve satisfying Eq. (f) is now as in Fig. 9.10c. Hence Eq. (g) at $z = \frac{1}{3}$ (by setting $v_n = v_1$) and at $z = \frac{2}{3}$ (by setting $v_n = v_2$) leads to

$$v_1 + 6\frac{3^4 + 1}{3^4}v_1 - 4v_2 = \frac{1.6}{3^4}$$

$$-4v_1 + 6\frac{3^4 + 1}{3^4}v_2 - v_2 = \frac{1.6}{3^4}$$

from which $v_1 = 9$ mm and $v_2 = 11$ mm.

For $m = 4$, a similar procedure yields $v_1 = 5.3$ mm, $v_2 = 9.7$ mm, and $v_3 = 7.5$ mm.

9.8 Applications

The theory of beams on elastic foundation is applicable to many problems of practical importance, of which one is discussed below.

Grid Configurations of Beams

The ability of a floor to sustain extreme loads without undue deflection, as in a machine shop, is significantly enhanced by combining the floor beams in a particular array or grid configuration. Such a design is illustrated in the ensuing problem.

Example 9.6. A single concentrated load P acts at the center of a machine room floor composed of 79 transverse beams (spaced $a = 0.3$ m apart) and one longitudinal beam, as shown in Fig. 9.11. If all beams have the same modulus of rigidity EI,

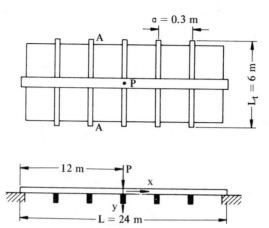

Figure 9.11

determine the deflection and the distribution of load over the various transverse beams supporting the longitudinal beam. Assume that the transverse and longitudinal beams are attached so that they deform together.

SOLUTION. The spring constant K of an individual elastic support such as beam AA is

$$K = \frac{R_C}{v_c} = \frac{R_C}{R_C L_t^3 / 48 EI} = \frac{48 EI}{L_t^3}$$

where v_C is the central deflection of a simply supported beam of length L_t carrying a center load R_C. From Eq. (9.16), the modulus k of the equivalent continuous elastic foundation is found to be

$$k = \frac{K}{a} = \frac{48 EI}{a L_t^3}$$

Thus

$$\beta = \sqrt[4]{\frac{k}{4 EI}} = \sqrt[4]{\frac{12}{a L_t^3}} = \frac{3.936}{L_t}$$

and

$$\beta L = \frac{3.936}{L_t}(4 L_t) = 15.744, \qquad \beta a = \frac{3.936}{L_t} \frac{L_t}{20} = 0.1968$$

In accordance with the criteria discussed in Secs. 9.4 and 9.5, the longitudinal beam may be classified as a long beam resting on a continuous elastic support of modulus k. Consequently, from Eqs. (9.8), the deflection at midspan is

$$v_P = \frac{P\beta}{2k}(1) = \frac{3.936 P}{2 L_t} \frac{a L_t^3}{48 EI} = \frac{1.968 P L_t^2 a}{48 EI}$$

The deflection of a transverse beam depends upon its distance x from the center of the longitudinal beam, as shown in the following tabulation:

x	0	$2a$	$4a$	$6a$	$8a$	$10a$	$12a$	$16a$	$20a$	$24a$
$f_1(\beta x)$	1	0.881	0.643	0.401	0.207	0.084	-0.0002	-0.043	-0.028	-0.009
v		v_P	$0.881v_P$	$0.643v_P$	$0.401v_P$	$0.207v_P$	$0.084v_P$	$-0.0002v_P$	$-0.043v_P$	$-0.028v_P$	$-0.009v_P$

We are now in a position to calculate the load R_{CC} supported by the central transverse beam. Since the midspan deflection v_M of the central transverse beam is equal to v_P, we have

$$v_M = \frac{R_{CC}L_t^3}{48EI} = \frac{1.968PL_t^2a}{48EI}$$

and

$$R_{CC} = 1.968P\frac{a}{L_t} = 0.0984P$$

The remaining transverse beam loads are now readily calculated on the basis of the deflections in the above tabulation, recalling that the loads are linearly proportional to the deflections.

We observe that beyond beam 11, it is possible for a transverse beam to be pulled up as a result of the central loading. This is indicated by the negative value of the deflection. The longitudinal beam thus serves to decrease transverse beam deflection only if it is sufficiently rigid.

Chapter 9—Problems

Secs. 9.1 to 9.3

9.1.   A very long S127×15 steel I-beam, 0.127 m deep, resting on a foundation for which $k = 1.4$ MPa, is subjected to a concentrated load at midlength. The flange is 0.0762 m wide, and the cross-sectional moment of inertia is 5.04×10^{-6} m^4. What is the maximum load that can be applied to the beam without causing the elastic limit to be exceeded? Assume that $E = 200$ GPa and $\sigma_{yp} = 210$ MPa.

9.2.   A long beam on an elastic foundation is subjected to a sinusoidal loading $p = p_1 \sin(2\pi x/L)$, where p_1 and L are the peak intensity and wavelength of loading, respectively. Determine the equation of the elastic deflection curve in terms of k and β.

9.3.   If point A is taken to the right of the loaded portion of the beam shown in Fig. 9.3, what is the deflection at this point?

9.4.   A single train wheel exerts a load of 135 KN upon a rail, assumed to be supported by an elastic foundation. For a modulus of foundation $k = 16.8$

MPa, determine the maximum deflection and maximum bending stress in the rail. The respective values of the section modulus and modulus of rigidity are $Z = 3.9 \times 10^{-4}$ m^3 and $EI = 8.437$ MN·m^2.

9.5.   Calculate the maximum resultant bending moment and deflection in the rail of Problem 9.4 if two wheel loads spaced 1.66 m apart act on the rail. The remaining conditions of the problem are unchanged.

9.6.   Determine the deflection at any point A under the triangular loading acting on an infinite beam on an elastic foundation (Fig. P9.6).

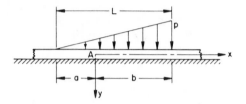

Figure P9.6

9.7.   What are the reactions acting on a semi-infinite beam built in at the left end and subjected to a uniformly distributed loading p? Use the method of superposition. [Hint: At a large distance from the left end, the deflection is p/k.]

9.8.   A semi-infinite beam on an elastic foundation is hinged at the left end and subjected to a moment M_L at that end. Determine the equation of the deflection curve, slope, moment, and shear force.

Secs. 9.4 to 9.8

9.9.   A machine base consists partly of an 5.4-m long S127×15 steel I-beam supported by coil springs spaced $a = 0.625$ m apart. The constant for each spring is $K = 180$ kN/m. The moment of inertia of the I-section is 5.4×10^{-6} m^4, the depth is 0.127 m, and the flange width is 0.0762. Assuming that a concentrated force of 6.75 kN transmitted from the machine acts at midspan, determine the maximum deflection, maximum bending moment, and maximum stress in the beam.

9.10.   A steel beam of 0.75-m length and 0.05-m square cross section is supported on three coil springs spaced $a = 0.375$ m apart. For each spring, $K = 18$ kN/m. Determine (a) the deflection of the beam if a load $P = 540$ N is applied at midspan; (b) the deflection at the ends of the beam if a load $P = 540$ N acts 0.25 m from the left end.

9.11.   A finite beam with $EI = 8.4$ MN·m^2 rests on an elastic foundation for which $k = 14$ MPa. The length L of the beam is 0.6 m. If the beam is subjected to a concentrated load $P=4.5$ kN at its midpoint, determine the maximum deflection and slope.

9.12. A finite beam is subjected to a concentrated force $P=9$ kN at its midlength and a uniform loading $p=750$ N/m. Determine the maximum deflection and slope if $L=0.15$ m, $EI=8.4$ MN·m^2, and $k=14$ MPa.

9.13. Redo Example 9.5 for the case in which both ends of the beam are simply supported.

9.14. Assume that all the data of Example 9.6 are unchanged except that a uniformly distributed load p replaces the concentrated force on the longitudinal beam. Compute the load R_{CC} supported by the central transverse beam.

Chapter 10

Energy Methods

10.1 Introduction

As an alternative to the methods based upon differential equations as outlined in Sec. 3.1, the analysis of stress and deformation can be accomplished through the use of energy methods. The latter are predicated upon the fact that the equations governing a given stress or strain configuration are derivable from consideration of the minimization of energy associated with deformation, stress, or deformation and stress. Applications of energy methods are effective in situations involving a variety of shapes and variable cross sections, and in complex problems involving elastic stability and multielement structures. In particular, strain energy methods offer concise and relatively simple approaches for computation of the displacements of slender structural and machine elements subjected to combined loading.

We shall deal with two principal energy methods.* The first is concerned with the finite deformation experienced by an element under load (Secs. 10.2 to 10.7). The second relies upon a hypothetical or *virtual* variation in stress or deformation, and represents one of the so-called variational methods (Secs. 10.8 to 10.10).

10.2 Work Done in Deformation

Consider a set of forces (applied forces and reactions) $P_k(k=1,2,\ldots,m)$, acting on an elastic body (Fig. 10.1). Let the displacement in the direction of P_k of the point at which the force P_k is applied be designated e_k. It is clear that e_k is attributable to the action of the entire force set, and not to P_k alone. Suppose that all the forces are applied statically, and let the final

---

*For more details, see H. L. Langhaar, *Energy Methods in Applied Mechanics*, New York: Wiley, 1962; I. S. Sokolnikoff, *Mathematical Theory of Elasticity*, New York: McGraw-Hill, 1956, Chapter 7.

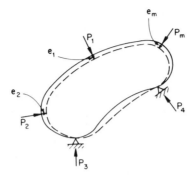

Figure 10.1

values of load and displacement be designated P_k and e_k. Based upon the linear relationship of load and deflection, the work W done by the external force system in deforming the body is given by $\frac{1}{2}\Sigma P_k e_k$. If no energy is dissipated during loading (which is certainly true of a conservative system), we may equate the work done on the body to the *strain energy* U gained by the body:

$$U = W = \frac{1}{2} \sum_{k=1}^{m} P_k e_k \tag{10.1}$$

While the force set P_k $(k = 1, 2, \ldots, m)$ includes applied forces and reactions, it is noted that the support displacements are zero, and therefore the support reactions do no work and do not contribute to the above summation. Equation (10.1) states simply that the work done by the forces acting on the body manifests itself as elastic strain energy. To further explore this concept, consider the body as a combination of small cubic elements. Owing to surface loading, the faces of an element are displaced, and stresses acting on these faces do work equal to the strain energy stored in the element. Consider two adjacent elements within the body. The work done by the stresses acting on two contiguous internal faces is equal but of opposite algebraic sign. We conclude therefore that the work done on all adjacent faces of the elements will cancel. All that remains is the work done by the stresses acting on the faces that lie on the surface of the body. As the internal stresses balance the external forces at the boundary, the work, whether expressed in terms of external forces (W) or internal stresses (U), is the same.

10.3   The Reciprocity Theorem

Consider now two sets of applied forces and reactions: P'_k $(k = 1, 2, \ldots, m)$, set 1; P''_j $(j = 1, 2, \ldots, n)$, set 2. If *only* the first set is applied, the strain energy is, from Eq. (10.1),

$$U_1 = \frac{1}{2} \sum_{k=1}^{m} P_k' e_k' \tag{a}$$

where e_k' are the displacements corresponding to the set P_k'. Application of *only* set 2 results in the strain energy

$$U_2 = \frac{1}{2} \sum_{j=1}^{n} P_j'' e_j'' \tag{10.2}$$

where e_j'' corresponds to the set P_j''.

Suppose that the first force system P_k' is applied, *followed* by the second force system P_j''. The total strain energy is

$$U = U_1 + U_2 + U_{1,2} \tag{b}$$

where $U_{1,2}$ is the strain energy attributable to the work done by the first force system as a result of deformations associated with the application of the second force system. Because the forces comprising the first set are unaffected by the action of the second set, we may write

$$U_{1,2} = \sum_{k=1}^{m} P_k' e_k'' \tag{10.3}$$

Here e_k'' represents the displacements caused by the forces of the second set at the points of application of P_k', the first set. If now the forces are applied in reverse order, we have

$$U = U_2 + U_1 + U_{2,1} \tag{c}$$

where

$$U_{2,1} = \sum_{j=1}^{n} P_j'' e_j' \tag{10.4}$$

Here e_j' represents the displacements caused by the forces of set 1 at the points of application of the forces P_j'', set 2.

The loading processes described above must, according to the principle of superposition, cause identical stresses within the body. The strain energy must therefore be independent of the order of loading, and it is concluded from Eqs. (b) and (c) that $U_{1,2} = U_{2,1}$. We thus have

$$\sum_{k=1}^{m} P_k' e_k'' = \sum_{j=1}^{n} P_j'' e_j' \tag{10.5}$$

The above expression is the *reciprocity* or *reciprocal theorem* due to E. Betti and Lord Rayleigh: the work done by one set of forces owing to displacements due to a second set, is equal to the work done by the second system of forces owing to displacements due to the first.

The utility of the reciprocal theorem lies principally in its application to the derivation of various approaches rather than as a method in itself.

10.4 Castigliano's Theorem

First formulated in 1879, Castigliano's theorem is in widespread use because of the ease with which it is applied to a variety of problems involving the deformation of structural elements, especially those classed as statically indeterminate. There are two theorems due to Castigliano. In this section we discuss the one restricted to structures composed of *linearly elastic* materials, i.e., those obeying Hooke's law. For these materials, the strain energy is equal to the complementary energy: $U = U^*$. In Sec. 10.8, another form of Castigliano's theorem is introduced which is appropriate to structures that behave *nonlinearly* as well as linearly. Both theorems are valid for the cases where any change in structure geometry owing to loading is so small that the action of the loads is not affected.

Refer again to Fig. 10.1, which shows an elastic body subjected to applied forces and reactions, P_k ($k = 1, 2, \ldots, m$). This set of forces will be designated set 1. Now let one force of set 1, P_i, experience an infinitesimal increment ΔP_i. We designate as set 2 the increment ΔP_i. According to the reciprocity theorem, Eq. (10.5), we may write

$$\sum_{k=1}^{m} P_k \Delta e_k = \Delta P_i e_i \qquad (a)$$

where Δe_k is the displacement in the direction and, at the point of application, of P_k attributable to the forces of set 2, and e_i is the displacement in the direction and, at the point of application, of P_i due to the forces of set 1.

The incremental strain energy $\Delta U = \Delta U_2 + \Delta U_{1,2}$ associated with the application of ΔP_i is, from Eqs. (10.2) and (10.3), $\Delta U = \frac{1}{2} \Delta P_i \Delta e_i + \Sigma P_k \Delta e_k$. Substituting Eq. (a) into the above, we have $\Delta U = \frac{1}{2} \Delta P_i \Delta e_i + \Delta P_i e_i$. Now divide this expression by ΔP_i and take the limit as the force ΔP_i approaches zero. In the limit, the displacement Δe_i produced by ΔP_i vanishes, leaving

$$\frac{\partial U}{\partial P_i} = e_i \qquad (10.6)$$

The above is known as *Castigliano's second theorem*: For a linear structure, the partial derivative of the strain energy with respect to an applied force is equal to that component of displacement at the point of application of the force which is in the direction of the force.

It can similarly be demonstrated that

$$\frac{\partial U}{\partial C_i} = \theta_i \qquad (10.7)$$

where C_i and θ_i are, respectively, the couple (bending or twisting) moment and the associated angular rotation at a point.

In applying Castigliano's theorem, the strain energy must be expressed as a function of the load. For example, the expression for the strain energy in a straight or curved slender bar (Sec. 5.11) subjected to a number of common loads (axial force N, bending moment M, shearing force V, and twisting moment M_t) is, from Eqs. (2.30), (2.34), (5.45), and (5.47),

$$U = \int \frac{N^2 \, dx}{2EA} + \int \frac{M^2 \, dx}{2EI} + \int \frac{f_s V^2 \, dx}{2GA} + \int \frac{M_t^2 \, dx}{2GJ} \tag{b}$$

where the integrations are carried out over the length of the bar. Recall that the term given by the last integral is valid only for a circular cross-sectional area. The displacement at any point in the bar may then readily be found by applying Castigliano's theorem. Inasmuch as the force P is not a function of x, we can perform the differentiation of U with respect to P under the integral. In so doing the displacement is obtained in the following convenient form:

$$e = \frac{1}{EA} \int N \frac{\partial N}{\partial P} dx + \frac{1}{EI} \int M \frac{\partial M}{\partial P} dx + \frac{1}{GA} \int f_s V \frac{\partial V}{\partial P} dx$$

$$+ \frac{1}{GJ} \int M_t \frac{\partial M_t}{\partial P} dx \tag{c}$$

Similarly, an expression may be written for $\theta = \partial U / \partial C$. For a slender beam, as observed in Sec. 5.4, the contribution of the shear force V to the displacement is negligible.

When it is necessary to determine the deflection at a point at which no load acts, the problem is treated as follows. A fictitious load P (or C) is introduced at the point in question in the direction of the desired displacement e (or θ). The displacement is then found by applying Castigliano's theorem, setting $P = 0$ in the final result.

Example 10.1. Determine the slope of the elastic curve at the left support of the uniformly loaded beam shown in Fig. 10.2.

SOLUTION. As a slope is sought, a fictitious couple moment C is introduced at point A. Applying the equations of statics, the reactions are found to be

$$R_A = \frac{15}{32} pL - \frac{C}{L}, \qquad R_B = \frac{25}{32} pL + \frac{C}{L}$$

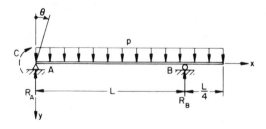

Figure 10.2

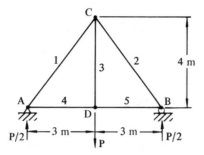

Figure 10.3

The following expressions for the moments are thus available:

$$M_1 = -\left(\frac{15}{32}pL - \frac{C}{L}\right)x + \frac{px^2}{2} - C \qquad (0 \leqslant x \leqslant L)$$

$$M_2 = -\left(\frac{15}{32}pL - \frac{C}{L}\right)x + \frac{px^2}{2} - C - \left(\frac{25}{32}pL + \frac{C}{L}\right)(x - L) \qquad \left(L \leqslant x \leqslant \frac{5}{4}L\right)$$

The slope at A is now found from Eq. (10.7):

$$\theta_A = \frac{1}{EI}\int_0^L \left(-\frac{15}{32}pLx + \frac{Cx}{L} + \frac{px^2}{2} - C\right)\left(\frac{x}{L} - 1\right)dx + \frac{1}{EI}\int_L^{5L/4} M_2(0)\,dx$$

Integrating and setting $C = 0$, we obtain $\theta_A = 7pL^3/192\,EI$.

Example 10.2. The simple pin connected truss shown in Fig. 10.3 supports a force P. If all members are of equal rigidity EA, what is the deflection of point D?

SOLUTION.   Applying the method of joints at points A and C and taking symmetry into account, we obtain $N_1 = N_2 = 5P/8$, $N_4 = N_5 = 3P/8$, and $N_3 = P$. Castigliano's theorem, $e_D = (1/EA)\Sigma N_i(\partial N_i/\partial P)L_i$, substituting the above values of axial forces in terms of applied load, leads to

$$e_D = \frac{1}{EA}\left[2\left(\frac{5}{8}P\right)\left(\frac{5}{8}\right)5 + P(1)4 + 2\left(\frac{3}{8}P\right)\left(\frac{3}{8}\right)3\right]$$

from which $e_D = 35p/4EA$.

Example 10.3 A piping system *expansion loop* is fabricated of pipe of constant size and subjected to a temperature differential (ΔT), Fig. 10.4. The overall length of the loop and the coefficient of thermal expansion of the tubing material are L and α, respectively. Determine, for each end of the loop, the restraining bending moment M and force N induced by the temperature change.

SOLUTION.   In labeling the end points for each segment the symmetry about a vertical axis through point A is taken into account, as shown in the figure. Expressions for the moments, associated with segments DC, CB, and BA, are

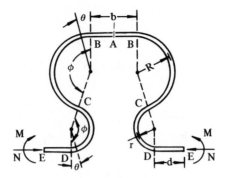

Figure 10.4

respectively,

$$M_1 = Nr(1 - \cos \theta) - M$$

$$M_2 = N[r + (r + R)(-\cos \phi) + R \cos \theta] - M \qquad \text{(d)}$$

$$M_3 = N[r + R + (r + R)(-\cos \phi)] - M$$

Upon application of Eqs. (10.6) and (10.7), the end deflection and end slope are found to be

$$e_{EN} = \frac{2}{EI}\left[\int_0^\phi M_1 \frac{\partial M_1}{\partial N} \cdot r\, d\theta + \int_0^\phi M_2 \frac{\partial M_2}{\partial N} \cdot R\, d\theta + M_3 \frac{\partial M_3}{\partial N} \cdot \frac{b}{2} \right] \qquad \text{(e)}$$

$$0 = \frac{2}{EI}\left[\int_0^\phi M_1 \frac{\partial m_1}{\partial M} \cdot r\, d\theta + \int_0^\phi M_2 \frac{\partial M_2}{\partial M} \cdot R\, d\theta + M_3 \frac{\partial M_3}{\partial M} \cdot \frac{b}{2} + M \frac{\partial M}{\partial M} \cdot a \right] \qquad \text{(f)}$$

Substitution of Eqs. (d) into Eq. (e) results in

$$e_{EN} = \frac{2N}{EI}\left\{ r^3\left(\frac{3}{2}\phi - 2\sin\phi + \frac{1}{4}\sin 2\phi \right) \right.$$

$$+ rR[r\phi - 2\phi\cos\phi(r + R) + 2R\sin\phi]$$

$$+ (r + R)^2\left[\frac{b}{2}(1 - \cos\phi) + R\phi\cos^2\phi \right]$$

$$\left. + R^2\left(\frac{1}{2}R\phi - \frac{3}{4}R\sin\phi - r\sin^2\phi \right) \right\}$$

$$- \frac{2M}{EI}\left\{ r^2(\phi - \sin\phi) + R^2[\sin\phi - \phi\cos\phi] \right.$$

$$\left. + rR\phi(-\cos\phi) + \frac{b}{2}(r + R)(1 - \cos\phi) \right\} \qquad \text{(g)}$$

Similarly, Eq. (f) and (d) lead to

$$M = \frac{1}{a + (b/2) + N(r + R)}\left\{ N\left[(r^2 + rR)\phi + \sin\phi(R^2 - r^2) \right.\right.$$

$$\left.\left. + \frac{b}{2}(r + R) - \cos\phi(r + R)\left(\frac{b}{2} + R\phi \right) \right] \right\} \qquad \text{(h)}$$

The deflections at E owing to the temperature variation and end restraints must be equal, i.e.,

$$e_{EN} = E_{ET} = \alpha(\Delta T)L \tag{i}$$

Expressions (g), (h), and (i) are then solved to yield the unknown reactions N and M in terms of the given material properties and loop dimensions.

10.5   The Unit or Dummy Load Method

Recall that the deformation at a point in an elastic body subjected to external loading P_i, expressed in terms of the moment produced by the force system, is, according to Castigliano's theorem,

$$e_i = \frac{\partial U}{\partial P_i} = \int \frac{1}{EI} M \frac{\partial M}{\partial P_i} dx \tag{a}$$

For small deformations of linearly elastic materials, the moment is linearly proportional to the external loads, and consequently we are justified in writing $M = mP$, m denoting a constant. It follows that $\partial M/\partial P_i = m$, the change in the bending moment per unit change in P_i, i.e., the moment caused by a unit load. The foregoing considerations lead to the so-called *unit* or *dummy load* approach, which finds extensive application in structural analysis. From Eq. (a),

$$e_i = \int \frac{Mm}{EI} dx \tag{10.8}$$

In a similar manner, the following expression is obtained for the change of slope:

$$\theta_i = \int \frac{Mm'}{EI} dx \tag{10.9}$$

Here $m' = \partial M/\partial C_i$ represents the change in the bending moment per unit change in C_i, i.e., the change in bending moment caused by a unit couple moment.

Example 10.4. Derive an expression for the deflection of point C of the simply supported beam shown in Fig. 10.5a.

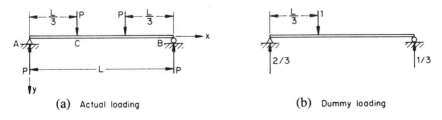

(a) Actual loading   (b) Dummy loading

Figure 10.5

SOLUTION. Figure 10.5b shows the dummy load of 1 N and the reactions it produces. Note that the unit load is applied at C because it is the deflection of C that is required. Referring to the figure, the following moment distributions are obtained:

$$M_1 = -Px, \qquad m_1 = -\tfrac{2}{3}x \qquad \left(0 \leqslant x \leqslant \tfrac{L}{3}\right)$$

$$M_2 = -Px + P\left(x - \frac{L}{3}\right) = -\frac{PL}{3}, \qquad m_2 = -\frac{2}{3}x + 1\left(x - \frac{L}{3}\right) = -\frac{L-x}{3}$$

$$\left(\frac{L}{3} \leqslant x \leqslant \frac{2L}{3}\right)$$

$$M_3 = -Px + P\left(x - \frac{L}{3}\right) + P\left(x - \frac{2L}{3}\right) = -P(L-x), \qquad m_3 = -\tfrac{1}{3}(L-x)$$

$$\left(\frac{2L}{3} \leqslant x \leqslant L\right)$$

In the above, the M's refer to Fig. 10.4a, and the m's to Fig. 10.4b. The vertical deflection at C is then, from Eq. (10.8),

$$e_C = \frac{1}{EI}\left[\int_0^{L/3} Px\left(\frac{2x}{3}\right) dx + \int_{L/3}^{2L/3} \frac{PL}{3}\left(\frac{L-x}{3}\right) dx \right.$$

$$\left. + \int_{2L/3}^{L} P(L-x)\left(\frac{L-x}{3}\right) dx \right]$$

The solution, after integration, is found to be $e_C = 5PL^3/162\,EI$.

10.6   The Crotti–Engesser Theorem

Consider a set of forces acting on a structure that behaves *nonlinearly*. Let the displacement of the point at which the force P_i is applied, in the direction of P_i, be designated e_i. This displacement is to be determined. The problem is the same as that stated in Sec. 10.4, but now it will be expressed in terms of P_i and the complementary energy U^* of the structure, the latter being given by Eq. (2.23). In deriving the theorem, a procedure is employed similar to that given in Sec. 10.4. Thus, U is replaced by U^* in Castigliano's second theorem, Eq. (10.6), to obtain

$$\frac{\partial U^*}{\partial P_i} = e_i \tag{10.10}$$

This equation is known as the *Crotti–Engesser theorem*: The partial derivative of the complementary energy with respect to an applied force is equal to that component of the displacement at the point of application of the force which is in the direction of the force. Obviously, here the complementary energy must be expressed in terms of the loads.

Example 10.5. A simple truss, constructed of pin-connected members 1 and 2, is subjected to a vertical force P at joint B, Fig. 10.6a. The bars are made of a nonlinearly elastic material displaying the stress–strain relation $\sigma = K\varepsilon^{1/2}$ equally in

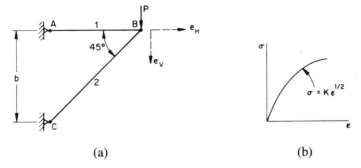

Figure 10.6

tension and compression (Fig. 10.6b). Here K is a constant. The cross-sectional area of each member is A. Determine the vertical deflection of joint B.

SOLUTION. The volume of member 1 is Ab, and that of member 2 is $Ab\sqrt{2}$. The total complementary energy of the structure is therefore

$$U^* = \left(U_{01}^* + U_{02}^* \sqrt{2} \right) Ab \tag{a}$$

The complementary energy densities are, from Eq. (2.23),

$$U_{01}^* = \int_0^{\sigma_1} \frac{\sigma^2}{K^2} d\sigma = \frac{\sigma_1^3}{3K^2}$$

$$U_{02}^* = \int_0^{\sigma_2} \frac{\sigma^2}{K^2} d\sigma = \frac{\sigma_2^3}{3K^2} \tag{b}$$

where σ_1 and σ_2 are the stresses in bars 1 and 2. Upon introduction of Eqs. (b) into (a), we have

$$U^* = \frac{Ab}{3K^2} \left(\sigma_1^3 + \sqrt{2}\, \sigma_2^3 \right) \tag{c}$$

From static equilibrium, the axial forces in 1 and 2 are found to be P and $P\sqrt{2}$, respectively. Thus, $\sigma_1 = P/A$ (tension) and $\sigma_2 = \sqrt{2}\, P/A$ (compression), which when introduced into Eq. (c) yields $U^* = 5P^3 b/3A^2 K^2$. Applying Eq. (10.10), the vertical deflection of B is found to be

$$e_V = \frac{5P^2 b}{A^2 K^2}$$

Another approach to the solution of this problem is given in Example 10.7.

10.7 Statically Indeterminate Systems

To supplement the discussion of statically indeterminate systems given in Sec. 5.9, energy methods are now applied to obtain the unknown, redundant forces (moments) in such systems. Consider, for example, the beam system of Fig. 10.2 rendered statically indeterminate by the addition of an extra or redundant support at the right end (not shown in the figure). The strain energy is, as before, written as a function of all external forces, including both applied loads and reactions. Castigliano's theorem may then be applied to derive an expression for the deflection at point B, which is clearly zero:

$$\frac{\partial U}{\partial R_b} = e_B = 0 \tag{a}$$

The above expression and the two equations of statics available for this force system provide the three equations required for determination of the three unknown reactions. Extending this reasoning to the case of a statically indeterminate beam with n redundant reactions, we write

$$\frac{\partial U}{\partial R_n} = 0 \tag{b}$$

The equations of statics together with the equations of the type given by Eq. (b) constitute a set sufficient for solution of all the reactions. This basic concept is fundamental to the analysis of structures of considerable complexity.

Example 10.6. The built-in beam shown in Fig. 10.7 is supported at one end by a spring of constant k. Determine the redundant reaction.

SOLUTION. The expressions for the moments are

$$M_1 = -R_D x \qquad \left(0 \leqslant x \leqslant \frac{L}{2}\right)$$

$$M_2 = -R_D x + P\left(x - \frac{L}{2}\right) \qquad \left(\frac{L}{2} \leqslant x \leqslant L\right)$$

Applying Castigliano's theorem to obtain the deflection at point D, $e_D = \partial U/\partial R_D$, we have

$$e_D = \frac{1}{EI}\int_0^{L/2} R_D x(x\,dx) + \frac{1}{EI}\int_{L/2}^{L}\left[R_D x - P\left(x - \frac{L}{2}\right)\right]x\,dx = \frac{R_D}{k}$$

from which $R_D = \frac{5}{16}P/(1 - 3EI/kL)$. Equilibrium of vertical forces yields

$$R_A = P - \frac{5P}{16(1 - 3EI/kL^3)}$$

Note that were the right end rigidly supported, e_D would be equated to zero.

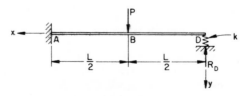

Figure 10.7

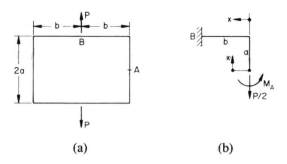

Figure 10.8

Example 10.7. A rectangular frame of constant EI is loaded as shown in Fig. 10.8a. Assuming the strain energy to be attributable to bending alone, determine the increase in distance between the points of application of the load.

SOLUTION. The situation described is statically indeterminate. For reasons of symmetry we need analyze only one quadrant. Because the slope B is zero before and after application of the load, the segment may be treated as fixed at B, Fig. 10.8b. The moment distributions are

$$M_1 = -M_A \qquad (0 \leqslant x \leqslant a)$$

$$M_2 = -M_A + \tfrac{1}{2}Px \qquad (0 \leqslant x \leqslant b)$$

Since the slope is zero at A, we have

$$\theta_A = \frac{1}{EI} \int M \frac{\partial M}{\partial M_A} dx = \frac{1}{EI}\left[\int_0^a M_A\, dx + \int_0^b \left(M_A - \frac{1}{2}Px \right) dx \right] = 0$$

from which $M_A = Pb^2/4(a+b)$. The relative displacement between the points of application of load is then, by applying Castigliano's theorem, found to be $e = Pb^3(4a+b)/12\,EI(a+b)$.

10.8   The Principle of Virtual Work

In this section a second type of energy approach is explored, based upon a hypothetical variation of deformation. This method, as is demonstrated, lends itself to the expeditious solution of a variety of important problems.

Consider a body in equilibrium under a system of forces. Accompanying a small displacement within the body we expect a change in the original force system. Suppose now that an *arbitrary* incremental displacement occurs, termed a *virtual displacement*. This displacement need not actually take place and need not be infinitesimal. If we assume the displacement to be infinitesimal, as is usually done, it is reasonable to regard the system of forces as unchanged.

Recall from *particle* mechanics that for a point mass, *unconstrained* and thereby free to experience arbitrary virtual displacements δu, δv, δw, the *virtual work* accompanying these displacements is $\Sigma F_x \delta u$, $\Sigma F_y \delta v$, $\Sigma F_z \delta w$, where ΣF_x, ΣF_y, ΣF_z are the force resultants. If the particle is in equilibrium, it follows that the *virtual work must vanish*, since $\Sigma F_x = \Sigma F_y = \Sigma F_z = 0$. This is the principle of virtual work.

For an elastic body, it is necessary to impose a number of restrictions on the arbitrary virtual displacements. To begin with, these displacements must be continuous and their derivatives must exist. In this way, material continuity is assured. Because certain displacements on the boundary may be dictated by the circumstances of a given situation (boundary conditions), the virtual displacements at such points on the boundary must be zero. A virtual displacement results in no alteration in the magnitude or direction of external and internal forces. The imposition of a virtual displacement field upon an elastic body does, however, result in the imposition of an increment in the strain field.

To determine the virtual strains, replace the displacements u, v, and w by virtual displacements δu, δv, and δw in the definition of the actual strains, Eq. (2.3):

$$\delta\varepsilon_x = \frac{\partial}{\partial x}\delta u, \ldots, \delta\gamma_{xy} = \frac{\partial}{\partial y}\delta u + \frac{\partial}{\partial x}\delta v, \ldots \qquad (10.11)$$

The strain energy δU acquired by a body of volume V as a result of virtual straining is, by application of Eq. (2.29) together with the second of Eqs. (2.28):

$$\delta U = \int_V \left(\sigma_x\delta\varepsilon_x + \sigma_y\delta\varepsilon_y + \sigma_z\delta\varepsilon_z + \tau_{xy}\delta\gamma_{xy} + \tau_{yz}\delta\gamma_{yz} + \tau_{xz}\delta\gamma_{xz}\right)dV \qquad (10.12)$$

Note the absence in the above equation of any term involving a variation in stress. This is attributable to the assumption that the *stress remains constant* during application of virtual displacement.

The variation in strain energy may be viewed as the work done against the mutual actions between the infinitesimal elements composing the body, owing to the virtual displacements (Sec. 10.2). The virtual work done in an elastic body by these mutual actions is therefore $-\delta U$.

Consider next the virtual work done by *external* forces. Again suppose that the body experiences virtual displacements δu, δv, δw. The virtual work done by a body force $\mathbf{F}$ per unit volume and a surface force $\mathbf{T}$ per unit area is

$$\delta W = \int_V \left(F_x\delta u + F_y\delta v + F_z\delta w\right)dV + \int_A \left(T_x\delta u + T_y\delta v + T_z\delta w\right)dA \qquad (10.13)$$

where A is the boundary surface.

We have already stated that the total work done during the virtual displacement is zero: $\delta W - \delta U = 0$. The *principle of virtual work* for an elastic body is therefore expressed as follows:

$$\delta W = \Delta U \qquad (10.14)$$

As the virtual displacements result in no geometric alteration of the body, and as the external forces are regarded as constants, Eq. (10.14) may be rewritten

$$\delta\Pi = \delta\left[U - \int_A (T_x u + T_y v + T_z w)\, dA - \int_V (F_x u + F_y v + F_z w)\, dV \right] = 0$$
(10.15a)

or briefly,

$$\delta\Pi = \delta(U - W) = 0 \qquad (10.15b)$$

where it is noted that δ has been removed from under the integral sign. The term $\Pi = U - W$ is called the *potential energy*, and Eq. (10.15) represents a condition of *stationary* potential energy of the system. It can be demonstrated that, for *stable equilibrium*, the potential energy is a minimum. Only for displacements which satisfy the boundary conditions and the equilibrium conditions will Π assume a minimum value. This is called the *principle of minimum potential energy*.

Consider now the case in which the loading system consists only of forces applied at points on the surface of the body, denoting each point force by P_i and the displacement in the direction of this force by e_i (corresponding to the equilibrium state). From Eq. (10.15), we have

$$\delta(U - P_i e_i) = 0 \qquad \text{or} \qquad \delta U = P_i \delta e_i$$

The principle of minimum potential energy thus leads to

$$\frac{\partial U}{\partial e_i} = P_i \qquad (10.16)$$

meaning that the partial derivative of the strain energy with respect to a displacement e_i equals the force acting in the direction of e_i at the point of application of P_i. Equation (10.16) is known as *Castigliano's first theorem*. This theorem, as with the Crotti–Engesser theorem, may be applied to any structure, linear or nonlinear.

Example 10.8. Determine the vertical displacement e_V and the horizontal displacement e_H of the joint B of the truss described in Example 10.5.

SOLUTION. First introduce the unknown vertical and horizontal displacements at the joint shown by dashed lines in Fig. 10.6a. Under the influence of e_V, member 1 does not deform, while member 2 is contracted by $e_V/2b$ per unit length. Under the influence of e_H, member 1 elongates by e_H/b, and member 2 by $e_H/2b$, per unit length. The strains produced in members 1 and 2 under the effect of both displacements are then calculated from

$$\varepsilon_1 = \frac{e_H}{b}, \qquad \varepsilon_2 = \frac{e_V - e_H}{2b} \qquad (a)$$

where ε_1 is an elongation and ε_2 is a shortening. Members 1 and 2 have volumes Ab and $Ab\sqrt{2}$, respectively. Next, the total strain energy of the truss, from Eq. (a) of

Sec. 2.8, is determined as follows:

$$U = Ab \int_0^{\varepsilon_1} \sigma \, d\varepsilon + Ab\sqrt{2} \int_0^{\varepsilon_2} \sigma \, d\varepsilon$$

Upon substituting $\sigma = K\varepsilon^{1/2}$ and Eqs. (a), and integrating, the above becomes

$$U = \frac{AK}{3\sqrt{b}} \left[2e_H^{3/2} + (e_V - e_H)^{3/2} \right]$$

Now we apply Castigliano's theorem in the horizontal and vertical directions at B, respectively:

$$\frac{\partial U}{\partial e_H} = \frac{AK}{2\sqrt{b}} \left[2e_H^{1/2} - (e_V - e_H)^{1/2} \right] = 0$$

$$\frac{\partial U}{\partial e_V} = \frac{AK}{2\sqrt{b}} (e_V - e_H)^{1/2} = P$$

Simplifying and solving these expressions simultaneously, the joint displacements are found to be

$$e_H = \frac{P^2 b}{A^2 K^2}, \qquad e_V = \frac{5P^2 b}{A^2 K^2} \tag{b}$$

The stress–strain law, together with Eqs. (a) and (b), yields the stresses in the members if required:

$$\sigma_1 = K\varepsilon_1^{1/2} = \frac{P}{A}, \qquad \sigma_2 = K\varepsilon_2^{1/2} = \frac{\sqrt{2}\,P}{A}$$

The axial forces are therefore

$$N_1 = \sigma_1 A = P, \qquad N_2 = \sigma_2 A = \sqrt{2}\,P$$

Here N_1 is tensile and N_2 is compressive. It is noted that for the statically determinate problem under consideration, the axial forces could readily be determined from static equilibrium. The solution procedure given above applies similarly to statically indeterminate structures as well as to linearly elastic structures.

10.9   Application of Trigonometric Series

Certain problems in the analysis of structural deformation, mechanical vibration, heat transfer, etc. are amenable to solution by means of trigonometric series. This approach offers as an important advantage the fact that a single expression may apply to the entire length of the member. The method is now illustrated using the case of a simply supported beam subjected to a moment at point A (Fig. 10.9a).

The deflection curve can be represented by a *Fourier* sine series:

$$v = a_1 \sin \frac{\pi x}{L} + a_2 \sin \frac{2\pi x}{L} + \cdots = \sum_{n=1}^{\infty} a_n \sin \frac{n\pi x}{L} \tag{a}$$

The end conditions of the beam ($v = 0$, $v'' = 0$ at $x = 0$, $x = L$) are observed to be satisfied by each term of this infinite series. The first and second

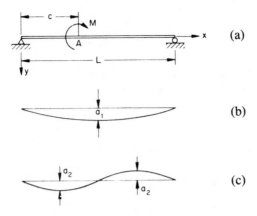

Figure 10.9

terms of the series are represented by the curves in Fig. 10.9b and c, respectively. As a physical interpretation of Eq. (a), consider the true deflection curve of the beam to be the superposition of sinusoidal curves of n different configurations. The coefficients a_n of the series are the maximum coordinates of the sine curves, and the n's indicate the number of half waves in the sine curves. It is demonstrable that, when the coefficients a_n are determined properly, the series given by Eq. (a) can be used to represent any deflection curve.* By increasing the number of terms in the series, the accuracy can, of course, be improved. To evaluate the coefficients, the principle of virtual work will be applied. The strain energy of the system, from Eqs. (5.45) and (a), is written

$$U = \frac{EI}{2} \int_0^L \left(\frac{d^2 v}{dx^2} \right)^2 dx = \frac{EI}{2} \int_0^L \left[\sum_{n=1}^{\infty} a_n \left(\frac{n\pi}{L} \right)^2 \sin \frac{n\pi x}{L} \right]^2 dx \qquad \text{(b)}$$

Expanding the term in brackets,

$$\left[\sum_{n=1}^{\infty} a_n \left(\frac{n\pi}{L} \right)^2 \sin \frac{n\pi x}{L} \right]^2 = \sum_{m=1}^{\infty} \sum_{n=1}^{\infty} a_m a_n \left(\frac{m\pi}{L} \right)^2 \left(\frac{n\pi}{L} \right)^2 \sin \frac{m\pi x}{L} \sin \frac{n\pi x}{L}$$

Since for the orthogonal functions $\sin (m\pi x/L)$ and $\sin (n\pi x/L)$ it can be shown by direct integration that

$$\int_0^L \sin \frac{m\pi x}{L} \sin \frac{n\pi x}{L} dx = \begin{cases} 0 & (m \neq n) \\ L/2 & (m = n) \end{cases} \qquad (10.17)$$

---

*See, for example, I. S. Sokolnikoff and R. M. Redheffer, *Mathematics of Physics and Modern Engineering*, New York: McGraw-Hill, 1966, Chapter 1.

Eq. (b) reduces to

$$U = \frac{\pi^4 EI}{4L^3} \sum_{n=1}^{\infty} n^4 a_n^2 \qquad (c)$$

The virtual work done by a moment M acting through a virtual rotation at A increases the strain energy of the beam by δU:

$$M\left(\delta \frac{\partial v}{\partial x}\right)_A = \delta U \qquad (d)$$

Therefore, from Eqs. (c) and (d), we have

$$M \sum_{n=1}^{\infty} \frac{n\pi}{L} \cos \frac{n\pi c}{L} \delta a_n = \frac{\pi^4 EI}{4L^3} \sum_{n=1}^{\infty} n^4 \delta a_n^2$$

which leads to

$$a_n = \frac{2ML^2}{\pi^3 EI} \sum_{n=1}^{\infty} \frac{1}{n^3} \cos \frac{n\pi c}{L}$$

Upon substitution of the above for a_n in the series given by Eq. (a), the equation for the deflection curve is obtained in the form

$$v = \frac{2ML^2}{\pi^3 EI} \sum_{n=1}^{\infty} \frac{1}{n^3} \cos \frac{n\pi c}{L} \sin \frac{n\pi x}{L}$$

Through the use of this infinite series, the deflection for any given value of x can be calculated.

Example 10.9. Derive an expression for the deflection of a cantilever beam of length L subjected to a concentrated force P at its free end.

SOLUTION. The origin of the coordinates is located at the fixed end. Let us represent the deflection by the infinite series

$$v = \sum_{n=1,3,5,\ldots}^{\infty} a_n\left(1 - \cos \frac{n\pi x}{2L}\right) \qquad (e)$$

It is clear that Eq. (e) satisfies the conditions related to the slope and deflection at $x=0$: $v=0$, $dv/dx=0$. The strain energy of the system is

$$U = \frac{EI}{2} \int_0^L \left(\frac{d^2 v}{dx^2}\right)^2 dx = \frac{EI}{2} \int_0^L \left[\sum_{n=1,3,5,\ldots}^{\infty} a_n\left(\frac{n\pi}{2L}\right)^2 \cos \frac{n\pi x}{2L}\right]^2 dx$$

Squaring the bracketed term above and noting that the orthogonality relationship yields

$$\int_0^L \cos \frac{m\pi x}{2L} \cos \frac{n\pi x}{2L} dx = \begin{cases} 0 & (m \neq n) \\ L/2 & (m = n) \end{cases} \qquad (10.18)$$

we obtain

$$U = \frac{\pi^4 EI}{64 L^3} \sum_{n=1,3,5,\ldots}^{\infty} n^4 a_n^2$$

Application of the principle of virtual work gives $P\delta v_E = \delta U$. Thus,

$$P \sum_{n=1,3,5,\ldots}^{\infty} \left(1 - \cos\frac{n\pi L}{2L}\right)\delta a_n = \frac{EI\pi^4}{64L^3} \sum_{n=1,3,5,\ldots}^{\infty} n^4 \,\delta a_n^2$$

from which $a_n = 32PL^3/n^4\pi^4 EI$. The beam deflection is obtained by substituting the value of a_n obtained above into Eq. (e). At $x = L$, disregarding terms beyond the first three, we obtain $v_{max} = PL^3/3.001EI$. The exact solution due to bending is $PL^3/3EI$ (Sec. 5.4).

10.10   The Rayleigh–Ritz Method

The so-called Rayleigh–Ritz method offers a convenient procedure for obtaining solutions by the principle of minimum potential energy. This method was originated by Lord Rayleigh, and a generalization contributed by W. Ritz. In this section, the application of the method to one-dimensional problems is discussed. The essentials of the method may be described as follows. First assume the solution in the form of a series containing unknown parameters a_n $(n = 1, 2, \ldots)$. This series function is such that it satisfies the *geometric* boundary conditions. These describe any end constraints pertaining to deflections and slopes. Another kind of condition, a *static* boundary condition, which equates the internal forces (and moments) at the edges of the member to prescribed external forces (and moments), need not be fulfilled. Next, using the assumed solution, determine the potential energy Π in terms of a_n. This indicates that the a_n's govern the variation of the potential energy. As the potential energy must be a minimum at equilibrium,

$$\frac{\partial \Pi}{\partial a_1} = 0, \ldots, \frac{\partial \Pi}{\partial a_n} = 0 \qquad (10.19)$$

The above represents a set of algebraic equations which are solved to yield the parameters a_n. Substituting these values into the assumed function, we obtain the solution for a given problem. In general, only a finite number of parameters can be employed, and the solution found is thus only approximate.

The method is illustrated in the following example. Application to buckling problems will be given in the next chapter.

Example 10.10. A simply supported beam of length L is subjected to uniform loading p per unit length. Determine the deflection $v(x)$ by employing (a) a power series and (b) a Fourier series.

SOLUTION.   Let the origin of the coordinates be placed at the left support.

(a) Assume a solution of polynomial form:

$$v = a_1 x(L - x) + a_2 x^2 (L - x)^2 + \cdots \qquad (a)$$

Note that this choice enables the deflection to vanish at either boundary. Consider now only the first term of the series:

$$v = a_1 x(L-x) \tag{b}$$

The corresponding potential energy, $\Pi = U - W$, is

$$\Pi = \int_0^L \left[\frac{EI}{2} \left(\frac{d^2 v}{dx^2} \right)^2 - pv \right] dx = \int_0^L \left[\frac{EI}{2} (-2a_1)^2 - a_1 px(L-x) \right] dx$$

From the minimizing condition, Eq. (10.19), we obtain $a_1 = pL^2/24EI$. The approximate displacement is therefore

$$v = \frac{pL^4}{24EI} \left(\frac{x}{L} - \frac{x^2}{L^2} \right)$$

which at midspan becomes $v_{max} = pL^4/96EI$. This result may be compared with the exact solution due to bending, $v_{max} = pL^4/76.8\,EI$, indicating an error in maximum deflection of roughly 17%. An improved approximation is obtained when two terms of the series given by Eq. (a) are retained. The same procedure as above now yields $a_1 = pL^2/24EI$ and $a_2 = p/24EI$, so that

$$v = \frac{pL^2}{24EI} [x(L-x)] + \frac{p}{24EI} \left[x^2(L-x)^2 \right]$$

At midspan, the above expression provides the exact solution. The foregoing is laborious and not considered practical when compared with the approach given below.

(b) Now suppose a solution of the form

$$v = \sum_{n=1}^{\infty} a_n \sin \frac{n\pi x}{L} \tag{c}$$

The boundary conditions are satisfied inasmuch as v and v'' both vanish at either end of the beam. We now substitute v and its derivatives into $\Pi = U - W$. Employing Eq. (10.17), we obtain, after integration,

$$\Pi = \frac{EI}{2} \left(\frac{L}{2} \right)^2 \sum_{n=1}^{\infty} a_n^2 \left(\frac{n\pi}{L} \right)^4 + p \left(\frac{L}{\pi} \right) \sum_{n=1}^{\infty} \frac{1}{n} a_n \cos \frac{n\pi x}{L} \Big|_0^L$$

Observe that if n is even, the second term vanishes. Thus

$$\Pi = \frac{\pi^4 EI}{4L^3} \sum_{n=1}^{\infty} a_n^2 n^4 - \frac{2pL}{\pi} \sum_{n=1,3,5,\ldots}^{\infty} \frac{a_n}{n}$$

and Eq. (10.19) yields $a_n = 4pL^4/EI(n\pi)^5$, $n = 1,3,5,\ldots$. The deflection at midspan is, from Eq. (c),

$$v_{max} = \frac{4pL^4}{EI\pi^5} \left(1 - \frac{1}{3^5} + \frac{1}{5^5} - \cdots \right)$$

Dropping all but the first term, $v_{max} = PL^4/76.6EI$. The exact solution is obtained when all terms in the series (c) are retained. Evaluation of all terms in the series may not always be possible, however.

It should be noted that the results obtained in this example, based upon only one or two terms of the series, are remarkably accurate. So few terms will not, in general, result in such accuracy when applying the Rayleigh–Ritz method.

Chapter 10—Problems

Secs. 10.1 to 10.7

10.1. A cantilever beam of constant flexural rigidity EI is loaded by forces P_H and P_V as shown in Fig. P10.1. Determine the vertical and horizontal deflections and the angular rotation of the free end. Employ Castigliano's theorem.

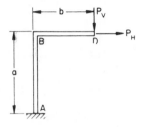

Figure P10.1

10.2. The truss shown in Fig. P10.2 supports concentrated forces of $P_1 = P_2 = P_3 = 45$ kN. Assuming all members are of the same cross section and material, find the vertical deflection of point B in terms of EI. Take $L = 3$ m. Use Castigliano's theorem.

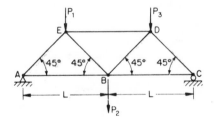

Figure P10.2

10.3. The moments of inertia of the tapered and constant area segments of the cantilever beam shown in Fig. P10.3 are given by $I_1 = (c_1 x + c_2)^{-1}$ and I_2, respectively. Determine the deflection of the beam under a load P. Use Castigliano's theorem.

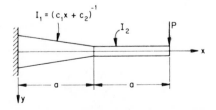

Figure P10.3

10.4. It is required to determine the horizontal deflection of point D of the frame shown in Fig. P10.4, subject to downward load F, applied at the top. The moment of inertia of segment BC is twice that of the remaining sections. Use Castigliano's theorem.

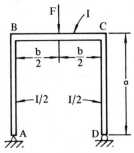

Figure P10.4

10.5. A steel spring of constant flexural rigidity is described by Fig. P10.5. If a force P is applied, determine the increase in the distance between the ends. Use Castigliano's theorem.

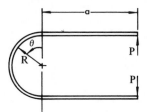

Figure P10.5

10.6. A cylindrical circular rod in the form of a quarter ring of radius R is fixed at one end (Fig. P10.6a). At the free end, a concentrated force P is applied in a diametral plane perpendicular to the plane of the ring. What is the deflection of the free end? Use unit load method. [Hint: At any section, $M_\theta = PR \sin \theta$ and $M_{t\theta} = PR(1 - \cos \theta)$.]

Figure P10.6

10.7. Redo Problem 10.6 if the curved bar is a *split circular ring* as shown in Fig. P10.6b.

10.8. Determine the deflection and slope at midspan of the beam described in Example 10.6.

10.9. A cantilever beam of length L subject to a linearly varying loading per unit length, having the value zero at the free end and p at the fixed end, is supported on a roller at its free end. Find the reactions using Castigliano's theorem.

10.10. The symmetrical frame shown in Fig. P.10.10 supports a uniform loading of p per unit length. Assume that each horizontal and vertical member has the modulus of rigidity E_1I_1 and E_2I_2, respectively. Determine the resultant reaction R_A at the left support, employing Castigliano's theorem.

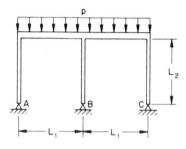

Figure P10.10

10.11. Forces P_m are applied to a *compound loop* or *link* of constant flexural rigidity EI (Fig. P10.11). Assuming that the dimension perpendicular to the plane of the page is small in comparison with radius R, and taking into account only the strain energy due to bending, determine the maximum moment.

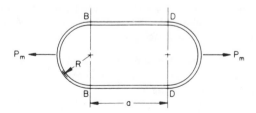

Figure P10.11

10.12. A large ring is loaded as shown in Fig. P10.12. Taking into account only the strain energy associated with bending, determine the bending moment and the force within the ring at the point of application of P. Employ the unit load method.

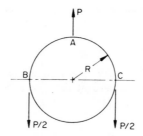

Figure P10.12

Secs. 10.8 to 10.10

10.13. Apply Castigliano's first theorem to compute the force P required to cause a vertical displacement of 5 mm in the hinge connected structure of Fig. P10.13. Let $\alpha = 45°$, $L_0 = 3$ m, and $E = 200$ GPa. The area of each member is 6.25×10^{-4} m^2.

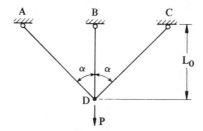

Figure P10.13

10.14. A hinge ended beam of length L rests on an elastic foundation and is subjected to a concentrated load P, located a distance c from the left end. Derive the equation of the deflection curve by applying the principle of virtual work.

10.15. A simply supported beam carries a load P at a distance of c away from its left end. Obtain the beam deflection at the point where P is applied. Use the Rayleigh–Ritz method. Assume a deflection curve of the form $v = ax(L - x)$, where a is to be determined.

10.16. Determine the midspan deflection for the fixed ended symmetrical beam of stepped section shown in Fig. P10.16. Take $v = a_1x^3 + a_2x^2 + a_3x + a_4$. Employ the Rayleigh–Ritz method.

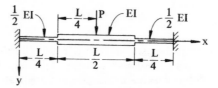

Figure P10.16

Chapter 11

Elastic Stability

11.1 Introduction

We have up to now dealt primarily with the prediction of stress and deformation in structural elements subject to various load configurations. Failure criteria have been based upon a number of theories relying on the attainment of a particular stress, strain, or energy level within the body. In this chapter it is demonstrated that the beginnings of structural failure may occur prior to the onset of any seriously high levels of stress. We thus explore failure owing to *elastic instability*, seeking to determine those conditions of load and geometry that lead to a compromise of structural integrity. In so doing, we shall deal only with beam and slender members subject to axial compression. The problem is essentially one of ascertaining those configurations of the system that lead to sustainable patterns of deformation. The principal difference between the theories of linear elasticity and linear stability is that in the former, equilibrium is based upon the undeformed geometry, whereas in the latter, the deformed geometry must be considered.

11.2 Critical Load

To demonstrate the concepts of stability and critical load, consider a rigid, weightless bar AB, shown in Fig. 11.1a. This member is pinned at B and acted upon by a force P. In the *absence* of restoring influences such as the spring shown, any small lateral disturbance (causing a displacement δ) will result in rotation of the bar, indicated by the dashed lines, with no possibility of return to the original configuration. *Without* the spring, therefore, the situation depicted in the figure is one of *unstable equilibrium*. With a spring present, different possibilities arise. A small *momentary* disturbance δ can now be sustained by the system (provided that P is also small) because the disturbing moment $P\delta$ is smaller than the restoring moment $k\delta L$ (where k represents the linear spring constant, force per unit of deformation). For a small enough value of P, the moment $k\delta L$ will thus

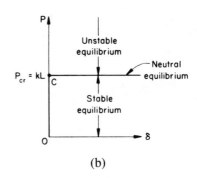

(a)   (b)

Figure 11.1

be sufficient to return the bar to $\delta=0$. Since the system reacts to a small disturbance by creating a counterbalancing effect acting to diminish the disturbance, the configuration is in *stable equilibrium*.

If now the load is increased to the point where

$$P\delta = kL\delta \qquad \text{(a)}$$

it is clear that any small disturbance δ will be neither diminished nor amplified. The system is now in *neutral equilibrium* at any small value of δ. The above expression defines the *critical load*:

$$P_{cr} = kL \qquad \text{(b)}$$

If $P > P_{cr}$, the net moment acting will be such as to increase δ, tending to further increase the disturbing moment $P\delta$, and so on. For $P > P_{cr}$, the system is in *unstable equilibrium* because *any* lateral disturbance will be amplified, as in the springless case discussed earlier.

The equilibrium regimes are shown in Fig. 11.1b. Note that C, termed the *bifurcation point*, marks the two branches of the equilibrium solution. One is the vertical branch ($P \leqslant P_{cr}, \delta = 0$), the other is the horizontal ($P = P_{cr}, \delta > 0$).

The buckling analysis of compression members usually follows in essentially the same manner as above. Stability may also be interpreted in terms of energy concepts, however. Referring again to Fig. 11.1a, the work done by P as it acts through a distance $L(1 - \cos\theta)$ is

$$\Delta W = PL(1 - \cos\theta)$$

$$= PL\left(1 - 1 + \frac{\theta^2}{2} + \cdots\right) \approx \frac{PL\theta^2}{2}$$

The elastic energy acquired as a result of the corresponding spring elongation $L\theta$ is

$$\Delta U = \tfrac{1}{2} k (L\theta)^2$$

If $\Delta U > \Delta W$, the configuration is *stable*; the work done is insufficient to result in a displacement which grows subsequent to a lateral disturbance. However, if $\Delta W > \Delta U$, the system is *unstable* because now the work is large enough to cause the displacement to grow following a disturbance. The boundary between stable and unstable configurations corresponds to $\Delta W = \Delta U$,

$$P_{cr} = kL$$

as before.

Either the static equilibrium or the energy approach may be used for determination of the critical load. The choice depends upon the particulars of the situation under analysis. While the static equilibrium method leads to exact solutions, the results offered by the energy approach (sometimes approximate) are often preferable because of the physical insights which may be more readily gained.

11.3   Buckling of a Column

Consideration is now given to a relatively slender straight bar subject to axial compression. This member, a *column*, is similar to the element shown in Fig. 11.1a, in that it too can experience unstable behavior. In the case of a column, the restoring force or moment is provided by elastic forces established within the member rather than by springs external to it.

Refer to Fig. 11.2a, in which is shown a straight, homogeneous, pin-ended column. For such a column, the end moments are zero. This is regarded as the fundamental or most basic case. It is reasonable to suppose that the column can be held in a deformed configuration by a load P while remaining in the elastic range. Note that the requisite axial motion is permitted by the movable end support. In Fig. 11.2b the postulated deflection is shown, having been caused by collinear forces P acting at the centroid of the cross section. The bending moment at any section, $M = -Pv$, when inserted into the equation for the elastic behavior of a beam, $EIv'' = M$, yields

$$EI \frac{d^2v}{dx^2} + Pv = 0 \tag{11.1}$$

The solution of this differential equation is

$$v = c_1 \sin \sqrt{\frac{P}{EI}}\, x + c_2 \cos \sqrt{\frac{P}{EI}}\, x \tag{a}$$

where the constants of integration, c_1 and c_2, are determined from the end

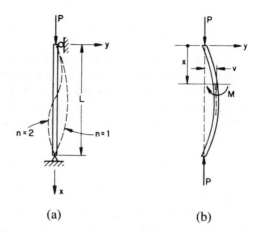

Figure 11.2

conditions: $v(0)=v(L)=0$. From $v(0)=0$, we find that $c_2=0$. Substituting the second condition into Eq. (a) we obtain

$$c_1 \sin\sqrt{\frac{P}{EI}}\ L=0 \qquad\qquad (b)$$

It must be concluded that either $c_1=0$, in which case $v=0$ for all x and the column remains straight regardless of load, or $\sin\sqrt{P/EI}\ L=0$. The case of $c_1=0$ corresponds to a condition of no buckling and yields a trivial solution [the energy approach (Sec. 11.9) sheds further light upon this case]. The latter is the acceptable alternative because it is consistent with column deflection. It is satisfied if

$$\sqrt{\frac{P}{EI}}\ L=n\pi \qquad (n=1,2,\dots) \qquad\qquad (c)$$

The value of P ascertained from Eq. (c), i.e., that load for which the column may be maintained in a deflected shape, is the *critical load*,

$$(P_{cr})_n = \frac{n^2\pi^2 EI}{L^2} \qquad\qquad (11.2)$$

where L represents the original length of the column. Assuming that column deflection is in no way restricted to a particular plane, the deflection may be expected to occur about an axis through the centroid for which the second moment of area is a minimum. The lowest critical load or *Euler buckling load* of the pin-ended column is of greatest interest; for $n=1$,

$$P_{cr} = \frac{\pi^2 EI}{L^2} \qquad\qquad (11.3)$$

The deflection is found by combining Eqs. (a) and (c) and inserting the values of c_1 and c_2:

$$v = c_1 \sin \frac{n \pi x}{L} \qquad (11.4)$$

Inasmuch as c_1, the amplitude of the elastic curve, is undetermined (and independent of P_{cr}), we are led to conclude that the critical load and deflection are independent, and that P_{cr} will sustain any small lateral deflection.

The foregoing conclusions are predicated upon the linearized beam theory with which the analysis began. Recall that in Eq. (11.1) the term d^2v/dx^2 is actually an approximation to the curvature, Eq. (5.7). Were the exact curvature used, the differential equation derived would apply to large deformations within the elastic range, and the results would be less restricted. For this case it is found that P is dependent upon the magnitude of the deflection or c_1. The exact or large amplitude analysis also reveals values of P exceeding P_{cr}. Because of practical considerations, large deflections are generally not permitted to occur in structures. The bending stresses accompanying large deflection could carry the material into the inelastic regime, this leading to diminished buckling loads.* In most applications P_{cr} is usually regarded as the maximum load sustainable by a column.

Returning to Eq. (11.4), we note that while theoretically any buckling mode is possible, the column will ordinarily deflect into the first mode ($n = 1$). The first two modes of buckling are indicated by the dashed lines in Fig. 11.2a. To initiate second mode buckling ($n = 2$) requires one lateral support at the nodal point, at midlength. As n increases, the number of nodal points also increases. One is therefore interested primarily in the lowest buckling mode.

We conclude this section by recalling that the boundary conditions employed in the solution of the differential equation led to an infinite set of discrete values of load, $(P_{cr})_n$. These solutions, typical of many engineering problems, are termed *eigenvalues*, and the corresponding deflections v are the *eigenfunctions*.

11.4   End Conditions

It is evident from the foregoing derivation that P_{cr} is dependent upon the end conditions of the column. For other than the pin-ended, fundamental case discussed, one need only substitute the appropriate conditions into Eq. (a) of Sec. 11.3 and proceed as before.

---

*S. P. Timoshenko and J. M. Gere, *Theory of Elastic Stability*, New York: McGraw-Hill, 1961, pp. 76–82.

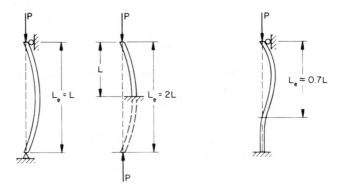

Figure 11.3

Consider an alternative approach, beginning with the following revised form of the Euler buckling formula for a pin-ended column, applicable to a variety of end conditions:

$$P_{cr} = \frac{\pi^2 EI}{L_e^2} \qquad (11.5)$$

Here L_e denotes the *effective* column length, which for the *pin-ended* column is the column length L. The effective length, shown in Fig. 11.3 for several end conditions, is determined by noting the length of a segment corresponding to a pin-ended column. In so doing one seeks the distance between points of inflection on the elastic curve or the distance between hinges, if any exist.

Regardless of end condition, it is observed that the critical load is dependent not on material strength but rather upon the *flexural rigidity*, *EI*. Buckling resistance can thus be enhanced by deploying material so as to maximize the moment of inertia but not to the point where the section thickness is so small as to result in local buckling or wrinkling.

11.5 Critical Stress in a Column

Assuming that at buckling the column material is stressed within the elastic regime, the stress corresponding to P_{cr} is simply

$$\sigma_{cr} = \frac{P_{cr}}{A} = \frac{\pi^2 E}{(L_e/r)^2} \qquad (11.6)$$

where A is the cross-sectional area, and r the *radius of gyration* defined by $I = Ar^2$. We seek the minimum value of σ_{cr}, and consequently the minimum radius of gyration should be used in Eq. (11.6). The quotient L_e/r, known

Figure 11.4

as the *slenderness ratio*, is an important parameter in the classification of compression members.

For a member of sufficiently large slenderness ratio, a *long* column, buckling occurs at a stress lower than the proportional limit. The Euler formula (11.6) is appropriate to this case. For very short members, on the other hand, failure occurs by compression, without appreciable buckling, at stresses exceeding the proportional limit. Between these extremes lies a column classified as *intermediate*,* characterized by slenderness ratios that are neither large nor small, which does not fail by direct compression. The failure of intermediate columns occurs *subsequent* to the onset of inelastic behavior. Presented below is one approach to the determination of the *inelastic buckling load*, referred to as the *tangent modulus theory*.

Consider the concentric compression of an intermediate column, and imagine the loading to occur in small increments until such time as the buckling load P_t is achieved. As one would expect, the column does not remain perfectly straight, but displays slight curvature as the increments in load are imposed. It is fundamental to the tangent modulus theory to assume that accompanying the increasing loads and curvature is a *continuous increase, or no decrease*, in the longitudinal stress and strain in *every* fiber of the column. That this should happen is not at all obvious, for it is reasonable to suppose that fibers on the convex side of the member might *elongate*, thereby reducing the stress. Accepting the former assumption, the stress distribution is as shown in Fig. 11.4. The increment of stress, $\Delta\sigma$, is attributable to bending effects; σ_{cr} is the value of stress associated with the

---

*The range of L_e/r depends upon the material under consideration. In the case of structural steel, for example, long columns are those for which $L_e/r > 100$; for intermediate columns, $30 < L_e/r < 100$, and for short struts, $L_e/r < 30$.

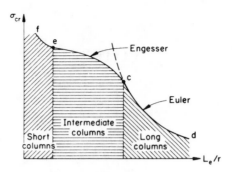

Figure 11.5

attainment of the critical load P_t. The distribution of strain will display a pattern similar to that shown in Fig. 11.4.

For small deformations Δv, the increments of stress and strain will likewise be small, and as in the case of elastic bending, sections originally plane are assumed to *remain plane* subsequent to bending. The change of stress $\Delta\sigma$ is thus assumed proportional to the increment of strain, $\Delta\epsilon$; i.e., $\Delta\sigma = E_t \Delta\epsilon$. The constant of proportionality E_t is the *slope* of the stress–strain diagram, termed the *tangent modulus*. Note that within the linearly elastic range, $E_t = E$. The stress–strain relationship beyond the proportional limit is thus assumed linear, as in the case of elastic buckling. The critical or so-called *Engesser stress* may, on the basis of the foregoing rationale, be expressed by means of a modification of Eq. (11.6) in which E_t replaces E:

$$\sigma_{\mathrm{cr}} = \frac{P_t}{A} = \frac{\pi^2 E_t}{(L_e/r)^2} \tag{11.7}$$

The buckling load P_t predicted by Eq. (11.7) is in good agreement with test results. This expression is therefore recommended for design purposes.

Figure 11.5 is a plot of critical stress as a function of slenderness ratio. Portions cd and ce are represented by Eqs. (11.6) and (11.7), respectively. Point c, the yield stress (approximately equal to the proportional stress), is denoted by σ_{yp}. The segment denoted ef represents the range of failure by axial compression.

When the critical stress is known and the slenderness ratio required, the application of Eq. (11.7) is straightforward. The value of E_t corresponding to σ_{cr} is read from the stress–strain curve obtained from a simple tension test, following which L_e/r is calculated using Eq. (11.7). If, however, L_e/r is known, and σ_{cr} is to be ascertained, a trial-and-error approach is necessary (see Problem 11.7).

11.6 Allowable Stress

The foregoing discussion and analysis have related to ideal, homogeneous, concentrically loaded columns. Inasmuch as such columns are not likely candidates for application in structures, actual design requires the use of *empirical formulas* based upon a strong background of test and experience. Care must be exercised in applying such special purpose formulas. The designer should be prepared to respond to the following questions:

a. To what material does the formula apply?
b. Does the formula already contain a factor of safety, or is the factor of safety separately applied?
c. For what range of slenderness ratios is the formula valid?

Included among the many special purpose relationships developed are the following, recommended by the American Institute of Steel Construction, and valid for a structural steel column*:

$$\sigma_w = \frac{1-(L_e/r)^2/2C^2}{N}\sigma_{yp} \qquad \left(0 < \frac{L_e}{r} < C\right)$$

$$\sigma_w = \frac{1027 \times 10^9}{(L_e/r)^2} \qquad \left(C < \frac{L_e}{r} < 200\right)$$

(11.8)

Here σ_w, σ_{yp}, C, and N denote, respectively, the allowable and yield stresses, a material constant, and the factor of safety. The values of C and N are given by

$$C = \sqrt{\frac{2\pi^2 E}{\sigma_{yp}}}, \qquad N = \frac{5}{3} + \frac{3(L_e/r)}{8C} - \frac{(L_e/r)^3}{8C^3}$$

This relationship provides a smaller N for a short strut than for a column of higher L_e/r, recognizing the fact that the former fails by yielding and the latter by buckling. The use of a variable factor of safety provides a consistent buckling formula for various ranges of L_e/r. The second equation of (11.8) includes a constant factor of safety and gives the value of allowable stress in pascals. Both formulas apply to principal load carrying (main) members.

Example 11.1. The boom of a crane, shown in Fig. 11.6, is constructed of steel, $E = 210$ GPa; the yield point stress is 210 MPa. The cross section is rectangular with a depth of 100 mm and a thickness of 50 mm. Determine the buckling load of the column.

---

*See *AISC Steel Construction Manual*, New York: AISC Inc., 1970.

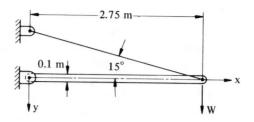

Figure 11.6

SOLUTION. The moments of inertia of the section are $I_z = 0.05(0.1)^3/12 = 4.17 \times 10^{-6}$ m^4, $I_y = 0.1(0.05)^3/12 = 1.04 \times 10^{-6}$ m^4. The least radius of gyration is thus $r = \sqrt{I_y/A} = 14$ mm, and the slenderness ratio, $L/r = 194$. The Euler formula is applicable in this range. From statics, the axial force in terms of W is $P = W/\tan 15° = 3.732W$. Applying the formula for a hinged-end column, Eq. (11.5), for buckling in the yx plane, we have

$$P_{cr} = \frac{\pi^2 EI_z}{L^2} = \frac{9.86 \times 210 \times 10^9 \times 4.17 \times 10^{-6}}{(2.75)^2} = 1141.739 = 3.732W$$

or

$$W = 305.932 \text{ kN}$$

To calculate the load required for buckling in the xz plane, we must take note of the fact that the line of action of the compressive force passes through the joint and thus causes no moment about the y axis at the fixed end. Therefore Eq. (11.5) may again be applied:

$$P_{cr} = \frac{\pi^2 EI_y}{L^2} = \frac{9.86 \times 210 \times 10^9 \times 1.04 \times 10^{-6}}{(2.75)^2} = 284.75 = 3.732W$$

or

$$W = 76.3 \text{ kN}$$

The member will thus fail by lateral buckling when the load W exceeds 76.3 kN. Note that the critical stress $P_{cr}/A = 76.3/0.005 = 15.26$ MPa. This, compared with the yield strength of 210 MPa, indicates the importance of buckling analysis in predicting the safe working load.

11.7   Initially Curved Members

As might be expected, the load-carrying capacity and deformation under load of a column are significantly affected by even a small initial curvature. In order to ascertain the extent of this influence, consider a pin-ended column for which the unloaded shape is described by

$$v_0 = a_0 \sin \frac{\pi x}{L} \tag{a}$$

as shown by the dashed lines in Fig. 11.7. Here a_0 is the maximum initial

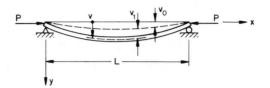

Figure 11.7

deflection. An additional deflection v_1 will accompany a subsequently applied load P, so that the total deflection is

$$v = v_0 + v_1 \tag{b}$$

The differential equation of the column is thus

$$EI\frac{d^2 v_1}{dx^2} = -P(v_0 + v_1) = -Pv \tag{c}$$

which together with Eq. (a) becomes

$$\frac{d^2 v_1}{dx^2} + \frac{P}{EI}v_1 = -\frac{P}{EI}a_0 \sin\frac{\pi x}{L}$$

When the trial particular solution

$$v_{1P} = B \sin\frac{\pi x}{L}$$

is substituted into Eq. (c), it is found that

$$B = \frac{Pa_0}{(EI\pi^2/L^2) - P} = \frac{a_0}{(P_{cr}/P) - 1} \tag{11.9}$$

The general solution of Eq. (c) is therefore

$$v_1 = c_1 \sin\sqrt{\frac{P}{EI}}\, x + c_2 \cos\sqrt{\frac{P}{EI}}\, x + B \sin\frac{\pi x}{L} \tag{d}$$

The constants c_1 and c_2 are evaluated upon consideration of the end conditions $v_1(0) = v_1(L) = 0$. The result of substituting these conditions is $c_1 = c_2 = 0$; the column deflection is thus

$$v = v_0 + v_1 = a_0 \sin\frac{\pi x}{L} + B \sin\frac{\pi x}{L} = \frac{a_0}{1 - (P/P_{cr})} \sin\frac{\pi x}{L} \tag{11.10}$$

As for the critical stress, we begin with the expression applicable to combined axial loading and bending: $\sigma_x = (P/A) \pm (My/I)$, where A and I are the cross-sectional area and the moment of inertia. On substitution of Eqs. (c) and (11.10) into the above expression, the maximum compressive stress at midspan is found to be

$$\sigma_{max} = \frac{P}{A}\left[1 + \frac{a_0 A}{Z}\frac{1}{1-(P/P_{cr})}\right] \tag{11.11}$$

Here Z is the section modulus I/c, where c represents the distance measured in the y direction from the centroid of the cross section to the extreme fibers. In Eq. (11.11), σ_{max} is limited to the proportional or yield stress of the column material. Thus, setting $\sigma_{max} = \sigma_{yp}$ and $P = P_L$, we rewrite Eq. (11.11) as follows:

$$\sigma_{yp} = \frac{P_L}{A}\left[1 + \frac{a_0 A}{Z}\frac{1}{1-(P_L/P_{cr})}\right] \tag{11.12}$$

where P_L is the *limit load* which results in impending yielding and subsequent failure. Given σ_{yp}, a_0, E, and the column dimensions, the above equation may be solved by trial-and-error for P_L. The allowable load P_w can then be found by dividing P_L by an appropriate factor of safety, N.

11.8 Eccentrically Loaded Columns

In contrast with the cases considered up to this point, we now analyze columns which are loaded eccentrically, i.e., those in which the load is not applied at the centroid of the cross section. This situation is clearly of great practical importance, since we ordinarily have no assurance that loads are truly concentric. As seen in Fig. 11.8a, the bending moment at any section is $-P(v+e)$, where e is the eccentricity, defined in the figure. It follows that the beam equation is given by

$$EI\frac{d^2v}{dx^2} + P(v+e) = 0 \tag{a}$$

or

$$EI\frac{d^2v}{dx^2} + Pv = -Pe$$

The general solution is

$$v = c_1 \sin\sqrt{\frac{P}{EI}}\ x + c_2 \cos\sqrt{\frac{P}{EI}}\ x - e \tag{b}$$

To determine the constants c_1 and c_2, the end conditions $v(L/2) = v(-L/2) = 0$ are applied, with the result

$$c_1 = 0, \qquad c_2 = \frac{e}{\cos\left[\sqrt{P/EI}\ (L/2)\right]}$$

Substituting these values into Eq. (b) provides an expression for the

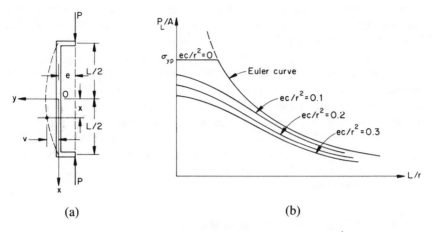

(a)   (b)

Figure 11.8

column deflection:

$$v = \frac{e}{\cos\sqrt{PL^2/(4EI)}} \cos\sqrt{\frac{P}{EI}}\, x - e \qquad (c)$$

In terms of the critical load $P_{cr} = \pi^2 EI/L^2$, the midspan deflection is

$$v_{max} = e\left[\sec\left(\frac{\pi}{2}\sqrt{\frac{P}{P_{cr}}}\right) - 1\right] \qquad (11.13)$$

As P approaches P_{cr}, the maximum deflection is thus observed to approach infinity.

The maximum compressive stress, $(P/A) + (Mc/I)$, occurs at $x = 0$, on the concave side of the column; it is given by

$$\sigma_{max} = \frac{P}{A}\left[1 + \frac{ec}{r^2}\sec\left(\frac{\pi}{2}\sqrt{\frac{P}{P_{cr}}}\right)\right] \qquad (11.14)$$

where r represents the radius of gyration and c is the distance from the centroid of the cross section to the extreme fibers, both in the direction of eccentricity. The above expression is referred to as the *secant formula*, giving σ_{max} in the column as a function of the average stress (P/A), the *eccentricity ratio* (ec/r^2), and the *slenderness ratio* (L/r). As in case of initially curved columns, if we let $\sigma_{max} = \sigma_{yp}$ and the limit load $P = P_L$, Eq. (11.14) becomes

$$\sigma_{yp} = \frac{P_L}{A}\left[1 + \frac{ec}{r^2}\sec\left(\frac{\pi}{2}\sqrt{\frac{P_L}{P_{cr}}}\right)\right] \qquad (11.15)$$

For any prescribed yield stress and eccentricity ratio, Eq. (11.15) can be solved by trial-and-error and P_L/A plotted as a function of L/r with E held constant (Fig. 11.8b). The allowable value of the average compressive load is found from $P_w = P_L/N$. The development on which Eq. (11.14) is based assumes buckling to occur in the xy plane. It is also necessary to investigate buckling in the xz plane, for which Eq. (11.14) does not apply. This possibility relates especially to narrow columns.

The behavior of a beam subjected to simultaneous axial and lateral loading, the so-called *beam-column*, is analogous to that of the bar shown in Fig. 11.1a, with an additional force acting, transverse to the bar. For problems of this type, energy methods are usually more efficient than the equilibrium approach previously employed.

11.9   Energy Methods Applied to Buckling

Energy techniques usually offer considerable ease of solution compared with equilibrium approaches to the analysis of elastic stability and the determination of critical loads. Recall from Chapter 10 that energy methods are especially useful in treating members of variable cross section, where the variation can be expressed as a function of the beam axial coordinate. The following examples serve to illustrate the method.

Example 11.2. Apply the principle of virtual work in analyzing the stability of a straight, pin-ended column. Locate the origin of coordinates at the stationary end.

SOLUTION.   Recall from Chapter 10 that the principle of virtual work may be stated

$$\delta W = \delta U \tag{a}$$

where W and U are the virtual work and strain energy, respectively. Consider the configuration of the column in the first buckling mode, denoting the arc length of a column segment by ds. We shall require the displacement, $\delta u = ds - dx$, experienced by the column in the direction of applied load P, in order to determine the work done:

$$\delta u = \int_0^L (\sqrt{dx^2 + dv^2} - dx) = \int_0^L \left\{ \left[1 + (dv/dx)^2\right]^{1/2} - 1 \right\} dx$$

Noting that for $(dv/dx)^2 \ll 1$,

$$\left[1 + \left(\frac{dv}{dx}\right)^2\right]^{1/2} = \left[1 + \frac{1}{2}\left(\frac{dv}{dx}\right)^2 + \cdots \right]$$

we obtain

$$\delta u \approx \frac{1}{2} \int_0^L \left(\frac{dv}{dx}\right)^2 dx \tag{b}$$

Since the load remains constant, the work done is

$$\delta W = \frac{1}{2} P \int_0^L \left(\frac{dv}{dx}\right)^2 dx \tag{11.16}$$

Next, the strain energy must be evaluated. There are components of strain energy associated with column bending, compression, and shear. We shall neglect the last. From Eq. (5.45), the bending component is

$$U_1 = \int_0^L \frac{M^2}{2EI}\, dx = \int_0^L \frac{EI}{2}\left(\frac{d^2 v}{dx^2}\right)^2 dx \qquad (11.17)$$

The energy due to a uniform compressive loading P is, according to Eq. (2.34),

$$U_2 = \frac{P^2 L}{2AE} \qquad (11.18)$$

Inasmuch as U_2 is constant, it plays no role in the analysis. The change in the strain energy as the column proceeds from its original to its buckled configuration is therefore

$$\delta U = \int_0^L \frac{EI}{2}\left(\frac{d^2 v}{dx^2}\right)^2 dx \qquad (c)$$

since the initial strain energy is zero. Substituting Eqs. (11.16) and (c) into Eq. (a), we have

$$\frac{1}{2}\int_0^L P\left(\frac{dv}{dx}\right)^2 dx = \frac{1}{2}\int_0^L EI\left(\frac{d^2 v}{dx^2}\right)^2 dx \qquad (11.19a)$$

From which

$$P_{cr} = \frac{\displaystyle\int_0^L EI(v'')^2\, dx}{\displaystyle\int_0^L (v')^2\, dx} \qquad (11.19b)$$

The result given above applies to a column with any end condition. The end conditions specific to this problem will be satisfied by a solution

$$v = a_1 \sin\frac{n\pi x}{L}$$

where a_1 is a constant. After substituting this assumed deflection into Eq. (11.19b) and integrating, we obtain

$$(P_{cr})_n = \left(\frac{n\pi}{L}\right)^2 EI$$

The minimum critical load and the deflection to which this corresponds are

$$P_{cr} = \frac{\pi^2 EI}{L^2}, \qquad v = a_1 \sin\frac{\pi x}{L} \qquad (d)$$

It is apparent from Eq. (11.19a) that for $P > P_{cr}$, the work done by P exceeds the strain energy stored in the column. The assertion can therefore be made that a straight column is unstable when $P > P_{cr}$. This point, with regard to stability, corresponds to $c_1 = 0$ in Eq. (b) of Sec. 11.3; it could not be obtained as readily from the equilibrium approach. In the event that $P = P_{cr}$, the column exists in neutral equilibrium. For $P < P_{cr}$, a straight column is in stable equilibrium.

Example 11.3. A simply supported beam is subjected to a moment M at point A and axial loading P as shown in Fig. 11.9. Determine the equation of the elastic curve.

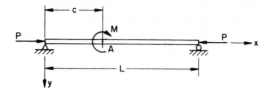

Figure 11.9

SOLUTION.   The displacement of the right end, which occurs during the deformation of the beam from its initially straight configuration to the equilibrium curve, is
given by Eq. (b). The total work done is evaluated by adding to Eq. (11.16) the
work due to the moment. In Sec. 10.9, we have already solved this problem for
$P=0$ by using the following Fourier series for displacement:

$$v= \sum_{n=1}^{\infty} a_n \sin\frac{n\pi x}{L} \tag{e}$$

Proceeding in the same manner, Eq. (d) of Sec. 10.9, representing $\Delta U = \delta W$, now
takes the form

$$\frac{\pi^4 EI}{4L^3} \sum_{n=1}^{\infty} n^4\delta(a_n^2) = \frac{\pi^2 P}{4L} \sum_{n=1}^{\infty} n^2\delta(a_n^2) + \frac{\pi M}{L} \sum_{n=1}^{\infty} n\cos\frac{n\pi c}{L}\delta(a_n)$$

From the above expression,

$$a_n = \frac{2M}{\pi} \sum_{n=1}^{\infty} \frac{1}{n} \frac{\cos(n\pi c/L)}{(n^2\pi^2 EI/L^2) - P} \tag{f}$$

For purposes of simplification let b denote the ratio of the axial force to its critical
value:

$$b = \frac{PL^2}{\pi^2 EI} \tag{11.20}$$

Then by substituting Eqs. (11.20) and (f) into Eq. (e), the following expression for
deflection results:

$$v = \frac{2ML^2}{\pi^3 EI} \sum_{n=1,3,5,\ldots}^{\infty} \frac{\cos(n\pi c/L)}{n(n^2 - b)} \sin\frac{n\pi x}{L} \qquad (0 < x < L) \tag{11.21}$$

Note that when P approaches its critical value in Eq. (11.20), $b \to 1$. The first term
in Eq. (11.21) is then

$$v = \frac{2ML^2}{\pi^3 EI} \frac{1}{1-b} \cos\frac{\pi c}{L} \sin\frac{\pi x}{L} \tag{11.22}$$

indicating that the deflection becomes infinite, as expected.

Comparison of Eq. (11.22) with the solution found in Sec. 10.9 (corresponding to
$P=0$ and $n=1$) indicates that the axial force P serves to increase the deflection
produced by the lateral load (moment M) by a factor of $1/(1-b)$.

In general, if we have a beam subjected to several moments or lateral loads in
addition to an axial load P, the deflections owing to the lateral moments or forces

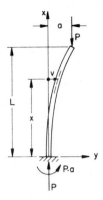

Figure 11.10

are found for the $P=0$ case. This usually involves superposition. The resulting deflection is then multiplied by the factor $1/(1-b)$ to account for the deflection effect due to P. This procedure is valid for any lateral load configuration composed of moments, concentrated forces, and distributed forces.

Example 11.4. Apply the Rayleigh–Ritz method to determine the buckling load of a straight, uniform cantilever column carrying a vertical load (Fig. 11.10).

SOLUTION. The analysis begins with an assumed parabolic deflection curve,

$$v = \frac{ax^2}{L^2} \tag{g}$$

where a represents the deflection of the free end, and L the column length. (The parabola is actually a very poor approximation to the true curve, since it describes a beam of constant curvature, whereas the curvature of the actual beam is zero at the top and a maximum at the bottom.) The assumed deflection satisfies the geometric boundary conditions pertaining to deflection and slope: $v(0)=0$, $v'(0)=0$. In accordance with the Rayleigh–Ritz procedure (see Sec. 10.10), it may therefore be used as a trial solution. The static boundary conditions, such as $v''(0) \neq 0$ or $M \neq 0$, need not be satisfied.

The work done by the load P and the strain energy gained are given by Eqs. (11.16) and (11.17). The potential energy function Π is thus given by

$$\Pi = U - W = \int_0^L \frac{EI}{2} \left(\frac{d^2 v}{dx^2} \right)^2 dx - \int_0^L \frac{P}{2} \left(\frac{dv}{dx} \right)^2 dx \tag{h}$$

Substituting Eq. (g) into this expression and integrating, we have

$$\Pi = \frac{2EIa^2}{L^3} - \frac{2Pa^2}{3L}$$

Applying Eq. (10.19), $\partial \Pi / \partial a = 0$, we find that

$$P_{cr} = 3 \frac{EI}{L^2} \tag{i}$$

Let us rework this problem by replacing the strain-energy expression due to bending,

$$U_1 = \int_0^L \frac{EI}{2} \left(\frac{d^2v}{dx^2} \right)^2 dx \tag{j}$$

with one containing the moment deduced from Fig. 11.10, $M = P(a-v)$:

$$U_1 = \int_0^L \frac{M^2 \, dx}{2EI} = \int_0^L \frac{P^2(a-v)^2}{2EI} dx \tag{k}$$

Equation (h) becomes

$$\Pi = \int_0^L \frac{P^2(a-v)^2}{2EI} dx - \int_0^L \frac{P}{2} \left(\frac{dv}{dx} \right)^2 dx$$

Substituting Eq. (g) into the above and integrating, we obtain

$$\Pi = \frac{8}{30} \frac{P^2 a^2 L}{EI} - \frac{2Pa^2}{3L}$$

Now $\partial \Pi / \partial a = 0$ yields

$$P_{cr} = 2.50 \frac{EI}{L^2} \tag{l}$$

Comparison with the exact solution, $2.4674 \, EI/L^2$, reveals errors for the solutions (i) and (l) of about 22% and 1.3%, respectively. The latter result is satisfactory, although it is predicated upon an assumed deflection curve differing considerably in shape from the true curve.

It is apparent from the foregoing example and a knowledge of the exact solution that one solution is quite a bit more accurate than the other. It can be shown that the expressions (j) and (k) will be identical only when the true deflection is initially assumed. Otherwise Eq. (k) will give better accuracy. This is because when one chooses Eq. (k), the accuracy of the solution depends upon the closeness of the assumed deflection to the actual deflection; with Eq. (j), the accuracy depends instead upon the rate of change of slope, d^2v/dx^2.

An additional point of interest relates to the consequences, in terms of the critical load, of selecting a deflection that departs from the actual curve. When other than the true deflection is used, not every beam element is in equilibrium. The approximate beam curve can be maintained only through the introduction of additional constraints that tend to render the beam more rigid. The critical loads thus obtained will be higher than those derived from exact analysis. It may be concluded that *energy methods always yield buckling loads higher than the exact values if the assumed beam deflection differs from the true curve.*

More efficient application of energy techniques may be realized by selecting a series approximation for the deflection, as in Example 11.3. Inasmuch as a series involves a number of parameters, as for example in Eq. (e) of Example 11.3, the approximation can be varied by appropriate manipulation of these parameters, i.e., by changing the number of terms in the series.

Example 11.5. A pin-ended, tapered bar of constant thickness is subjected to axial compression (Fig. 11.11). Determine the critical load. The variation of the moment of inertia is given by

$$I(x) = I_1\left(1 + \frac{3x}{L}\right) \qquad \left(0 < x < \frac{L}{2}\right)$$

$$= I_1\left(4 - \frac{3x}{L}\right) \qquad \left(\frac{L}{2} < x < L\right)$$

where I_1 is the constant moment of inertia at $x=0$ and $x=L$.

SOLUTION. As before, we begin by representing the deflection by

$$v = a_1 \sin\frac{\pi x}{L}$$

Taking symmetry into account, the variation of strain energy and the work done [Eqs. (11.16) and (11.17)] are expressed by

$$\delta U = \delta \int_0^L \frac{EI}{2}\left(\frac{d^2v}{dx^2}\right)^2 dx = 2\delta \int_0^{L/2} \frac{EI_1}{2}\left(1 + \frac{3x}{L}\right)a_1^2 \frac{\pi^4}{L^4}\sin^2\frac{\pi x}{L}dx$$

$$= \frac{\pi^4 EI_1}{4L^3}\delta(a_1^2) + \frac{4.215\pi^4 EI_1}{16L^3}\delta(a_1^2) = \frac{8.215\pi^4 EI_1}{16L^3}\delta(a_1^2)$$

$$\delta W = \delta \int_0^L \frac{P}{2}\left(\frac{dv}{dx}\right)^2 dx = 2\delta \int_0^{L/2}\frac{P}{2}a_1^2\frac{\pi^2}{L^2}\cos^2\frac{\pi x}{L}dx = \frac{\pi^2 P}{4L}\delta(a_1^2)$$

From the principle of virtual work, $\delta W = \delta U$, we have

$$\frac{8.215\pi^4 EI_1}{16L^3} = \frac{\pi^2 P}{4L}$$

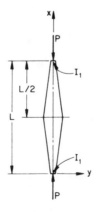

Figure 11.11

or

$$P_{cr} = 20.25 \frac{EI_1}{L^2}$$

In Example 11.7, this problem is solved by numerical analysis, revealing that the above solution overestimates the buckling load.

11.10 Solution by Finite Differences

The equilibrium analysis of buckling often leads to differential equations which resist solution. Even energy methods are limited in that they require that the moment of inertia be expressible as a function of the column length coordinate. Therefore, in order to enable the analyst to cope with the numerous and varied columns of practical interest, numerical techniques must be relied upon. The examples which follow apply the method of finite differences to the differential equation of a column.

Example 11.6. Determine the buckling load of a pin-ended column of length L and constant cross section. Use four subdivisions of equal length. Denote the nodal points by $0, 1, 2, 3, 4$, with 0 and 4 located at the ends. Locate the origin of coordinates at the stationary end.

SOLUTION.   The governing differential equation (11.1) may be put into the form

$$\frac{d^2v}{dx^2} + \lambda^2 v = 0$$

by substituting $\lambda^2 = P/EI$. The boundary conditions are $v(0) = v(L) = 0$. The finite difference equation corresponding to the above is, according to Eq. (7.9),

$$v_{m+1} + (\lambda^2 h^2 - 2)v_m + v_{m-1} = 0 \tag{11.23}$$

valid at every node along the length. Here the integer m denotes the nodal points, and h the segment length. Applying Eq. (11.23) at points $1, 2, 3$, we have

$$(\lambda^2 h^2 - 2)v_1 + v_2 = 0$$
$$v_1 + (\lambda^2 h^2 - 2)v_2 + v_3 = 0 \tag{a}$$
$$v_2 + (\lambda^2 h^2 - 2)v_3 = 0$$

or in convenient matrix form,

$$\begin{bmatrix} \lambda^2 h^2 - 2 & 1 & 0 \\ 1 & \lambda^2 h^2 - 2 & 1 \\ 0 & 1 & \lambda^2 h^2 - 2 \end{bmatrix} \begin{Bmatrix} v_1 \\ v_2 \\ v_3 \end{Bmatrix} = 0 \tag{b}$$

This set of simultaneous equations has a nontrivial solution for v_1, v_2, v_3 only if the determinant of the coefficients vanishes:

$$\begin{bmatrix} \lambda^2 h^2 - 2 & 1 & 0 \\ 1 & \lambda^2 h^2 - 2 & 1 \\ 0 & 1 & \lambda^2 h^2 - 2 \end{bmatrix} = 0 \tag{c}$$

The solution of the buckling problem is thus reduced to the determination of the roots (the λ's of the characteristic equation) resulting from the expansion of the determinant.

To expedite the solution, we take into account the symmetry of deflection. For the lowest critical load, the buckling configuration is given by the first mode, shown by the dashed lines in Fig. 11.2a. Thus $v_1 = v_3$, and the equations (a) become

$$(\lambda^2 h^2 - 2)v_1 + v_2 = 0$$

$$2v_1 + (\lambda^2 h^2 - 2)v_2 = 0$$

Setting equal to zero the characteristic determinant of the above set, we find that $(\lambda^2 h^2 - 2)^2 - 2 = 0$, which has the solution $\lambda^2 h^2 = 2 \pm \sqrt{2}$. Selecting $2 - 2\sqrt{2}$ to obtain a minimum critical value and letting $h = L/4$, we obtain $\lambda^2 = (2 - \sqrt{2})16/L^2$. Thus,

$$P_{\mathrm{cr}} = 9.373 \frac{EI}{L^2}$$

This result differs from the exact solution by approximately 5%. By increasing the number of segments, the accuracy may be improved.

If required, the critical load corresponding to second mode buckling, indicated by the dashed line in Fig. 11.2a, may be determined by recognizing that for this case $v_0 = v_2 = v_4 = 0$ and $v_1 = -v_3$. We then proceed from Eqs. (a) as above. For buckling of higher than second mode, a similar procedure is followed, in which the number of segments is increased and the appropriate conditions of symmetry satisfied.

Example 11.7. What load will cause buckling of the tapered pin-ended column shown in Fig. 11.11 of Example 11.5?

SOLUTION.   The finite difference equation is given by Eq. (11.23):

$$v_{m+1} + \left(\frac{Ph^2}{EI(x)} - 2 \right) v_m + v_{m-1} = 0 \tag{d}$$

The foregoing becomes, after substitution of $I(x)$,

$$v_{m+1} + \left[\frac{\lambda^2 h^2}{1 + (3x/L)} - 2 \right] v_m + v_{m-1} = 0 \qquad \left(0 \leqslant x \leqslant \frac{L}{2} \right) \tag{e}$$

where $\lambda^2 = P/EI_1$. Note that the coefficient of v_m is a variable, dependent upon x. This introduces no additional difficulties, however.

Dividing the beam into two segments, we have $h = L/2$ (Fig. 11.2a). Applying Eq. (e) at $x = L/2$,

$$\left[\frac{\lambda^2 L^2}{4(1 + 3/2)} - 2 \right] v_1 = 0$$

The nontrivial solution corresponds to $v_1 \neq 0$; then $(\lambda^2 L^2/10) - 2 = 0$ or, by letting $\lambda^2 = P/EI$,

$$P_{\mathrm{cr}} = \frac{20 EI_1}{L^2} \tag{f}$$

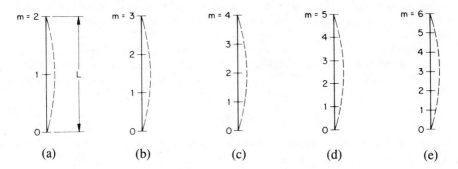

Figure 11.12

Similarly, for three segments $h = L/3$ (Fig. 11.12b). From symmetry we have $v_1 = v_2$, and $v_0 = v_3 = 0$. Thus Eq. (e) applied at $x = L/3$ yields

$$v_1 + \left[\frac{\lambda^2 L^2}{9(1+1)} - 2 \right] v_1 = 0 \quad \text{or} \quad \left(\frac{\lambda^2 L^2}{18} - 1 \right) v_1 = 0$$

The nontrivial solution is

$$P_{\text{cr}} = \frac{18 E I_1}{L^2} \tag{g}$$

For $h = L/4$, referring to Fig. 11.12c, Eq. (e) leads to

$$\left(\frac{\lambda^2 L^2}{28} - 2 \right) v_1 + v_2 = 0, \qquad 2v_1 + \left(\frac{\lambda^2 L^2}{40} - 2 \right) v_2 = 0$$

For a nontrivial solution, the characteristic determinant is zero:

$$\begin{bmatrix} \lambda^2 L^2/28 - 2 & 1 \\ 2 & \lambda^2 L^2/40 - 2 \end{bmatrix} = 0$$

Expansion of the above yields

$$\lambda^4 L^4 - 136 \lambda^2 L^2 + 2240 = 0$$

from which

$$P_{\text{cr}} = \frac{19.55 E I_1}{L^2} \tag{h}$$

Similar procedures considering the symmetry shown in Fig. 11.12d and e lead to the following results:
For $h = L/5$,

$$P_{\text{cr}} = \frac{18.96 E I_1}{L^2} \tag{i}$$

For $h = L/6$,

$$P_{\text{cr}} = \frac{19.22 E I_1}{L^2} \tag{j}$$

Results (f) through (j) indicate that for columns of variable moment of inertia,

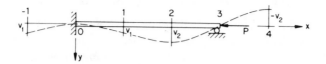

Figure 11.13

increasing the number of segments does not necessarily lead to improved P_{cr}. An energy approach to this problem, Example 11.5, gives the result $P_{cr} = 20.25 EI_1/L^2$. Because this value is higher than those obtained above, we conclude that the column does not deflect into the half sine curve assumed in Example 11.5.

Example 11.8. In Fig. 11.13 is shown a column of constant moment of inertia I and of length L, fixed at the left end and simply supported at the right end, subjected to an axial compressive load P. Determine the critical value of P, using $m = 3$.

SOLUTION. The characteristic value problem is defined by

$$\frac{d^4v}{dx^4} + \frac{P}{EI}\frac{d^2v}{dx^2} = 0$$

$$v(0) = v'(0) = v(L) = v''(L) = 0$$

(k)

where the first equation is found from Eq. (P11.15) by setting $p = 0$; the second expression represents the end conditions related to deflection, slope, and moment. Equations (k), referring to Sec. 7.3 and letting $\lambda^2 = P/EI$, may be written in the finite difference form as follows:

$$v_{m+2} + (\lambda^2 h^2 - 4)v_{m+1} + (6 - 2\lambda^2 h^2)v_m + (\lambda^2 h^2 - 4)v_{m-1} + v_{m-2} = 0 \qquad (l)$$

and

$$v_0 = 0, \qquad v_{-1} = v_1, \qquad v_m = 0, \qquad v_{m+1} = -v_{m-1} \qquad (m)$$

The quantities v_{-1} and v_{m+1} represent the deflections at the nodal points of the column prolonged by h beyond the supports. By dividing the column into three subintervals, the pattern of the deflection curve and the conditions (m) are represented in the figure by dashed lines. Now, Eq. (l) is applied at nodes 1 and 2 to yield, respectively.

$$v_1 + (6 - 2\lambda^2 h^2)v_1 + (\lambda^2 h^2 - 4)v_2 = 0$$

$$(\lambda^2 h^2 - 4)v_1 + (6 - 2\lambda^2 h^2)v_2 - v_2 = 0$$

We have a nonzero solution if the determinant of the coefficients of these equations vanishes:

$$\begin{vmatrix} 7 - 2\lambda^2 h^2 & \lambda^2 h^2 - 4 \\ \lambda^2 h^2 - 4 & 5 - 2\lambda^2 h^2 \end{vmatrix} = 3\lambda^4 h^4 - 16\lambda^2 h^2 + 19 = 0$$

From the above, setting $h = L/3$, we obtain $\lambda^2 = 16.063/L$. Thus,

$$P_{cr} = 16.063\frac{EI}{L^2}$$

(n)

The exact solution is $20.187 EI/L$. By increasing the number of segments and by employing an extrapolation technique, the results may be improved.

Finite Difference Solutions for Unevenly Spaced Nodes

It is often advantageous, primarily because of geometrical considerations, to divide a structural element so as to produce uneven spacing between nodal points. In some problems, uneven spacing provides more than a saving in time and effort. As seen in Sec. 7.6, some situations cannot be solved without resort to this approach.

Consider the problem of the buckling of a straight pin-ended column governed by

$$\frac{d^2v}{dx^2} + \frac{P}{EI} v = 0, \qquad v(0) = v(L) = 0$$

Upon substitution from Eq. (7.25), the following corresponding finite difference equation is obtained:

$$\frac{2}{\alpha(\alpha+1)} [\alpha v_{m-1} - (1+\alpha)v_m + v_{m+1}] + \frac{Ph^2}{EI} v_m = 0 \qquad (11.24)$$

where

$$h = x_m - x_{m-1}, \qquad \alpha = \frac{x_{m+1} - x_m}{x_m - x_{m-1}} \qquad (11.25)$$

Equation (11.24), valid throughout the length of the column, is illustrated in the example following.

Example 11.9. Determine the buckling load of a stepped pin-ended column (Fig. 11.14a). The variation of the moment of inertia is indicated in the figure.

SOLUTION.   The nodal points are shown in Fig. 11.14b and are numbered in a manner consistent with the symmetry of the beam. Note that the nodes are unevenly spaced. From Eq. (11.25), we have $\alpha_1 = 2$ and $\alpha_2 = 1$. Application of Eq. (11.24) at points 1 and 2 leads to

$$\frac{2}{2(2+1)} [2(0) - (1+2)v_1 + v_2] + \frac{P}{EI_1} \frac{L^2}{36} v_1 = 0$$

$$\frac{2}{1(1+1)} [v_1 - (1+1)v_2 + v_1] + \frac{P}{3EI_1} \frac{4L^2}{36} v_2 = 0$$

or

$$\left(\frac{PL^2}{36EI_1} - 1\right)v_1 + \frac{1}{3}v_2 = 0$$

$$2v_1 + \left(\frac{PL^2}{27EI_1} - 2\right)v_2 = 0$$

For a nontrivial solution, it is required that

$$\begin{bmatrix} (PL^2/36EI_1) - 1 & 1/3 \\ 2 & (PL^2/27EI_1) - 2 \end{bmatrix} = 0$$

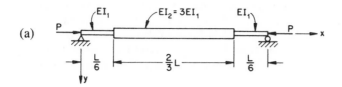

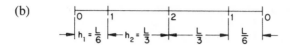

Figure 11.14

Solving, we find that the root corresponding to minimum P is

$$P_{cr} = 18\frac{EI}{L^2}$$

Employing additional nodal points may result in greater accuracy. The above procedure lends itself to columns of arbitrarily varying section and various end conditions.

We conclude our discussion by noting that column buckling represents but one case of structural instability. Other examples include the lateral buckling of a narrow beam; the buckling of a flat plate compressed in its plane; the buckling of a circular ring subject to radial compression; the buckling of a cylinder loaded by torsion, compression, bending, or internal pressure, and the snap buckling of arches. Buckling analyses for these cases are often not performed as readily as in the examples presented in this chapter. The solutions more often involve considerable difficulty and subtlety.*

Chapter 11—Problems

Secs. 11.1 to 11.6.

11.1. A column of length $3L$ is approximated by three bars of equal length connected by a torsional spring of appropriate stiffness k at each joint. The column is supported by a torsional spring of stiffness k at one end and is free at the other end. Derive an expression for determining the critical load of the system. Generalize the problem to the case of n connected bars.

---

*For a complete discussion of this subject, see S. P. Timoshenko and J. M. Gere, loc. cit. See also D. O. Brush and B. O. Almroth, *Buckling of Bars, Plates, and Shells*, New York: McGraw-Hill, 1975.

11.2. A uniform steel column, with fixed and hinge connected ends, is subjected to a vertical load $P = 450$ kN. The cross section of the column is 0.05 by 0.075 m and the length is 3.6 m. Taking $\sigma_{yp} = 28$ MPa and $E = 210$ GPa, calculate (a) the critical load and critical Euler stress, assuming a factor of safety of 2; (b) the allowable stress according to the AISC formula, Eq. (11.8).

11.3. In Fig. P11.3 is shown a square frame. Determine the critical value of the compressive forces P. All members are of equal length L and of equal modulus of rigidity EI. Assume that symmetrical buckling, indicated by the dashed lines in the figure, occurs.

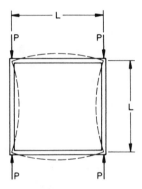

Figure P11.3

11.4. A rigid block of weight W is to be supported by three identical steel bars. The bars are fixed at each end (Fig. P11.4). Assume that sidesway is not prevented and that when an additional downward force of $2W$ is applied at the middle of the block, buckling will take place as indicated by the dashed lines in the figure. Find the effective lengths of the columns by solving the differential equation for deflection of the column axis.

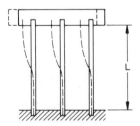

Figure P11.4

11.5. A simply supported beam of flexural rigidity EI_b is propped up at its center by a column of flexural rigidity EI_c (Fig P11.5). Determine the midspan deflection of the beam if it is subjected to a uniform load p per unit length.

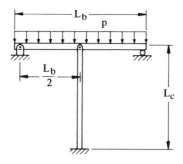

Figure P11.5

11.6. Two in-line identical cantilevers of cross-sectional area A, rigidity EI, and coefficient of thermal expansion α are separated by a small gap δ. What temperature rise will cause the beams to (a) just touch and (b) buckle elastically?

11.7. A W203×25 column fixed at both ends has a minimum radius of gyration $r = 29.4$ mm, cross-sectional area $A = 3230$ mm², and length 1.94 m. It is made of a material whose compression stress–strain diagram is given in Fig. P11.7 by dashed lines. Find the critical load. The stress–strain diagram may be approximated by a series of tangentlike segments, the accuracy improving as the number of segments increases. For simplicity, use four segments as indicated in the figure. The modulus of elasticity and various tangent moduli (the slopes) are labeled.

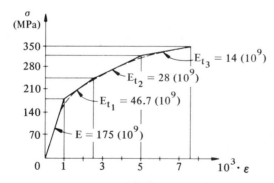

Figure P11.7

11.8. The pin-jointed structure shown in Fig. P11.8. is constructed of two 0.025-m diameter tubes, having the following properties: $A = 5.4 \times 10^{-5}$ m², $I = 3.91 \times 10^{-9}$ m⁴, $E = 210$ GPa. The stress–strain curve for the tube material can be accurately approximated by three straight lines as shown. If the load P is increased until the structure fails, which tube buckles first? Describe the nature of the failure and determine the critical load.

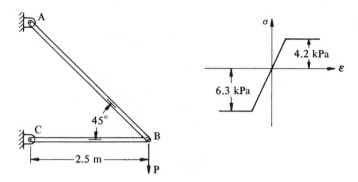

Figure P11.8

11.9. Two 0.075- by 0.075-m equal leg angles, positioned with the legs 0.025 m apart back to back, as shown in Fig. P11.9, are used as a column. The angles are made of structural steel with $\sigma_{yp} = 203$ MPa and $E = 210$ GPa. The area properties of an angle are thickness $t = 0.0125$ m, $A = 1.719 \times 10^{-3}$ m², $I_c = 8.6 \times 10^{-7}$ m⁴, $I/c = Z = 1.719 \times 10^{-5}$ m³, $r_c = 0.0225$ m, and $\bar{z} = \bar{y} = 0.02325$ m. Assume that the columns are connected by lacing bars which cause them to act as a unit. Determine the critical stress of the column by using the AISC formula, Eq. (11.8), for effective column lengths (a) 2.1 m, (b) 4.2 m.

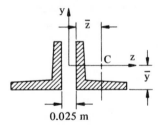

Figure P11.9

11.10. Redo Problem 11.9(b) by the use of the Euler formula.

11.11. A 1.2-m long, 0.025- by 0.05-m rectangular column with rounded ends fits perfectly between a rigid ceiling and a rigid floor. Compute the change in temperature that will cause the column to buckle. Let $\alpha = 10 \times 10^{-6}/°C$, $E = 140$ GPa, $\sigma_{yp} = 280$ MPa.

Secs. 11.7 and 11.8

11.12. A 0.05-m square, horizontal steel bar, 9 m long, is simply supported at each end. The only force acting is the weight of the bar. (a) Find the maximum stress and deflection. (b) Assume that an axial compressive load of 4.5 kN is also applied at each end through the centroid of cross-sectional area. Determine the stress and deflection under this combined loading. For steel, the specific weight is 77 kN/m³, $E = 210$ GPa, and $\nu = 0.3$.

11.13. The properties of a W203×46 steel link are $A = 5880$ mm², $I_z \approx 45.66 \times 10^6$ mm⁴, $I_y \approx 15.4 \times 10^6$ mm⁴, depth $= 203.2$ mm, width of flange $= 203.2$ mm, and $E = 210$ GPa. What maximum end load P can be applied at both ends, given an eccentricity of 0.05 m along axis yy? A stress of 210 MPa is not to be exceeded. Assume that the effective column length of the link is 4.5 m.

11.14. A hinge-ended bar of length L and stiffness EI has an initial curvature expressed by $v_0 = a_1 \sin(\pi x/L) + 5a_1 \sin(2\pi x/L)$. If this bar is subjected to an increasing axial load P, what value of the load P, expressed in terms of L, E, and I, will result in zero deflection at $x = 3L/4$?

11.15. Employing the equilibrium approach, derive the following differential equation for a simply supported beam-column subjected to an arbitrary distributed transverse loading $p(x)$ and axial force P:

$$\frac{d^4v}{dx^4} + \frac{P}{EI}\frac{d^2v}{dx^2} = \frac{p}{EI} \qquad \text{(P11.15)}$$

Demonstrate that the homogeneous solution of the above is

$$v = c_1 \sin\sqrt{\frac{P}{EI}}\, x + c_2 \cos\sqrt{\frac{P}{EI}}\, x + c_3 x + c_4$$

where the four constants of integration will require, for evaluation, four boundary conditions.

Secs. 11.9

11.16. Assuming $v = a_0[1 - (2x/L)^2]$, determine the buckling load of a pin-ended column. Employ the Rayleigh–Ritz method, placing the origin at midspan.

11.17. The cross section of a pin-ended column varies as in Fig. P11.17. Determine the critical load using an energy approach.

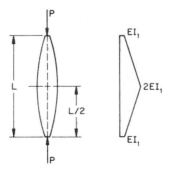

Figure P11.17

11.18. A cantilever column has a moment of inertia varying as $I=I_1(1-x/2L)$, where I_1 is the constant moment of inertia at the fixed end $(x=0)$. find the buckling load by choosing $v=v_1(x/L)^2$. Here v_1 is the deflection of the free end.

11.19. Derive an expression for the buckling load of the uniform pin-ended beam-column of length L, subjected to a uniform transverse load p and axial compressive force P. Use an energy approach.

11.20. A simply supported beam-column of length L is subjected to compression forces P at both ends and lateral loads F and $2F$ at quarter length and midlength, respectively. Employ the Rayleigh–Ritz method to determine the beam deflection.

11.21. Determine the critical compressive load P that can be carried by a cantilever at its free end $(x=L)$. Use the Rayleigh–Ritz method and let $v=x^2(L-x)(a+bx,)$, where a and b are constants.

Secs. 1.10

11.22. A stepped cantilever beam with a hinged end, subjected to the axial compressive load P, is shown in Fig. P11.22. Determine the critical value of P, applying the method of finite differences. Let $m=3$ and $L_1=L_2=L/2$.

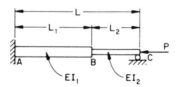

Figure P11.22

11.23. A uniform cantilever column is subjected to axial compression at the free end $(x=L)$. Determine the critical load. Employ finite differences, using $m=2$. [Hint: the boundary conditions are $v(0)=v'(0)=v''(L)=v'''(L)=0$.]

11.24. The cross section of a pin-ended column varies as in Fig. P11.17. Determine the critical load using the method of finite differences. Let $m=4$.

11.25. Find the critical value of the load P in Fig. P11.22 if both ends of the beam are simply supported. Let $L_1=L/4$ and $L_2=3L/4$. Employ the method of finite differences by taking the nodes at $x=0$, $x=L/4$, $x=L/2$, and $x=L$.

Chapter 12

Plastic Behavior of Solids

12.1 Introduction

The subject of plasticity is perhaps best introduced by recalling the principal characteristics of elastic behavior. First, a material subjected to stressing within the elastic regime will return to its original state upon the removal of those external influences causing application of load or displacement. Second, the deformation corresponding to a given stress is dependent solely upon that stress and not upon the history of strain or load. In plastic behavior, opposite characteristics are observed. The permanent distortion that takes place in the plastic range of a material can assume considerable proportions. This distortion depends not only upon the final state of stress, but upon the stress states existing from the start of the loading process as well. The equations of equilibrium (1.5), the conditions of compatibility (2.8), and the strain-displacement relationships (2.3) are all valid in plastic theory. New relationships must, however, be derived to connect stress and strain. The various yield criteria, which strictly speaking are not required in solving a problem in elasticity, play a direct and important role in plasticity.

This chapter can provide only an introduction to what is an active area of contemporary design and research in the mechanics of solids.* The basics presented can, however, indicate the potential of the field as well as its complexities.

12.2 Plastic Deformation

We shall here deal with the permanent alteration in the shape of a polycrystalline solid subject to external loading. The crystals are assumed to be randomly oriented. As has been demonstrated in Sec. 2.9, the stresses

---

*See, for example, A. Mendelson, *Plasticity*, New York: Macmillan, 1968; J. H. Faupel, *Engineering Design*, New York: Wiley, 1964.

acting on an elemental cube can be resolved into those associated with change of volume (dilatation) and those causing distortion or change of shape. The distortional stresses are usually referred to as the *deviator stresses*.

The dilatational stresses, such as hydrostatic pressures, can clearly decrease the volume while they are applied. The volume change is recoverable, however, upon removal of external load. This is because the material cannot be compelled to assume, in the absence of external loading, interatomic distances different from their initial values. When the dilatational stresses are removed, therefore, the atoms revert to their original position. Under the conditions described, no plastic behavior is noted, and the volume is essentially unchanged upon removal of load.

In contrast with the situation described above, during change of shape, the atoms within a crystal of a polycrystalline solid slide over one another. This slip action, referred to as *dislocation*, is a complex phenomenon. Dislocation can occur only by the shearing of atomic layers, and consequently it is primarily the shear component of the deviator stresses that controls plastic deformation. There is experimental evidence to support the assumption that associated with plastic deformation essentially *no volume change* occurs, i.e., the material is incompressible (Sec. 2.6):

$$\varepsilon_x + \varepsilon_y + \varepsilon_z = \varepsilon_1 + \varepsilon_2 + \varepsilon_3 = 0 \tag{12.1}$$

Therefore from Eq. (2.21), *Poisson's ratio* $\nu = \frac{1}{2}$ *for a plastic material*.

Slip begins at an imperfection in the lattice, e.g., along a plane separating two regions, one having one more atom per row than the other. Because slip does not occur simultaneously along every atomic plane, the deformation appears discontinuous on the microscopic level of the crystal grains. The overall effect, however, is plastic shear along certain slip planes, and the behavior described is approximately that of the ideal plastic solid. As the deformation continues, a locking of the dislocations takes place, resulting in *strain hardening*.

In performing engineering analyses of stresses in the plastic range, one does not usually need to consider dislocation theory, and the explanation offered above, while overly simple, will suffice. What is of great importance to the analyst, however, is the experimentally determined curves of stress and strain.

12.3 True Stress–True Strain Curve in Simple Tension

The bulk of present day analysis in plasticity is predicated upon materials displaying idealized stress–strain curves as in Fig. 12.1a. Such materials are referred to as *perfectly plastic*. Examples include mild steel and nylon, which exhibit negligible elastic strains in comparison with large plastic deformations at practically constant stress. A more realistic portrayal

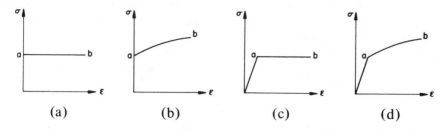

Figure 12.1

including strain hardening is given in Fig. 12.1b for what is called a *rigid plastic* solid. In the curves, *a* and *b* designate the tensile yield and ultimate stresses, σ_{yp} and σ_u, respectively. The curves of Fig. 12.1c and d do not ignore the elastic strain, which must be included in a more general stress–strain depiction. The latter figures thus represent idealized *elastic-plastic* diagrams for the perfectly plastic and rigid plastic materials, respectively.

A general *true stress–true strain curve* may be represented by the *empirical expression*

$$\sigma = \sigma_{yp} + k\varepsilon^n \tag{a}$$

where *n* and *k* are termed the *strain-hardening index* and the *strength coefficient*. The true stress σ and the true strain ε are defined by Eqs. (2.12) and (2.13). For $0 \leqslant n \leqslant 1$, the above form describes the material idealized in Fig. 12.1b. For $\sigma_{yp} = E\varepsilon$, Eq. (a) represents an elastic-plastic stress as shown in the stress diagram of Fig. 12.1d. Clearly, for $k = 0$, the above expression is represented by Fig. 12.1a.

When no yielding occurs prior to plastic deformation (i.e., $\sigma_{yp} = 0$), the true stress and the true strain are connected by a parabola,

$$\sigma = k\varepsilon^n \tag{12.2}$$

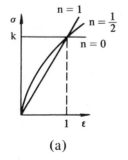

(a)

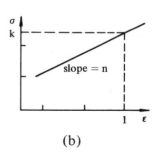

(b)

Figure 12.2

and the curves are as in Fig. 12.2a. We observe from the figure that the slope $d\sigma/d\varepsilon$ grows without limit as ε approaches zero for $n \neq 1$. Hence, Eq. (12.2) should not be used for small strains.

For a particular material, true stress–true strain data available, k and n are readily evaluated inasmuch as Eq. (12.2) plots as a straight line on *logarithmic coordinates*. We can thus rewrite Eq. (12.2) in the form

$$\log \sigma = \log k + n \log \varepsilon \tag{b}$$

Here n is the *slope* of the line and k the true stress associated with the true strain at 1.0 on the *log–log plot*, Fig. 12.2b. The strain-hardening coefficient n for commercially used materials falls between 0.2 and 0.5.

We now describe an *instability phenomenon* in uniaxial tension of practical importance in predicting the maximum allowable plastic stress in a rigid-plastic material. At the ultimate stress in a tensile test (Fig. 2.6), an unstable flow results from the effects of strain hardening and the decreasing cross-sectional area of the specimen. These tend to weaken the material. When the rate of the former effect is less than the latter, an instability occurs. This point corresponds to the *maximum tensile load* and is defined by

$$dP = 0 \tag{c}$$

Since axial load P is a function of both the true stress and the area (i.e., $P = \sigma A$), the above is rewritten

$$\sigma \, dA + A \, d\sigma = 0 \tag{d}$$

The condition of incompressibility, $A_0 L_0 = AL$, also yields

$$L \, dA + A \, dL = 0 \tag{e}$$

as the original volume $A_0 L_0$ is constant. Expressions (d) and (e) result in

$$\frac{d\sigma}{\sigma} = \frac{dL}{L} = d\varepsilon \tag{f}$$

From Eqs. (2.13) and (f), we thus obtain the relationships

$$\frac{d\sigma}{d\varepsilon} = \sigma \quad \text{or} \quad \frac{d\sigma}{d\varepsilon_0} = \frac{\sigma}{1 + \varepsilon_0} = \sigma_0 \tag{12.3}$$

for the *instability of a tensile member*. Here the subscript o denotes the engineering strain and stress (Sec. 2.5).

Introduction of Eq. (12.2) into Eq. (12.3) results in

$$\sigma = k\varepsilon^n = \frac{d}{d\varepsilon}(k\varepsilon^n) = nk\varepsilon^{n-1}$$

or

$$\varepsilon = n \tag{12.4}$$

That is, at the instant of instability of flow *in tension*, the *true strain ε has the same numerical value as the strain-hardening index*. The state of true stress and the true strain under uniaxial tension are therefore

$$\sigma_1 = kn^n, \qquad \sigma_2 = \sigma_3 = 0$$

$$\varepsilon_1 = n, \qquad \varepsilon_2 = \varepsilon_3 = -\frac{n}{2} \tag{g}$$

The problem of instability under *simple compression* or *plastic buckling* is discussed in Sec. 11.5. The instability condition for cases involving *biaxial tension* is derived in Secs. 12.9 and 12.10.

Example 12.1. Determine the maximum allowable plastic stress and strain in the frame sustaining a vertical load P, shown in Fig. P10.13. Assume that $\alpha = 45°$ and that each element is constructed of an aluminum alloy with the following properties:

$$\sigma_{yp} = 350 \text{ MPa}, \qquad k = 840 \text{ MPa}, \qquad n = 0.2$$

$$A_{AD} = A_{CD} = 10 \times 10^{-5} \text{ m}^2, \qquad A_{BD} = 15 \times 10^{-5} \text{ m}^2, \qquad L_0 = 3 \text{ m}$$

SOLUTION. The frame is elastically statically indeterminate and the solution may readily be obtained on applying Castigliano's theorem (Sec. 10.7). Plastic yielding begins upon loading:

$$P = \sigma_{yp} A_{BD} + 2\sigma_{yp} A_{AD} \cos \alpha$$

$$= 350 \times 10^6 [15 + 2 \times 10 \cos 45°] 10^{-5} = 101{,}990 \; N$$

On Applying Eqs. (g), the maximum allowable stress

$$\sigma = kn^n = 840 \times 10^6 (0.2)^{0.2} = 608.9 \text{ MPa}$$

occurs at the following axial and transverse strains

$$\varepsilon_1 = n = 0.2, \qquad \varepsilon_2 = \varepsilon_3 = -0.1$$

For example, the total elongation for instability of the central bar is thus $3(0.2) = 0.6$ m.

Example 12.2. A tube of original mean diameter d_0 and thickness t_0 is subjected to axial tensile loading. Assume a true stress–engineering strain relation of the form $\sigma = k_1 \varepsilon_0^n$ and derive expressions for thickness and diameter at the instant of instability. Let $n = 0.3$.

SOLUTION.   Differentiating the given expression for stress

$$\frac{d\sigma}{d\varepsilon_0} = nk_1 \varepsilon_0^{n-1} = \frac{n\sigma}{\varepsilon_0}$$

This result and Eq. (12.3) yield the engineering axial strain at instability:

$$\varepsilon_0 = \frac{n}{1-n}$$

The transverse strains are $-\varepsilon_0/2$, and hence the decrease of wall thickness equals $nt_0/2(1-n)$. The thickness at instability is thus

$$t = t_0 - \frac{nt_0}{2(1-n)} = \frac{2-3n}{2(1-n)} t_0 \tag{h}$$

Similarly, the diameter at instability is

$$d=d_0-\frac{nd_0}{2(1-n)}=\frac{2-3n}{2(1-n)}d_0 \qquad (i)$$

From Eqs. (h) and (i), with $n=0.3$, we have $t=0.79t_0$ and $d=0.79d_0$. Thus, for the tube under axial tension, the diameter and thickness decreases approximately 21% at the instant of instability.

12.4 Theory of Plastic Bending

In this section we treat the deflection of a rigid-plastic beam, employing the mechanics of materials approach. Consider a beam of rectangular section, as in Fig. 12.3a, wherein the bending moment M produces a radius of curvature r. The longitudinal strain of any fiber located a distance y from the neutral surface, from Eq. (5.9), is given by

$$\varepsilon=\frac{y}{r} \qquad (12.5)$$

Assume the beam material to possess equal properties in tension and compression. Owing to the distribution of stress shown, the longitudinal tensile and compressive forces cancel, and the equilibrium of axial forces is satisfied. The following describes the equilibrium of moments about the z axis (Fig. 12.3a):

$$\int_A y\sigma\,dA=b\int_{-h}^{h} y\sigma\,dy=M \qquad (a)$$

For any specific distribution of stress, as for example that shown in Fig. 12.3b, Eq. (a) provides M and then the deflection, as is demonstrated below.

Consider the true stress–true strain relationship of the form $\sigma=k\varepsilon^n$. Introducing the above, together with Eq. (12.5), into Eq. (a), we obtain

$$M=b\int_{-h}^{h}\frac{1}{r^n}ky^{n+1}\,dy=\frac{1}{r^n}kI_n \qquad (b)$$

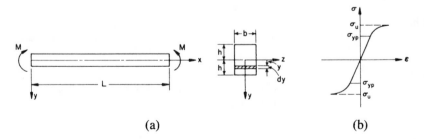

(a)   (b)

Figure 12.3

where

$$I_n = b \int_{-h}^{h} y^{n+1} dy \tag{12.6}$$

From Eqs. (12.5), $\sigma = k\varepsilon^n$, and (b) the following is derived:

$$\frac{\sigma}{y^n} = \frac{k}{r^n} = \frac{M}{I_n} \tag{c}$$

In addition, on the basis of the elementary beam theory, we have, from Eq. (5.7),

$$\frac{1}{r} = \frac{d^2 v}{dx^2} \tag{d}$$

Upon substituting Eq. (c) into Eq. (d), we obtain the following equation for a rigid plastic beam

$$\frac{d^2 v}{dx^2} = \left(\frac{M}{kI_n} \right)^{1/n} \tag{12.7}$$

It is noted that when $n = 1$ (and hence $k = E$), this expression, as expected, reduces to that of an elastic beam, Eq. (5.10).

Example 12.3. Determine the deflection of a rigid plastic simply supported beam subjected to a downward concentrated force P at its midlength. The beam has a rectangular cross section of depth $2h$ and width b. The span length is L.

SOLUTION. The bending moment at any section is given by

$$M = -\tfrac{1}{2} Px \tag{e}$$

where the minus sign is due to the sign convention of Sec. 5.2.

Substituting Eq. (e) into Eq. (12.7) and integrating, we have

$$\frac{dv}{dx} = - \frac{\lambda x^{(1/n)+1}}{(1/n)+1} + c_1$$

$$v = - \frac{\lambda x^{(1/n)+2}}{[(1/n)+1][(1/n)+2]} + c_1 x + c_2 \tag{f}$$

where

$$\lambda = \left(\frac{P}{2kI_n} \right)^{1/n} \tag{g}$$

The constants of integration c_1 and c_2 depend upon the boundary conditions $v(0) = dv/dx(L/2) = 0$:

$$c_2 = 0, \qquad c_1 = \frac{\lambda (L/2)^{(1/n)+1}}{(1/n)+1}$$

Upon introduction of c_1 and c_2 into Eq. (f), the beam deflection is found to be

$$v = - \frac{\lambda}{(1/n)+1} \left[\frac{x^{(1/n)+2}}{(1/n)+2} - \left(\frac{L}{2} \right)^{(1/n)+1} x \right] \tag{12.8}$$

Interestingly, in the case of an elastic beam, the above becomes

$$v = -\frac{P}{4EI_1}\left(\frac{x^3}{3} - \frac{L^2}{4}x\right)$$

For $x = L/2$, the familiar result is

$$v = v_{max} = \frac{PL^3}{48EI_1}$$

The foregoing procedure is applicable to the determination of the deflection of beams subject to a variety of end conditions and load configurations. It is clear, however, that owing to the nonlinearity of the stress law, $\sigma = k\varepsilon^n$, the principle of superposition cannot validly be applied.

12.5 Analysis of Perfectly Plastic Beams

By neglecting strain hardening, that is, by assuming a perfectly plastic material, considerable simplification can be realized. We shall, in this section, focus our attention on the analysis of a perfectly plastic straight beam of rectangular section subject to *pure bending* (Fig. 12.3a).

The bending moment at which plastic deformation impends, M_{yp}, may be found directly from the flexure relationship:

$$M_{yp} = \frac{\sigma_{yp}I}{h} = \tfrac{2}{3}bh^2\sigma_{yp} \tag{12.9}$$

Here σ_{yp} represents the stress at which yielding begins (and at which deformation continues in a perfectly plastic material). The stress distribution corresponding to M_{yp}, assuming identical material properties in tension and compression, is shown in Fig. 12.4a. As the bending moment is increased, the region of the beam which has yielded progresses in toward the neutral surface (Fig. 12.4b). The distance from the neutral surface to the point at which yielding begins is denoted by the symbol e as shown.

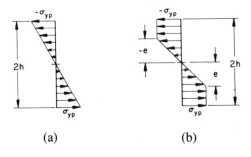

(a)   (b)

Figure 12.4

It is clear, upon examining Fig. 12.4b, that the normal stress varies in accordance with the relations

$$\sigma_x = \frac{\sigma_{yp} y}{e} \qquad (-e \leqslant y \leqslant e) \tag{a}$$

and

$$\sigma_x = \sigma_{yp} \qquad (e \leqslant y \leqslant h)$$
$$\sigma_x = -\sigma_{yp} \qquad (-e \geqslant y \geqslant -h) \tag{b}$$

It will be useful to determine the manner in which the bending moment M relates to the distance e. To do this, we begin with a statement of the x equilibrium of forces:

$$\int_{-h}^{-e} -\sigma_{yp} b \, dy + \int_{-e}^{e} \sigma_x b \, dy + \int_{e}^{h} \sigma_{yp} b \, dy = 0$$

Canceling the first and third integrals and combining the remaining integral with Eq. (a), we have

$$\frac{\sigma_{yp}}{e} \int_{-e}^{e} yb \, dy = 0$$

The above expression indicates that the *neutral and centroidal axes of the cross section coincide*, as in the case of an entirely elastic distribution of stress. Next, the equilibrium of moments about the neutral axis provides the following relation:

$$\int_{-h}^{-e} -\sigma_{yp} yb \, dy + \int_{-e}^{e} \sigma_x yb \, dy + \int_{e}^{h} \sigma_{yp} yb \, dy = M$$

Substituting σ_x from Eq. (a) into the above gives, after integration

$$M = b\sigma_{yp}\left(h^2 - \frac{e^2}{3}\right) \tag{12.10}$$

The general stress distribution is thus defined in terms of the applied moment, inasmuch as e is connected to σ_x by Eq. (a). For the case in which $e = h$, Eq. (12.10) reduces to Eq. (12.9) and $M = M_{yp}$. For $e = 0$, which applies to a totally plastic beam, Eq. (12.10) becomes

$$M_u = bh^2\sigma_{yp} \tag{12.11}$$

where M_u is the *ultimate moment*.

Through application of the foregoing analysis, similar relationships can be derived for other cross-sectional shapes. In general, for any cross section the plastic or ultimate resisting moment for a beam is

$$M_u = \sigma_{yp} Z \tag{12.12}$$

where Z is the *plastic section modulus*. Clearly, for the rectangular beam analyzed above, $Z = bh^2$. The *Steel Construction Manual* (Sec. 11.6) lists plastic section moduli for many common geometries.

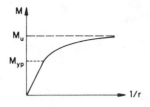

Figure 12.5

To express the beam curvature in terms of the yield stress, we begin by noting that the longitudinal strain given by Eq. (12.5), combined with Hooke's law, leads to

$$\sigma_{yp} = E\varepsilon_{yp} = E\frac{e}{r}$$

The curvature and yield stress are thus related by

$$\frac{1}{r} = \frac{\sigma_{yp}}{Ee} \tag{12.13}$$

A plot of Eq. (12.10) combined with Eq. (12.13) (Fig. 12.5) reveals that after the yielding moment M_{yp} is achieved, M and $1/r$ are connected by a nonlinear relationship. As M approaches M_u, the curvature grows without limit. The ultimate moment is also referred to as the *plastic hinge moment*. The rationale for the term *hinge* becomes apparent upon describing the behavior of a beam under a concentrated loading, discussed next.

Consider a simply supported beam of rectangular cross section, subjected to a load P at its midspan, (Fig. 12.6a). The corresponding bending moment diagram is shown in Fig. 12.6b. Clearly, when $M_{yp} < |PL/4| < M_u$,

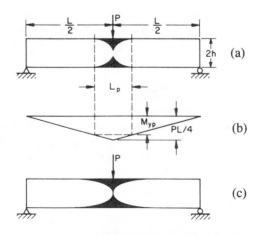

Figure 12.6

a region of plastic deformation occurs, as indicated in the figure by the shaded areas. The depth of penetration of these zones can be found from $h-e$, where e is determined using Eq. (12.10), as M at midspan is known. The length of the middle portion of the beam where plastic deformation occurs can readily be determined with reference to the figure. The magnitude of bending moment at the edge of the plastic zone is $M_{yp}=(P/2)(L-L_p)/2$, from which

$$L_p=L-\frac{4M_{yp}}{P} \tag{c}$$

With the increase of P, $M_{max}\to M_u$, and the plastic region extends further inward. When the magnitude of the maximum moment $PL/4$ is equal to M_u, the cross section at the midspan becomes fully plastic (Fig. 12.6c). Then, as in case of pure bending, the curvature at the center of the beam grows without limit, and the beam fails. The beam halves on either side of the midspan experience rotation in the manner of a rigid body about the neutral axis, as about a *plastic hinge*, under the influence of the constant ultimate moment M_u. For a plastic hinge, $P=4M_u/L$ is substituted into Eq. (c), leading to $L_p=L(1-M_{yp}/M_u)$.

The capacity of a beam to resist collapse is revealed by comparing Eqs. (12.9) and (12.11). Note that the M_u is 1.5 times as large as M_{yp}. Elastic design is thus conservative. Considerations such as this lead to concepts of *limit design* in structures, discussed in the next section.

Example 12.4. A link of rectangular cross section is subjected to a load N (Fig. 12.7a). Derive general relationships involving N and M which govern first the case of initial yielding, and then fully plastic deformation, for the straight part of the link of length L.

SOLUTION. Suppose N and M are such that the state of stress is as shown in Fig. 12.7b at any straight beam section. The maximum stress in the beam is then, by superposition of the axial and bending stresses,

$$\sigma_{yp}=\frac{N_1}{2hb}+\frac{3}{2}\frac{M_1}{bh^2} \tag{d}$$

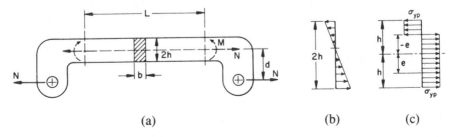

(a)   (b)   (c)

Figure 12.7

The upper limits on N ($M=0$) and M ($N=0$), corresponding to the condition of yielding, are

$$N_{\text{yp}}=2hb\,\sigma_{\text{yp}}, \qquad M_{\text{yp}}=\frac{I}{h}\sigma_{\text{yp}}=\tfrac{2}{3}bh^2\sigma_{\text{yp}} \tag{e}$$

Substituting $2hb$ and I/h from the above into Eq. (d) and rearranging terms, we have

$$\frac{N_1}{N_{\text{yp}}}+\frac{M_1}{M_{\text{yp}}}=1 \tag{12.14}$$

If N_1 is zero, then M_1 must achieve its maximum value M_{yp} for yielding to impend. Similarly, for $M_1=0$, it is necessary for N_1 to equal N_{yp} to initiate yielding. Between these extremes, Eq. (12.14) provides the infinity of combinations of N_1 and M_1 which will result in σ_{yp}.

For the fully plastic case (Fig. 12.7c), we shall denote the state of loading by N_2 and M_2. It is apparent that the stresses acting within the range $-e<y<e$ contribute pure axial load only. The stresses within the range $e<y<h$ and $-e>y>-h$ form a couple, however. For the total load system described, we may write

$$N_2=2eb\,\sigma_{\text{yp}}, \qquad e=\frac{N_2}{2b\sigma_{\text{yp}}} \tag{f}$$

$$M_2=(h-e)b\sigma_{\text{yp}}\cdot 2\!\left(e+\frac{h-e}{2}\right)=b(h^2-e^2)\sigma_{\text{yp}} \tag{g}$$

Introducing Eqs. (12.11) and (f) into the above expression, one has

$$M_2=M_{\text{u}}-\frac{N_2^2}{4b\sigma_{\text{yp}}}$$

Finally, dividing by M_{u} and noting that $M_{\text{u}}=\tfrac{3}{2}M_{\text{yp}}=bh^2\sigma_{\text{yp}}$, $N_{\text{yp}}=2bh\sigma_{\text{yp}}$, we obtain

$$\frac{2}{3}\frac{M_2}{M_{\text{yp}}}+\left(\frac{N_2}{N_{\text{yp}}}\right)^2=1 \tag{12.15}$$

Figure 12.8 is a plot of Eqs. (12.14) and (12.15). By employing these *interaction curves*, any combination of limiting values of bending moment and axial force is easily arrived at.

Let, for instance, $d=h$, $h=2b=24$ mm (Fig. 12.7a), and $\sigma_{\text{yp}}=280$ MPa. Then the value of $N/M=\frac{1}{24}$, and from Eq. (e), $M_{\text{yp}}/N_{\text{yp}}=h/3=8$. The radial line representing $(N/M)(M_{\text{yp}}/N_{\text{yp}})=\frac{1}{3}$ is indicated by the dashed line in the figure. This line intersects the interaction curves at $A(0.75, 0.25)$ and $B(1.24, 0.41)$. Thus yielding impends for $N_1=0.25N_{\text{yp}}=0.25(2bh\cdot\sigma_{\text{yp}})=40.32$ kN, and for fully plastic deformation, $N_2=0.41N_{\text{yp}}=66.125$ kN.

Note that the distance d is assumed constant and the values of N found are conservative. If the link deflection were taken into account, d would be smaller and the above calculations would yield larger N.

Example 12.5. An I-beam (Fig. 12.9a) is subjected to pure bending resulting from end couples. Determine the moment causing initial yielding and that resulting in complete plastic deformation.

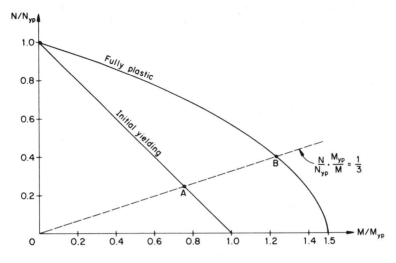

Figure 12.8

SOLUTION. The moment corresponding to σ_{yp} is, from Eq. (12.9),

$$M_{yp} = \frac{I}{h}\sigma_{yp} = \frac{1}{h}\left(\frac{2}{3}bh^3 - \frac{2}{3}b_1h_1^3\right)\sigma_{yp} \tag{h}$$

Refer now to the completely plastic stress distribution of Fig. 12.9b. The moments of force owing to σ_{yp}, taken about the neutral axis, provide

$$M_u = (bh^2 - b_1h_1^2)\sigma_{yp} \tag{i}$$

Combining Eqs. (h) and (i), we have

$$\frac{M_u}{M_{yp}} = \frac{3}{2}\frac{1 - (b_1h_1^2/bh^2)}{1 - (b_1h_1^3/bh^3)}$$

From the above expression, it is seen that $M_u/M_{yp} < \frac{3}{2}$, while it is $\frac{3}{2}$ for a beam of rectangular section ($h_1 = 0$). We conclude therefore that if a rectangular beam and an I-beam are designed plastically, the former will be more resistant to complete plastic failure.

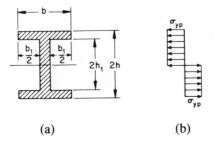

(a)   (b)

Figure 12.9

Example 12.6. Determine the maximum deflection due to an applied force P acting on the simply supported rectangular beam shown in Fig. 12.6.

SOLUTION. The center deflection in the *elastic range* is given by

$$v_{max} = \frac{PL}{48EI} \tag{j}$$

At the *start of yielding*,

$$P = P_{yp}, \qquad M_{max} = M_{yp} = \frac{P_{yp}L}{4} \tag{k}$$

Expression (j), together with Eqs. (k) and (12.9), leads to

$$v_{max} = \frac{M_{yp}L^2}{12EI} = \frac{1}{12}\left(\frac{L^2}{Eh}\right)\sigma_{yp} \tag{l}$$

In a like manner, we obtain

$$v_{max} = \frac{M_u L^2}{12EI} = \frac{1}{8}\left(\frac{L^2}{Eh}\right)\sigma_{yp} \tag{m}$$

for the center deflection at the instant of *plastic collapse*.

12.6 The Collapse Load of Structures

On the basis of the simple examples in the previous section, it may be deduced that structures may withstand loads in excess of those that lead to initial yielding. We recognize that while such loads need not cause structural collapse, they will result in some amount of permanent deformation. If no permanent deformation is to be permitted, the load configuration must be such that the stress does not attain the yield point anywhere in the structure. This is, of course, the basis of elastic design.

When a limited amount of permanent deformation may be tolerated in a structure, the design can be predicated upon higher loads than correspond to initiation of yielding. On the basis of the ultimate or plastic load determination, safe dimensions can be determined in what is termed *limit design*. Clearly such design requires higher than usual factors of safety. Examples of ultimate load determination are presented below.

Consider first a built-in beam subjected to a concentrated load at midspan (Fig. 12.10a). The general bending moment variation is sketched in Fig. 12.10b. As the load is progressively increased, we may anticipate plastic hinges at points 1, 2, and 3, as these are the points at which maximum bending moments are found (Fig. 12.10c). The configuration indicating the assumed location of the plastic hinges (Fig. 12.10c) is the *mechanism of collapse*. At every hinge, the hinge moment must clearly be the same.

The equilibrium and the energy approaches are available for determination of the ultimate loading. Electing the latter, we refer to Fig. 12.10c and note that the change in energy associated with rotation at points 1 and 2 is $M_u \cdot \delta\theta$, while at point 3, it is $M_u(2\,\delta\theta)$. The work done by the concentrated

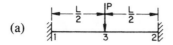

(a)

(b)

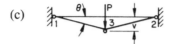

(c)

Figure 12.10

force is $P \cdot \delta v$. According to the principle of virtual work, we may write

$$P_u(\delta v) = M_u(\delta \theta) + M_u(\delta \theta) + M_u(2\,\delta \theta) = 4M_u(\delta \theta)$$

where P_u represents the *ultimate load*. Because the deformations are limited to small values, it may be stated that $v \approx \frac{1}{2} L\theta$ and $\delta v = \frac{1}{2} L\,\delta \theta$. Substituting in the above expression for δv, it is found that

$$P_u = \frac{8M_u}{L} \tag{a}$$

where M_u is calculated for a given beam using Eq. (12.12). It is interesting that by introduction of the plastic hinges, the originally statically indeterminate beam is rendered determinate. The determination of P_u is thus simpler than that of P_{yp}, on which elastic analysis is based. An advantage of limit design may also be found in noting that a small rotation at either end of the beam or a slight lowering of a support will not influence the value of P_u. Moderate departures from the ideal case, such as these, will, however, have a pronounced effect upon the value of P_{yp} in a statically indeterminate system.

While the positioning of the plastic hinges in the above problem is limited to the single possibility shown in Fig. 12.10c, more than one possibility will exist for situations in which several forces act. Correspondingly, a number of collapse mechanisms may exist, and it is incumbent upon the designer to select from among them the one associated with the lowest load.

Example 12.7. Determine the collapse load of the continuous beam shown in Fig. 12.11a.

SOLUTION. The four possibilities of collapse are indicated in Fig. 12.11b through *d*. We first consider the mechanism of Fig. 12.11b. In this system, motion occurs

(a)
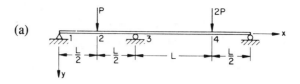

(b)
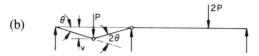

(c)
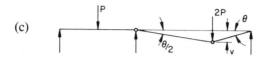

(d)

Figure 12.11

because of rotations at hinges 1, 2, and 3. The remainder of the beam remains rigid. Applying the principle of virtual work, noting that the moment at point 1 is zero, we have

$$P(\delta v) = M_u(2\,\delta\theta) + M_u(\delta\theta) = 3M_u(\delta\theta)$$

As

$$v \approx \tfrac{1}{2}L\theta, \qquad \delta v = \tfrac{1}{2}L\,\delta\theta$$

the above equation yields $P_u = 6M_u/L$.

For the collapse mode of Fig. 12.11c,

$$2P(\delta v) = M_u\left(\tfrac{1}{2}\delta\theta\right) + M_u\left(\tfrac{3}{2}\delta\theta\right) = 2M_u(\delta\theta)$$

and thus $P_u = 2M_u/L$.

The collapse mechanisms indicated by the solid and dashed lines of Fig. 12.11d are unacceptable because they imply a zero bending moment at section 3. We conclude that collapse will occur as in Fig. 12.11c, when $P \to 2M_u/L$.

Example 12.8. Determine the collapse load of the beam shown in Fig. 12.12a.

SOLUTION. There are a number of collapse possibilities, of which one is indicated in Fig. 12.12b. Let us suppose that there exists a hinge at point 2, a distance e from the left support. Then examination of the geometry leads to $\theta_1 = \theta_2(L-e)/e$ or $\theta_1 + \theta_2 = L\theta_2/e$. Applying the principle of virtual work,

$$\delta\int_0^e (\theta_1 x)p\,dx + \delta\int_e^L [\theta_1 e - \theta_2(x-e)]p\,dx = M_u\left(\frac{L}{e}\delta\theta_2\right) + M_u\,\delta\theta_2$$

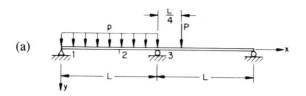

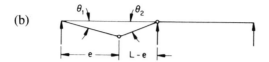

Figure 12.12

or

$$\frac{L(L-e)}{2}p = \left(\frac{L}{e}+1\right)M_u$$

from which

$$p = \frac{2(e+L)M_u}{e(L-e)L} \qquad \text{(b)}$$

The minimization condition for p in Eq. (b), $dp/de = 0$, results in

$$e = L\left(\frac{\sqrt{8}}{2}-1\right) \qquad \text{(c)}$$

Thus Eq. (b) together with (c) provides a possible collapse configuration. The remaining possibilities are similar to those discussed in the previous example and should be checked to ascertain the minimum collapse load.

Determination of the collapse load of frames involves much the same analysis as above. For complex frames, however, the approaches used in the foregoing examples would lead to extremely cumbersome calculations. For these, special purpose methods are available to provide approximate solutions.*

Example 12.9. Apply the method of virtual work to determine the collapse load of the structure shown in Fig. 12.13a. Assume that the rigidity of member BC is 1.2 times greater than that of the vertical members AB and CD.

SOLUTION. Of the several collapse modes, we consider only the two given in Fig. 12.13b and c. On the basis of Fig. 12.13b, plastic hinges will be formed at the ends of the vertical members. Thus from the principal of virtual work,

$$P\left(\tfrac{1}{2}\delta u\right) = 4M_u(\delta\theta)$$

Substituting $u = L\theta$, the above expression leads to $P_u = 8M_u/L$.

---

*See P. G. Hodge, *Plastic Design Analysis of Structures*, New York: McGraw-Hill, 1963.

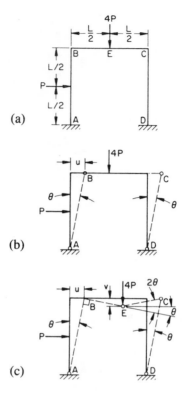

Figure 12.13

Referring to Fig. 12.13c, we have $M_{uE} = 1.2 M_u$, where M_u is the collapse moment of the vertical elements. Applying the principle of virtual work,

$$P\left(\tfrac{1}{2}\delta u\right) + 4P(\delta v) = 4M_u(\delta\theta) + 2(1.2 M_u \delta\theta)$$

Noting that $u = L\theta$ and $v = \tfrac{1}{2}L\theta$, the above equation provides the following expression for the collapse load: $P_u = 2.56 M_u / L$.

12.7 Elastic-Plastic Torsion

Recall from Chapter 6 that the maximum shearing stress in a slender bar of arbitrary section subject to pure torsion is always found on the boundary. As the applied torque is increased, we expect yielding to occur on the boundary and to move progressively toward the interior, as sketched in Fig. 12.14a for a bar of rectangular section. We now determine the ultimate torque $(M_t)_u$ that can be carried. This torque corresponds to the totally plastic state of the bar, as was the case of the beams previously discussed. Our analysis treats only perfectly plastic materials.

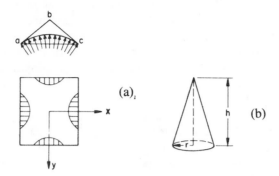

Figure 12.14

The stress distribution within the elastic region of the bar is governed by Eq. (6.5),

$$\frac{\partial^2\phi}{\partial x^2} + \frac{\partial^2\phi}{\partial y^2} = -2G\theta \tag{a}$$

where ϕ represents the stress function ($\phi = 0$ at the boundary) and θ is the angle of twist. The shearing stresses, in terms of ϕ, are

$$\tau_{zx} = \frac{\partial\phi}{\partial y}, \qquad \tau_{zy} = -\frac{\partial\phi}{\partial x} \tag{b}$$

Inasmuch as the bar is in a state of pure shear, the stress field in the plastic region is, according to the *Mises yield criterion*, expressed by

$$\left(\frac{\partial\phi}{\partial x}\right)^2 + \left(\frac{\partial\phi}{\partial y}\right)^2 = \tau_{yp}^2 \tag{c}$$

where τ_{yp} is the yield stress in shear. This expression indicates that the slope of the ϕ surface remains constant throughout the plastic region, and is equal to τ_{yp}.

Bearing in mind the condition imposed upon ϕ by Eq. (c), the membrane analogy (Sec. 6.3) may be extended from the purely elastic to the elastic-plastic case. As shown in Fig. 12.14a, a roof abc of constant slope is erected with the membrane as its base. Figure 12.14b shows such a roof for a circular section. As the pressure acting beneath the membrane increases, more and more contact is made between the membrane and the roof. In the fully plastic state, the membrane is in total contact with the roof, membrane and roof being of identical slope. Whether the membrane makes partial or complete contact with the roof clearly depends upon the pressure. The *membrane-roof analogy* thus permits solution of elastic-plastic torsion problems.

For the case of totally yielded bar, the membrane-roof analogy leads quite naturally to the so-called *sand hill analogy*. One need not construct a roof at all, using this method. Instead, sand is heaped upon a plate whose outline is cut into the shape of the cross section of the torsion member. The torque is, according to the membrane analogy, proportional to twice the volume of the sand figure so formed. The ultimate torque corresponding to the fully plastic state is thus found.

Referring to Fig. 12.14b, let us apply the sand hill analogy to determine the ultimate torque for a circular bar of radius r. The volume of the corresponding cone is $V = \frac{1}{3}\pi r^2 h$, where h is the height of the sand hill. The slope h/r represents the yield point stress τ_{yp}. The ultimate torque is therefore

$$(M_t)_u = \tfrac{2}{3}\pi r^3 \tau_{yp} \tag{12.16}$$

Note that the maximum elastic torque is $(M_t)_{yp} = (\pi r^3/2)\tau_{yp}$. We may thus form the ratio

$$\frac{(M_t)_u}{(M_t)_{yp}} = \frac{4}{3} \tag{12.17}$$

Other solid sections may be treated similarly.* Table 12.1 lists the ultimate torques for bars of various cross-sectional geometry.

The above procedure may also be applied to members having a symmetrically located hole. In this situation, the plate representing the cross section must contain the same hole as the actual cross section.

Table 12.1

Cross section	Radius or sides	Torque $(M_t)_u$ for full plasticity
Circular	r	$\frac{2}{3}\pi r^3 \tau_{yp}$
Equilateral triangle	a	$\frac{1}{12}a^3 \tau_{yp}$
Rectangle	a, b $(b>a)$	$\frac{1}{6}a^2(3b-a)\tau_{yp}$
Square	a	$\frac{1}{3}a^3 \tau_{yp}$
Thick-walled tube	b: outer a: inner	$\frac{2}{3}\pi(b^3-a^3)\tau_{yp}$

*See, for details, A. Nadai, *Theory of Flow and Fracture of Solids*, New York: McGraw-Hill, 1950, Chapter 35.

12.8   Elastic-Plastic Stresses in Rotating Disks

This section treats the stresses in a flat disk fabricated of a *perfectly plastic material*, rotating at constant angular velocity. The maximum elastic stresses for this geometry are, from Eqs. (8.26) and (8.25):

For the *solid disk* at $r=0$,

$$\sigma_\theta = \sigma_r = \frac{\rho\omega^2(3+\nu)b^2}{8} \tag{a}$$

For the *annular disk*, at $r=a$

$$\sigma_\theta = \frac{3+\nu}{4}\rho\omega^2\left(b^2 + \frac{1-\nu}{3+\nu}a^2\right) \tag{b}$$

Here a and b represent the inner and outer radii, respectively, ρ the mass density, and ω the angular speed. The following discussion relates to initial, partial, and complete yielding of an annular disk. Analysis of the solid disk is treated in a very similar manner.

Initial Yielding

According to the Tresca yield condition, yielding impends when the maximum stress is equal to the yield stress. Denoting the critical speed as ω_0 and using $\nu = \frac{1}{3}$, we have, from Eq. (b),

$$\omega_0 = \left(\frac{6}{5b^2+a^2}\frac{\sigma_{yp}}{\rho}\right)^{1/2} \tag{12.18}$$

Partial Yielding

For angular speeds in excess of ω_0, but lower than speeds resulting in total plasticity, the disk contains both an elastic and a plastic region, as shown in Fig. 12.15a. In the plastic range, the equation of radial equilibrium, Eq. (8.23), with σ_{yp} replacing the maximum stress σ_θ, becomes

$$r\frac{d\sigma_r}{dr} + \sigma_r - \sigma_{yp} + \rho\omega^2 r^2 = 0 \tag{12.19}$$

or

$$\frac{d}{dr}(r\sigma_r) - \sigma_{yp} + \rho\omega^2 r^2 = 0$$

The solution of the above is given by

$$r\sigma_r - \sigma_{yp}r + \frac{\rho\omega^2 r^3}{3} + c_1 = 0 \tag{c}$$

By satisfying the boundary condition $\sigma_r = 0$ at $r = a$, Eq. (c) provides an

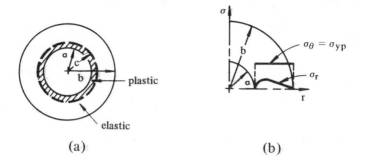

Figure 12.15

expression for the constant c_1, which when introduced above results in

$$r\sigma_r - \sigma_{yp}(r-a) + \frac{\rho\omega^2}{3}(r^3 - a^3) = 0 \qquad (d)$$

The stress within the plastic region is now determined by letting $r=c$ in Eq. (d):

$$\sigma_c = \frac{\rho\omega^2}{3}\frac{a^3 - c^3}{c} + \frac{c-a}{c}\sigma_{yp} \qquad (12.20)$$

Referring to the elastic region, the distribution of stress is determined from Eq. (8.24) with $\sigma_r = \sigma_c$ at $r=c$, and $\sigma_r = 0$ at $r=b$. Applying these conditions, we obtain

$$c_1 = -\frac{c^2(1-\nu)\sigma_c}{E(b^2 - c^2)} + \frac{\rho\omega^2(1-\nu)(3+\nu)}{8E}(b^2 + c^2)$$

$$c_2 = -\frac{b^2 c^2(1+\nu)\sigma_c}{E(b^2 - c^2)} + \frac{\rho\omega^2(1+\nu)(3+\nu)b^2 c^2}{8E} \qquad (e)$$

The stresses in the outer region are then obtained by substituting Eqs. (e) into Eq. (8.24)

$$\sigma_r = \frac{c^2}{b^2 - c^2}\left(-1 + \frac{b^2}{r^2}\right)\sigma_c + \frac{\rho\omega^2}{8}(3+\nu)\left(b^2 + c^2 - \frac{b^2 c^2}{r^2} - r^2\right)$$

$$\sigma_\theta = -\frac{c^2}{b^2 - c^2}\left(1 + \frac{b^2}{r^2}\right)\sigma_c + \frac{\rho\omega^2}{8}(3+\nu)\left(b^2 + c^2 - \frac{1+3\nu}{3+\nu}r^2 + \frac{b^2 c^2}{r^2}\right)$$

$$(12.21)$$

In order to determine that value of ω which causes yielding up to radius c, one need only substitute σ_θ for σ_{yp} above, and introduce σ_c as given by Eq. (12.20).

Complete Yielding

We turn finally to a determination of the speed ω_1 at which the disk becomes fully plastic. First Eq. (c) is rewritten

$$\sigma_r - \sigma_{yp} + \frac{\rho\omega^2 r^2}{3} + \frac{c_1}{r} = 0 \tag{f}$$

Applying the boundary conditions, $\sigma_r = 0$ at $r = a$ and $r = b$ in Eq. (f), we have

$$c_1 = a\sigma_{yp} - \frac{\rho^2\omega^2 a^3}{3} \tag{g}$$

and the critical speed ($\omega = \omega_1$) is given by

$$\omega_1 = \left(\frac{3\sigma_{yp}}{\rho} - \frac{b-a}{b^3 - a^3} \right)^{1/2} \tag{12.22}$$

Substitution of Eqs. (g) and (12.22) into Eq. (f) provides the radial stress in a fully plastic disk:

$$\sigma_r = \left(1 - \frac{a}{r} - \frac{1-a^2/b^2}{1-a^3/b^3} - \frac{r^2}{b^2}\frac{1-a/b}{1-a^3/b^3} \right)\sigma_{yp} \tag{12.23}$$

The distributions of radial and tangential stress are plotted in Fig. 12.15b.

12.9   Plastic Stress–Strain Relations

Consider an element subject to true stresses $\sigma_1, \sigma_2, \sigma_3$ with corresponding true straining. The true strain, which is plastic, is denoted $\varepsilon_1, \varepsilon_2, \varepsilon_3$. A simple way to derive expressions relating true stress and strain is to replace the elastic constants E and ν by E_s and $\frac{1}{2}$, respectively, in Eqs. (2.15). In so doing, we obtain equations of the *total strain theory* or the *deformational theory*, also known as *Hencky's plastic stress–strain relations*:

$$\varepsilon_1 = \frac{1}{E_s}\left[\sigma_1 - \tfrac{1}{2}(\sigma_2 + \sigma_3) \right]$$

$$\varepsilon_2 = \frac{1}{E_s}\left[\sigma_2 - \tfrac{1}{2}(\sigma_1 + \sigma_3) \right] \tag{12.24a}$$

$$\varepsilon_3 = \frac{1}{E_s}\left[\sigma_3 - \tfrac{1}{2}(\sigma_1 + \sigma_2) \right]$$

The foregoing may be restated

$$\frac{\varepsilon_1}{\sigma_1 - \tfrac{1}{2}(\sigma_2 + \sigma_3)} = \frac{\varepsilon_2}{\sigma_2 - \tfrac{1}{2}(\sigma_1 + \sigma_3)} = \frac{\varepsilon_3}{\sigma_3 - \tfrac{1}{2}(\sigma_1 + \sigma_2)} = \frac{1}{E_s} \tag{12.24b}$$

Here E_s, a function of the state of plastic stress, is termed the *modulus of*

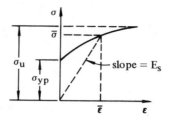

Figure 12.16

plasticity or *secant-modulus*. It is defined by (Fig. 12.16),

$$E_s = \frac{\bar{\sigma}}{\bar{\varepsilon}} \tag{12.25}$$

in which the quantities $\bar{\sigma}$ and $\bar{\varepsilon}$ are the *effective stress* and the *effective strain*, respectively.

While other yield theories may be employed to determine $\bar{\sigma}$, the maximum energy of distortion or Mises theory (Sec. 4.7) is most suitable. According to the Mises theory the following relationship connects the uniaxial yield stress to the general state of stress at a point:

$$\bar{\sigma} = \frac{1}{\sqrt{2}} \left[(\sigma_1 - \sigma_2)^2 + (\sigma_2 - \sigma_3)^2 + (\sigma_3 - \sigma_1)^2 \right]^{1/2} \tag{12.26}$$

It is assumed that the above expression applies not only to yielding or the beginning of inelastic action ($\bar{\sigma} = \sigma_{yp}$), but to any stage of plastic behavior. That is to say, $\bar{\sigma}$ has the value σ_{yp} at yielding, and as inelastic deformation progresses, $\bar{\sigma}$ increases in accordance with the right-hand side of Eq. (12.26). Equation (12.26) then represents the logical *extension* of the yield condition to describe plastic deformation after the yield stress is exceeded.

Collecting terms of Eqs. (12.24), we have

$$\left[\tfrac{2}{3} (\varepsilon_1^2 + \varepsilon_2^2 + \varepsilon_3^2) \right]^{1/2} = \frac{1}{E_s \sqrt{2}} \left[(\sigma_1 - \sigma_2)^2 + (\sigma_2 - \sigma_3)^2 + (\sigma_3 - \sigma_1)^2 \right]^{1/2} \tag{a}$$

The foregoing, together with Eqs. (12.25) and (12.26), leads to definition

$$\bar{\varepsilon} = \left[\tfrac{2}{3} (\varepsilon_1^2 + \varepsilon_2^2 + \varepsilon_3^2) \right]^{1/2} \tag{12.27a}$$

or (on the basis of $\varepsilon_1 + \varepsilon_2 + \varepsilon_3 = 0$) in different form

$$\bar{\varepsilon} = \frac{\sqrt{2}}{3} \left[(\varepsilon_1 - \varepsilon_2)^2 + (\varepsilon_2 - \varepsilon_3)^2 + (\varepsilon_3 - \varepsilon_1)^2 \right]^{1/2} \tag{12.27b}$$

relating the effective plastic strain and the true strain components. Note that, for simple tension $\sigma_2 = \sigma_3 = 0$, $\varepsilon_2 = \varepsilon_3 = -\varepsilon_1/2$, and Eqs. (12.26) and (12.27) result in

$$\bar{\sigma} = \sigma_1, \qquad \bar{\varepsilon} = \varepsilon_1 \tag{b}$$

Therefore, if we know $\bar{\sigma}, \bar{\varepsilon}$ can be read directly from true stress–strain diagram for uniaxial tension (Fig. 12.16).

Hencky's equations as they appear in Eqs. (12.24) have little utility. In order to give these expressions generality and convert them to a more convenient form, it is useful to employ the empirical relationship (12.2)

$$\bar{\sigma} = k(\bar{\varepsilon})^n$$

from which

$$\frac{\bar{\varepsilon}}{\bar{\sigma}} = \frac{(\bar{\sigma})^{1/n - 1}}{k^{1/n}} \tag{c}$$

The *true stress–strain relations*, upon substitution of Eqs. (12.26) and (c) into (12.24), then assume the following more useful form:

$$\varepsilon_1 = \left(\frac{\sigma_1}{k}\right)^{1/n} \left[\alpha^2 - \beta^2 - \alpha\beta - \alpha - \beta + 1\right]^{(1-n)/2n} \left(1 - \frac{\alpha}{2} - \frac{\beta}{2}\right) \tag{12.28a}$$

$$\varepsilon_2 = \left(\frac{\sigma_1}{k}\right)^{1/n} \left[\alpha^2 - \beta^2 - \alpha\beta - \alpha - \beta + 1\right]^{(1-n)/2n} \left(\alpha - \frac{\beta}{2} - \frac{1}{2}\right) \tag{12.28b}$$

$$\varepsilon_3 = \left(\frac{\sigma_1}{k}\right)^{1/n} \left[\alpha^2 - \beta^2 - \alpha\beta - \alpha - \beta + 1\right]^{(1-n)/2n} \left(\beta - \frac{\alpha}{2} - 1\right) \tag{12.28c}$$

where $\alpha = \sigma_2/\sigma_1$ and $\beta = \sigma_3/\sigma_1$.

In the case of an elastic material ($k = E$ and $n = 1$), it is observed that Eqs. (12.28) reduce to the familiar generalized Hooke's law.

Example 12.10. A thin-walled cylindrical tube of initial radius r_0 is subjected to internal pressure p. Assume that the values of r_0, p, and the material properties (k and n) are given. Apply Hencky's relations to determine (a) the maximum allowable stress and (b) the initial thickness t_0 for the cylinder to become unstable at internal pressure p.

SOLUTION. The current radius, thickness, and the length are denoted by r, t, and L, respectively. In the plastic range, the hoop, axial, and radial stresses are

$$\sigma_\theta = \sigma_1 = \frac{pr}{t}, \qquad \sigma_z = \sigma_2 = \frac{pr}{2t}, \qquad \sigma_r = \sigma_3 = 0 \tag{d}$$

We thus have $\alpha = \sigma_2/\sigma_1$ and $\beta = 0$ in Eqs. (12.28).

Corresponding to the above stresses, the components of true strain, are from Eq. (2.12):

$$\varepsilon_\theta = \varepsilon_1 = \ln\frac{r}{r_0}$$

$$\varepsilon_z = \varepsilon_2 = \ln\frac{L}{L_0} \tag{e}$$

$$\varepsilon_r = \varepsilon_3 = \ln\frac{t}{t_0}$$

Based upon the constancy of volume, Eq. (12.1), we then have

$$\varepsilon_3 = -(\varepsilon_1 + \varepsilon_2) = \ln\frac{t}{t_0}$$

or

$$t = t_0 e^{-\varepsilon_1 - \varepsilon_2} = t_0 \ln^{-1}\varepsilon_3 \tag{f}$$

The first of Eqs. (e) gives

$$r = r_0 e^{\varepsilon_1} = r_0 \ln^{-1}\varepsilon_1 \tag{g}$$

The tangential stress, the first of Eqs. (d), is therefore

$$\sigma_1 = pr_0 e^{\varepsilon_1}\frac{1}{t_0 e^{-\varepsilon_1 - \varepsilon_2}}$$

from which

$$p = \sigma_1\left(\frac{t_0}{r_0}\right)e^{-\varepsilon_1(2 + \varepsilon_2/\varepsilon_1)} \tag{h}$$

Simultaneous solution of Eqs. (12.28) leads readily to

$$\frac{\varepsilon_1}{\varepsilon_2} = \frac{2 - \alpha}{2\alpha - 1} \tag{i}$$

Equation (h) then appears as

$$p = \sigma_1\left(\frac{t_0}{r_0}\right)e^{-3\varepsilon_1/(2 - \alpha)} \tag{j}$$

For material instability,

$$dp = \frac{\partial p}{\partial\sigma_1}d\sigma_1 + \frac{\partial p}{\partial\varepsilon_1}d\varepsilon_1 = 0$$

which upon substitution of $\partial p/\partial\sigma_1$ and $\partial p/\partial\varepsilon_1$ derived from Eq. (j), becomes

$$dp = \left(\frac{t_0}{r_0}\right)e^{-3\varepsilon_1/(2 - \alpha)}d\sigma_1 + \sigma_1\left(\frac{t_0}{r_0}\right)e^{-3\varepsilon_1/(2 - \alpha)}\left(-\frac{3}{2 - \alpha}\right)d\varepsilon_1 = 0$$

or

$$\frac{d\sigma_1}{d\varepsilon_1} = \sigma_1\left(\frac{3}{2 - \alpha}\right) \tag{k}$$

In Eqs. (12.28) it is observed that σ_1 depends upon α and ε_1^n. That is

$$\sigma_1 = f(\alpha)\varepsilon_1^n \tag{l}$$

Differentiating the above, we have

$$\frac{d\sigma_1}{d\varepsilon_1} = nf(\alpha)\varepsilon_1^{n - 1} \tag{m}$$

Expressions (k), (l), and (m) lead to the *instability condition*:

$$\varepsilon_1 = \left(\frac{2 - \alpha}{3}\right)n \tag{12.29}$$

(a) Equating expressions (12.29) and (12.28a), we obtain

$$\left(\frac{2 - \alpha}{3}\right)n = \left(\frac{\sigma_1}{k}\right)^{1/n}(\alpha^2 - \alpha + 1)^{(1 - n)/2n}\left(\frac{2 - \alpha}{2}\right)$$

and the true tangential stress is thus

$$\sigma_1 = k\left(\frac{2n}{3}\right)^n\left(\frac{1}{\alpha^2-\alpha+1}\right)^{(1-n)/2} \tag{12.30}$$

(b) On the other hand, Eqs. (d), (f), (g), and (12.28) yield

$$\sigma_1 = \frac{pr_0}{t_0}\frac{\ln^{-1}(\sigma_1/k)^{1/n}(\alpha^2-\alpha+1)^{(1-n)/2n}[(2-\alpha)/2]}{\ln^{-1}(\sigma_1/k)^{1/n}(\alpha^2-\alpha+1)^{(1-n)/2n}(-1-\alpha)}$$

From this expression, the required original thickness is found to be

$$t_0 = \frac{pr_0}{\sigma_1}\ln^{-1}\frac{\alpha-2}{2(1+\alpha)} \tag{12.31}$$

wherein σ_1 is given by Eq. (12.30).

In the case under consideration, $\alpha=1/2$ and Eqs. (12.29), (12.30), and (12.31) thus become

$$\varepsilon_1 = \frac{n}{2}$$

$$\sigma_1 = k\left(\frac{2n}{3}\right)^n\left(\frac{4}{3}\right)^{(1-n)/2} = \frac{2k}{\sqrt{3}}\left(\frac{n}{\sqrt{3}}\right)^n \tag{12.32}$$

$$t_0 = \frac{0.606pr_0}{(2k/\sqrt{3})(n/\sqrt{3})^n}$$

For a thin-walled *spherical shell* under internal pressure, the two principal stresses are equal and hence $\alpha=1$. Equations (12.29), (12.30), and (12.31) then reduce to

$$\varepsilon_1 = \frac{n}{3}, \qquad \sigma_1 = k\left(\frac{2n}{3}\right)^n, \qquad t_0 = \frac{0.717pr_0}{k(2n/3)^n} \tag{12.33}$$

Based upon the relations derived in this example, the effective stress and the effective strain are determined readily. Table 12.2 furnishes the maximum true and effective stresses and strains in a thin-walled cylinder and a thin-walled sphere. For

Table 12.2

Member	Tensile bar	Cylindrical tube	Spherical tube
ε_1	n	$\dfrac{n}{2}$	$\dfrac{n}{3}$
ε_2	$-\dfrac{n}{2}$	0	$\dfrac{n}{3}$
ε_3	$-\dfrac{n}{2}$	$-\dfrac{n}{2}$	$-\dfrac{2n}{3}$
σ_1/k	$(n)^n$	$\dfrac{2}{\sqrt{3}}\left(\dfrac{n}{\sqrt{3}}\right)^n$	$\left(\dfrac{2n}{3}\right)^n$
σ_2/k	0	$\dfrac{1}{\sqrt{3}}\left(\dfrac{n}{\sqrt{3}}\right)^n$	$\left(\dfrac{2n}{3}\right)^n$

purposes of comparison, the table also lists the results (12.4) pertaining to simple tension. We observe that at instability the maximum true strains in a sphere and cylinder are much lower than the corresponding longitudinal strain in uniaxial tension.

It is significant that, for loading situation in which the components of stress do not increase continuously, Hencky's equations provide results which are somewhat in error and the incremental theory (Sec. 12.10) must be used. Under these circumstances, Eqs. (12.24) or (12.28) cannot describe the *complete plastic behavior* of the material. The latter is made clearer by considering the following. Suppose that subsequent to a given plastic deformation, the material is unloaded, either partially or completely, and then reloaded to a new state of stress that does not result in yielding. We expect no change to occur in the plastic strains; but Hencky's equations indicate different values of the plastic strain, because the stress components have changed. The latter cannot be valid because during the unloading–loading process the plastic strains have, in reality, not been affected.

12.10   Plastic Stress–Strain Increment Relations

We have already discussed the limitations of the deformational theory in connection with a situation in which the loading does not continuously increase. The *incremental theory* offers another approach, treating not the total strain associated with a state of stress, but rather the increment of strain.

Suppose now that the true stresses at a point experience very small changes in magnitude $d\sigma_1, d\sigma_2, d\sigma_3$. As a consequence of these increments, the effective stress $\bar{\sigma}$ will be altered by $d\bar{\sigma}$ and the effective strain $\bar{\varepsilon}$ by $d\bar{\varepsilon}$. The plastic strains thus suffer increments $d\varepsilon_1, d\varepsilon_2, d\varepsilon_3$.

The following modification of Hencky's equations, due to Lévy and Mises, describe the foregoing, giving good results in metals:

$$de_1 = \frac{d\bar{\varepsilon}}{\bar{\sigma}}\left[\sigma_1 - \frac{1}{2}(\sigma_2 + \sigma_3)\right]$$

$$de_2 = \frac{d\bar{\varepsilon}}{\bar{\sigma}}\left[\sigma_2 - \frac{1}{2}(\sigma_1 + \sigma_3)\right] \qquad (12.34a)$$

$$de_3 = \frac{d\bar{\varepsilon}}{\bar{\sigma}}\left[\sigma_3 - \frac{1}{2}(\sigma_1 + \sigma_2)\right]$$

An alternate form of the above is

$$\frac{d\varepsilon_1}{\sigma_1 - \frac{1}{2}(\sigma_2 + \sigma_3)} = \frac{d\varepsilon_2}{\sigma_2 - \frac{1}{2}(\sigma_1 + \sigma_3)} = \frac{d\varepsilon_3}{\sigma_3 - \frac{1}{2}(\sigma_1 + \sigma_2)} = \frac{d\bar{\varepsilon}}{\bar{\sigma}} \qquad (12.34b)$$

The plastic strain is, as before, assumed to occur at constant volume, i.e.,

$$d\varepsilon_1 + d\varepsilon_2 + d\varepsilon_3 = 0.$$

The *effective strain increment*, referring to Eq. (12.27b), may be written

$$d\bar{\varepsilon} = \frac{\sqrt{2}}{3}\left[(d\varepsilon_1 - d\varepsilon_2)^2 + (d\varepsilon_2 - d\varepsilon_3)^2 + (d\varepsilon_3 - d\varepsilon_1)^2\right]^{1/2} \quad (12.35)$$

The effective stress $\bar{\sigma}$ is given by Eq. (12.26). Alternately, in order to ascertain $d\bar{\varepsilon}$ from a uniaxial true stress–strain curve such as Fig. 12.16, it is necessary to know the increase of equivalent stress $d\bar{\sigma}$. Given $\bar{\sigma}$ and $d\bar{\varepsilon}$ it then follows that at any point in the loading process, application of Eqs. (12.34) provides the increment of strain as a unique function of the state of stress and the increment of the stress. According to the Lévy–Mises theory, therefore, the deformation suffered by an element varies in accordance with the specific loading path taken.

In a particular case of straining of sheet metal under *biaxial tension* the Lévy–Mises equations (12.34) become

$$\frac{d\varepsilon_1}{2-\alpha} = \frac{d\varepsilon_2}{2\alpha-1} = -\frac{d\varepsilon_3}{1+\alpha} \quad (12.36)$$

where $\alpha = \sigma_2/\sigma_1$ and σ_3 taken as zero. The effective stress and strain increment, Eqs. (12.26) and (12.35), is now written

$$\bar{\sigma} = \sigma_1(1 - \alpha + \alpha^2)^{1/2} \quad (12.37)$$

$$d\bar{\varepsilon} = d\varepsilon_1\left(\frac{2}{2-\alpha}\right)(1 - \alpha + \alpha^2)^{1/2} \quad (12.38)$$

Combining Eqs. (12.36) and (12.38) and integrating yields

$$\frac{\bar{\varepsilon}}{2(1-\alpha+\alpha^2)^{1/2}} = \frac{\varepsilon_1}{2-\alpha} = \frac{\varepsilon_2}{2\alpha-1} = -\frac{\varepsilon_3}{1+\alpha} \quad (12.39)$$

To generalize the above result, it is useful to employ Eq. (12.2), $\bar{\sigma} = k(\bar{\varepsilon})^n$, to include strain hardening characteristics. Differentiating this expression, we have

$$\frac{d\bar{\sigma}}{d\bar{\varepsilon}} = \frac{n\bar{\sigma}}{\bar{\varepsilon}} \quad (12.40)$$

Note that for simple tension, $n = \bar{\varepsilon} = \varepsilon$ and Eq. (12.40) reduces to Eq. (12.3).

The utility of the foregoing development is illustrated in Examples 12.11 and 12.12.

In closing, we note that the total (elastic-plastic) strains are determined by adding the elastic strains to the plastic strains. The elastic-plastic strain relations together with the equations of equilibrium and compatibility, and appropriate boundary conditions, completely describe a given situation. The general form of the Lévy–Mises relationships, including the elastic incremental components of strain, are referred to as the *Prandtl* and *Reuss* equations.

Example 12.11. Redo Example 12.10 employing the Lévy–Mises stress–strain increment relations.

SOLUTION. For the thin-walled cylinder under internal pressure, the plastic stresses are

$$\sigma_\theta = \sigma_1 = \frac{pr}{t}, \qquad \sigma_z = \sigma_2 = \frac{pr}{2t}, \qquad \sigma_r = \sigma_3 = 0 \tag{a}$$

where r and t are the current radius and the thickness. At instability,

$$dp = \frac{\partial p}{\partial \sigma_i} d\sigma_i + \frac{\partial p}{\partial r} dr + \frac{\partial p}{\partial t} dt = 0 \qquad (i = 1, 2) \tag{b}$$

As $\alpha = \frac{1}{2}$, introduction of Eqs. (a) into the above, provides

$$\frac{d\sigma_1}{\sigma_1} = \frac{d\sigma_2}{\sigma_2} = \frac{dr}{r} - \frac{dt}{t} = d\varepsilon_2 - d\varepsilon_3 \tag{12.41}$$

Clearly, dr/r is the hoop strain increment $d\varepsilon_2$, and dt/t is the incremental thickness strain or radial strain increment $d\varepsilon_3$. Equation (12.41) is the condition of instability for the cylinder material.

Upon application of Eqs. (12.37) and (12.38), the effective stress and the effective strains are found to be

$$\bar\sigma = \frac{\sqrt{3}}{2} \sigma_1$$

$$d\bar\varepsilon = \frac{2}{\sqrt{3}} d\varepsilon_1 = -\frac{2}{\sqrt{3}} d\varepsilon_3, \qquad d\varepsilon_2 = 0 \tag{c}$$

It is observed that axial strain does not occur and the situation is one of *plane strain*. The first of Eqs. (c) leads to $d\bar\sigma = (\sqrt{3}/2) d\sigma_1$, and condition (12.41) gives

$$d\bar\sigma = \frac{\sqrt{3}}{2} \sigma_1(d\varepsilon_1 - d\varepsilon_3) = \sqrt{3}\,\bar\sigma\,d\bar\varepsilon$$

from which

$$\frac{d\bar\sigma}{d\bar\varepsilon} = \sqrt{3}\,\bar\sigma$$

A comparison of this result with Eq. (12.40) shows that

$$\bar\varepsilon = \frac{n}{\sqrt{3}} \tag{12.42}$$

The true stresses and true strains are then obtained from Eqs. (c) and $\bar\sigma = k(\bar\varepsilon)^n$ and the results found to be identical with that obtained using Hencky's relations (Table 12.2).

For a *spherical shell* subjected to internal pressure $\sigma_1 = \sigma_2 = pr/2t$, $\alpha = 1$ and $d\varepsilon_1 = d\varepsilon_2 = -d\varepsilon_3/2$. At stability $dp = 0$, and we now have

$$\frac{d\sigma_1}{\sigma_1} = \frac{dr}{r} - \frac{dt}{t} = d\varepsilon_1 - d\varepsilon_3 \tag{d}$$

Equations (12.37) and (12.38) result in

$$\bar\sigma = \sigma_1$$

$$d\bar\varepsilon = 2 d\varepsilon_1 = 2 d\varepsilon_2 = -d\varepsilon_3 \tag{e}$$

Equations (d) and (e) are combined to yield

$$\frac{d\bar\sigma}{d\bar\varepsilon} = \frac{2}{3}\bar\sigma \tag{f}$$

From Eqs. (f) and (12.40) it is concluded that

$$\bar{\varepsilon} = \tfrac{2}{3}n \tag{12.43}$$

True stress and true strain are easily obtained, and are the same as the values obtained by a different method (Table 12.2).

Example 12.12. Determine the plastic stress distribution in a long, thick-walled cylinder, subjected only to internal pressure p (Fig. 8.2). Assume the cylinder to be fabricated of a rigid plastic material of yield strength σ_{yp} and ultimate strength σ_u, Fig. 12.16.

SOLUTION. The elastic stress distribution in the cylinder is described in Sec. 8.2. As the pressure is increased, it is clear that the entire cylinder will eventually yield. The limiting pressure as well as the stress distribution corresponding to this pressure is determined by application of the Lévy–Mises relations. In polar coordinates, the axial strain increment is

$$d\varepsilon_z = \frac{d\bar{\varepsilon}}{\bar{\sigma}}\left[\sigma_z - \frac{1}{2}(\sigma_r + \sigma_\theta)\right] \tag{12.44}$$

If the ends of the cylinder are restrained so that the axial displacement $w = 0$, the problem may be regarded as a case of *plane strain*, for which $\varepsilon_z = 0$. It follows that $d\varepsilon_z = 0$ and Eq. (12.44) gives

$$\sigma_z = \tfrac{1}{2}(\sigma_r + \sigma_\theta) \tag{g}$$

The equation of equilibrium is, from Eq. (8.2),

$$\frac{d\sigma_r}{dr} + \frac{\sigma_r - \sigma_\theta}{r} = 0 \tag{h}$$

subject to the following boundary conditions:

$$(\sigma_r)_{r=a} = -p, \qquad (\sigma_r)_{r=b} = 0 \tag{i}$$

Setting $\sigma_1 = \sigma_r$, $\sigma_2 = \sigma_\theta$, and $\sigma_z = \tfrac{1}{2}(\sigma_r + \sigma_\theta)$ in Eq. (12.26), the Mises yield criterion results in $(\bar{\sigma})^2 = (\tfrac{3}{4})(\sigma_r - \sigma_\theta)^2$. From this expression,

$$\sigma_r - \sigma_\theta = \pm \frac{2}{\sqrt{3}}\bar{\sigma} \tag{j}$$

Introducing Eq. (j) into Eq. (h), we obtain $d\sigma_r/dr = \pm 2\bar{\sigma}/(\sqrt{3}\,r)$, which has the solution

$$\sigma_r = \pm \frac{2}{\sqrt{3}}\bar{\sigma}\ln r + c_1 \tag{k}$$

The constant of integration is determined by applying the second of conditions (i)

$$c_1 = \pm \frac{2}{\sqrt{3}}\bar{\sigma}\ln b$$

Equation (k) is thus

$$\sigma_r = \pm \frac{2}{\sqrt{3}}\bar{\sigma}\ln\frac{b}{r} \tag{l}$$

The first of conditions (i) now leads to

$$p = \frac{2}{\sqrt{3}}\bar{\sigma}\ln\frac{b}{a} \qquad (r=a) \tag{12.45}$$

This pressure causes the *initial* plastic *yielding* when $\bar{\sigma} = \sigma_{yp}$, and the *collapse* of the cylinder when $\bar{\sigma} = \sigma_u$.

An expression for σ_θ can now be found by substituting Eq. (l) into Eq. (j). Consequently, Eq. (g) provides σ_z. The complete plastic stress distribution, for a specified $\bar{\sigma}$, is thus found to be

$$\sigma_r = -\frac{2}{\sqrt{3}}\bar{\sigma}\ln\frac{b}{r}$$

$$\sigma_\theta = \frac{2}{\sqrt{3}}\bar{\sigma}\left(1 - \ln\frac{b}{r}\right) \qquad\qquad (12.46)$$

$$\sigma_z = \frac{2}{\sqrt{3}}\bar{\sigma}\left(\frac{1}{2} - \ln\frac{b}{r}\right)$$

Following a procedure similar to that outlined above, a number of problems of practical importance, involving spherical or cylindrical symmetry, can be solved.

Chapter 2—Problems

Secs. 12.1 to 12.4

12.1. A solid circular cylinder of 0.1-m diameter is subjected to a bending moment $M = 3.375$ kN·m, an axial tensile force $P = 90$ kN, and a twisting end couple $M_t = 4.5$ kN·m. Determine the stress deviator tensor. [Hint: Refer to Sec. 2.9.]

12.2. In the pin-connected frame shown in Fig. P10.13, the true stress–engineering strain curves of the members are expressed by $\sigma_{AD} = \sigma_{DC} = k_1 \varepsilon_0^{n_1}$ *and* $\sigma_{BD} = k_2 \varepsilon_0^{n_2}$. Verify that, in order for all three bars to reach tensile instability simultaneously, they should be set initially at an angle described by

$$\cos\alpha = \frac{L_{BD}}{L_{AD}} = \frac{1-n_2}{1-n_1}\left[\frac{n_1(2-n_1)}{n_2(2-n_2)}\right]^{1/2} \qquad (P12.2)$$

Calculate the value of this initial angle for $n_1 = 0.2$ and $n_2 = 0.3$.

12.3. Determine the deflection of a uniformly loaded rigid plastic cantilever beam of length L. Locate the origin of coordinates at the fixed end, and denote the loading by p.

12.4. Redo Problem 12.3 for $p = 0$ and a concentrated load P applied at the free end.

12.5. Consider a beam of rectangular section, subjected to end moments as shown in Fig. 12.3a. Assuming that the relationship for tensile and compressive stress for the material is approximated by $\sigma = k\varepsilon^{1/4}$, determine the maximum stress.

12.6. A simply supported rigid plastic beam is described in Fig. P12.6. Compute the maximum deflection. Reduce the result to the case of a linearly elastic material, $EIv_{max}=(\frac{1}{8})PaL^2-(\frac{1}{6})Pa^3$. Let $P=8$ kN, $E=200$ GPa, $L=$ 1.2 m and $a=0.45$ m. Cross-sectional dimensions shown are in millimeters.

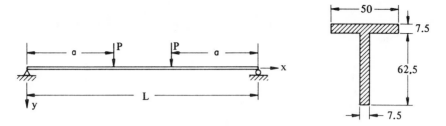

Figure P12.6

Secs. 12.5 to 12.6

12.7. A beam of rectangular cross section (width a, depth h) is subjected to bending moments M at its ends. The beam is constructed of a material displaying the stress–strain relationship shown in Fig. P12.7. What value of M can be carried by the beam?

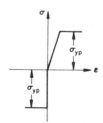

Figure P12.7

12.8. Consider a perfectly plastic cantilever beam of rectangular cross section and length L subjected to a concentrated downward force P at the free end. Determine the maximum deflection at the beginning of inelastic action and at the instant of plastic collapse.

12.9. Consider a uniform bar of solid circular cross section with radius r, subjected to axial tension and bending moments at both ends. Derive general relationships involving N and M which govern first the case of initial yielding, and then fully plastic deformation. Sketch the interaction curves.

12.10. In Fig. P5.16 is shown a hook made of steel with $\sigma_{yp}=280$ MPa, equal in tension and compression. What load P results in complete plastic deformation in section 1–2? Neglect the effect of curvature upon the stress distribution.

12.11. Obtain the interaction curves for the beam cross section shown in Fig. 12.9a. The beam is subjected to a bending moment M and an axial load N at both ends. Take $b=2h$, $b_1=1.8h$, $h_1=0.7h$.

12.12. Obtain the collapse load of the structure shown in Fig. P12.12. Assume that plastic hinges form at 1, 3, and 4.

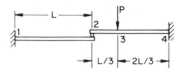

Figure P12.12

12.13. What is the collapse load of the beam shown in Fig. P12.13? Assume two possible modes of collapse such that plastic hinges form at 2, 3, and 4.

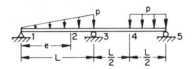

Figure P12.13

12.14. In Fig. P12.14 are shown two beam cross sections. Determine M_u/M_{yp} for each case.

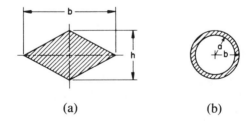

(a)   (b)

Figure P12.14

Secs. 12.7 to 12.10

12.15. Determine the elastic-plastic stresses in a rotating solid disk.

12.16. Determine, for a rectangular bar of sides a and b, the ultimate torque corresponding to the fully plastic state (Table 12.1). Use the sand hill analogy.

12.17. Determine, for an equilateral triangular bar of sides $2a$, (a) the ultimate torque corresponding to the fully plastic state (Table 12.1)(Use the sand hill analogy.) and (b) the maximum elastic torque by referring to the Table 6.1. (c) Compare the results found in (a) and (b).

12.18. An annular disk of equilateral hyperbolic profile with outer and inner radii b and a (Fig. 8.9) is shrunk onto a solid shaft so that the interfacial pressure is p_i. Demonstrate that, according to the Tresca yield criterion, when the disk becomes fully plastic

$$p_i = \frac{a(b-r)}{rb}\sigma_{yp} \qquad (P12.18)$$

Here σ_{yp} is the yield point stress and r represents any arbitrary radius.

12.19. Consider a thin-walled cylinder of original radius r_0, subjected to internal pressure p. Determine the value of the required original thickness at instability employing Hencky's relations. Use the following:

$$k = 900 \text{ MPa}, \qquad n = 0.2$$
$$r_0 = 0.5 \text{ m}, \qquad p = 14 \text{ MPa}$$

12.20. Redo Example 12.10 for the cylinder under uniform axial tensile load P and $p = 0$.

12.21. A thin-walled tube of original thickness t_0 and outer radius R_o just fits over a rigid rod of radius r_0. Employ the Lévy–Mises relations to verify that the axial load the tube can sustain before instability occurs is represented by

$$P = 2\pi r_0 t_0 e^{-\frac{2nk}{\sqrt{3}}} \left(\frac{2n}{\sqrt{3}}\right)^n \qquad (P12.21)$$

Assume the tube rod interface to be frictionless. Use as the true stress–true strain relationship of the tube material in simple tension $\sigma = k\varepsilon^n$.

Chapter 13

Introduction to Plates and Shells

PART 1—Bending of Thin Plates

13.1 Basic Definitions

Plates and shells are initially flat and curved structural elements, respectively, with thicknesses small compared with the remaining dimensions. In Part 1 of this chapter, we consider plates, for which it is usual to divide the thickness t into equal halves by a plane parallel to the faces. This plane is termed the *midsurface* of the plate. The plate thickness is measured in a direction normal to the midsurface at each point under consideration. We here treat the small deflection theory of homogeneous, uniform plates, leaving for the numerical approach of Sec. 13.7 a discussion of plates of nonuniform thickness and irregular shape.

Consider now a plate prior to deformation, shown in Fig. 13.1a, in which the xy plane coincides with the midsurface and hence the z deflection is zero. When, owing to external loading, deformation occurs, the midsurface at any point x_A, y_A suffers a deflection w. Referring to the coordinate system shown, the fundamental assumptions of the *small deflection* theory of bending for *isotropic, homogeneous, thin* plates may be summarized as follows:

1. The deflection of the midsurface is small in comparison with the thickness of the plate. The slope of the deflected surface is much less than unity.

2. Straight lines initially normal to the midsurface remain straight and normal to that surface subsequent to bending. This is equivalent to stating that the vertical shear strains γ_{xz} and γ_{yz} are negligible. The deflection of the plate is thus associated principally with bending strains, with the implication that the normal strain ε_z owing to vertical loading may also be neglected.

3. No midsurface straining or so-called in-plane straining, stretching, or contracting occurs as a result of bending.

4. The component of stress normal to the midsurface, σ_z, is negligible.

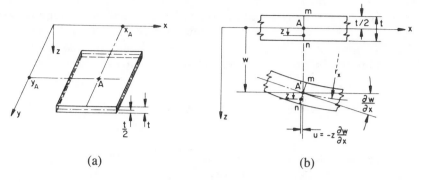

(a)   (b)

Figure 13.1

The above assumptions are analogous to those associated with the simple bending theory of beams. On the basis of assumption (2), the strain-displacement relations of Eq. (2.3) reduce to

$$\varepsilon_x = \frac{\partial u}{\partial x}, \qquad\qquad \varepsilon_z = \frac{\partial w}{\partial z} = 0$$

$$\varepsilon_y = \frac{\partial v}{\partial y}, \qquad\qquad \gamma_{xz} = \frac{\partial w}{\partial x} + \frac{\partial u}{\partial z} = 0 \qquad\qquad \text{(a)}$$

$$\gamma_{xy} = \frac{\partial u}{\partial y} + \frac{\partial v}{\partial x}, \qquad \gamma_{yz} = \frac{\partial w}{\partial y} + \frac{\partial v}{\partial z} = 0$$

Integration of $\varepsilon_z = \partial w / \partial z$ yields

$$w = f_1(x, y) \qquad\qquad (13.1)$$

indicating that the lateral deflection does not vary throughout the plate thickness. Similarly, integrating the expressions for γ_{xz} and γ_{yz}, we obtain

$$u = -z \frac{\partial w}{\partial x} + f_2(x, y),$$

$$\text{(b)}$$

$$v = -z \frac{\partial w}{\partial y} + f_3(x, y)$$

It is clear that $f_2(x, y)$ and $f_3(x, y)$ represent, respectively, the values of u and v corresponding to $z = 0$ (the midsurface). As assumption (3) precludes such in-plane straining, we conclude that $f_2 = f_3 = 0$, and therefore

$$u = -z \frac{\partial w}{\partial x}, \qquad v = -z \frac{\partial w}{\partial y} \qquad\qquad (13.2)$$

where $\partial w/\partial x$ and $\partial w/\partial y$ are the slopes of the midsurface. The above expression for u is represented in Fig. 13.1b at section mn passing through arbitrary point $A(x_A, y_A)$. A similar interpretation applies for v in the zy plane. It is observed that Eqs. (13.2) are consistent with assumption (2). Combining the first three equations of (a) with Eq. (13.2), we have

$$\varepsilon_x = -z\frac{\partial^2 w}{\partial x^2}, \qquad \varepsilon_y = -z\frac{\partial^2 w}{\partial y^2}, \qquad \gamma_{xy} = -2z\frac{\partial^2 w}{\partial x\, \partial y} \qquad (13.3a)$$

which provide the strains at any point.

Because in small deflection theory the square of a slope may be regarded as negligible, the partial derivatives of Eqs. (13.3a) represent the *curvatures* of the plate (see Eq. 5.7). Therefore the curvatures at the midsurface in planes parallel to the xz (Fig. 13.1b), yz, and xy planes are, respectively,

$$\frac{1}{r_x} = \frac{\partial}{\partial x}\left(\frac{\partial w}{\partial x}\right) = \frac{\partial^2 w}{\partial x^2}$$

$$\frac{1}{r_y} = \frac{\partial}{\partial y}\left(\frac{\partial w}{\partial y}\right) = \frac{\partial^2 w}{\partial y^2}$$

$$\frac{1}{r_{xy}} = \frac{1}{r_{yx}} = \frac{\partial}{\partial x}\left(\frac{\partial w}{\partial y}\right) \qquad\qquad (13.4)$$

$$= \frac{\partial}{\partial y}\left(\frac{\partial w}{\partial x}\right) = \frac{\partial^2 w}{\partial x\, \partial y}$$

The foregoing are simply the *rates* at which the slopes vary over the plate.

In terms of the radii of curvature, the strain-deflection relations (13.3a) may be written

$$\varepsilon_x = -z\frac{1}{r_x}, \qquad \varepsilon_y = -z\frac{1}{r_y}, \qquad \gamma_{xy} = -2z\frac{1}{r_{xy}} \qquad (13.3b)$$

Examining these equations, we are led to conclude that a circle of curvature can be constructed similarly to Mohr's circle of strain. The curvatures therefore transform in the same manner as the strains. It can be verified by employing Mohr's circle that $(1/r_x) + (1/r_y) = \nabla^2 w$. The sum of the curvatures in perpendicular directions, called the *average curvature*, is invariant with respect to rotation of the coordinate axes. This assertion is valid at any location on the midsurface.

13.2 Stress, Curvature, and Moment Relations

The stress components σ_x, σ_y, and $\tau_{xy} = \tau_{yx}$ are related to the strains by Hooke's law, which for a thin plate becomes

$$\sigma_x = \frac{E}{1-\nu^2}(\varepsilon_x + \nu\varepsilon_y) = -\frac{Ez}{1-\nu^2}\left(\frac{\partial^2 w}{\partial x^2} + \nu\frac{\partial^2 w}{\partial y^2}\right)$$

$$\sigma_y = \frac{E}{1-\nu^2}(\varepsilon_y + \nu\varepsilon_x) = -\frac{Ez}{1-\nu^2}\left(\frac{\partial^2 w}{\partial y^2} + \nu\frac{\partial^2 w}{\partial x^2}\right) \qquad (13.5)$$

$$\tau_{xy} = \frac{E}{2(1+\nu)}\gamma_{xy} = -\frac{Ez}{1+\nu}\frac{\partial^2 w}{\partial x\,\partial y}$$

These expressions demonstrate clearly that the stresses vanish at the midsurface and vary linearly over the thickness of the plate.

The stresses distributed over the side surfaces of the plate, while producing no net force, do result in bending and twisting moments. These moment resultants (per unit length—e.g., newtons times meters divided by meters, or simply newtons) are denoted M_x, M_y, M_{xy}. Referring to Fig. 13.2a,

$$\int_{-t/2}^{t/2} z\sigma_x\,dy\,dz = dy\int_{-t/2}^{t/2} z\sigma_x\,dz = M_x\,dy$$

Expressions involving M_y and $M_{xy} = M_{yx}$ are similarly derived. The bending and twisting moments per unit length are thus

$$M_x = \int_{-t/2}^{t/2} z\sigma_x\,dz$$

$$M_y = \int_{-t/2}^{t/2} z\sigma_y\,dz \qquad (13.6)$$

$$M_{xy} = \int_{-t/2}^{t/2} z\tau_{xy}\,dz$$

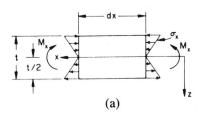

(a)

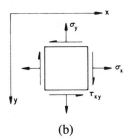

(b)

Figure 13.2

Introducing into the above the stresses given by Eqs. (13.5), and taking into account the fact that $w = w(x, y)$, we obtain

$$M_x = -D\left(\frac{\partial^2 w}{\partial x^2} + \nu \frac{\partial^2 w}{\partial y^2}\right)$$

$$M_y = -D\left(\frac{\partial^2 w}{\partial y^2} + \nu \frac{\partial^2 w}{\partial x^2}\right) \qquad (13.7)$$

$$M_{xy} = -D(1-\nu)\frac{\partial^2 w}{\partial x\, \partial y}$$

where

$$D = \frac{Et^3}{12(1-\nu^2)} \qquad (13.8)$$

is the *flexural rigidity* of the plate. Note that if a plate element of unit width were free to expand sidewise under the given loading, anticlastic curvature would not be prevented; the flexural rigidity would be $Et^3/12$. The remainder of the plate does not permit this action, however. Because of this, a plate manifests *greater stiffness* than a narrow beam by a factor $1/(1-\nu^2)$ or about 10%. Under the sign convention, a positive moment is one which results in positive stresses in the positive (bottom) half of the plate (Sec. 1.3), as shown in Fig. 13.2b.

Substitution of $z = t/2$ into Eq. (13.5), together with the use of Eq. (13.7), provides expressions for the maximum stresses (which occur on the surface of the plate):

$$\sigma_{x,\max} = \frac{6M_x}{t^2}, \qquad \sigma_{y,\max} = \frac{6M_y}{t^2}, \qquad \tau_{xy,\max} = \frac{6M_{xy}}{t^2} \qquad (13.9)$$

Employing a new set of coordinates in which x', y' replaces x, y and $z' = z$, we first transform $\sigma_x, \sigma_y, \tau_{xy}$ into $\sigma_{x'}, \sigma_{y'}, \tau_{x'y'}$ through the use of Eq. (1.7). These are then substituted into Eq. (13.6) to obtain the corresponding $M_{x'}, M_{y'}, M_{x'y'}$. Examination of Eqs. (13.6) and (13.9) indicates a direct correspondence between the moments and stresses. It is concluded, therefore, that the equation for transforming the stresses should be identical with that used for transforming the moments. Mohr's circle may thus be applied to moments as well as to stresses.

13.3   The Differential Equation of Plate Deflection

Consider now a plate element $dx\, dy$ subject to a uniformly distributed lateral load per unit area p (Fig. 13.3). In addition to the moments M_x, M_y, M_{xy} previously discussed, we now find vertical shearing forces Q_x and Q_y (force per unit length) acting on the sides of the element. These forces are related directly to the vertical shearing stresses:

$$Q_x = \int_{-t/2}^{t/2} \tau_{xz}\, dz, \qquad Q_y = \int_{-t/2}^{t/2} \tau_{yz}\, dz \tag{13.10}$$

The sign convention associated with Q_x and Q_y is identical with that for the shearing stresses τ_{xz} and τ_{yz}: a positive shearing force acts on a positive face in the positive z direction (or on a negative face in the negative z direction). The bending moment sign convention is as previously given. On this basis, all forces and moments shown in Fig. 13.3 are positive.

It is appropriate to emphasize that while the simple theory of thin plates neglects the effect on bending of σ_z, $\gamma_{xz} = \tau_{xz}/G$, and $\gamma_{yz} = \tau_{yz}/G$ (as discussed in Sec. 13.1), the vertical forces Q_x and Q_y resulting from τ_{xz} and τ_{yz} are not negligible. In fact, they are of the same order of magnitude as the lateral loading and moments.

It is our next task to obtain the equation of equilibrium for an element and eventually to reduce the system of equations to a single expression involving the deflection w. Referring to Fig. 13.3, we note that body forces are assumed negligible relative to the surface loading and that no horizontal shear and normal forces act on the sides of the element. The equilibrium of z-directed forces is governed by

$$\frac{\partial Q_x}{\partial x} dx\, dy + \frac{\partial Q_y}{\partial y} dy\, dx + p\, dx\, dy = 0$$

or

$$\frac{\partial Q_x}{\partial x} + \frac{\partial Q_y}{\partial y} + p = 0 \tag{a}$$

For the equilibrium of moments about the x axis,

$$\frac{\partial M_{xy}}{\partial x} dx\, dy + \frac{\partial M_y}{\partial y} dx\, dy - Q_y\, dx\, dy = 0$$

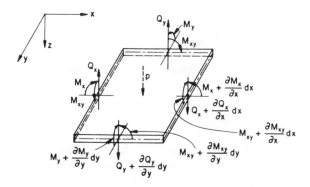

Figure 13.3

from which

$$\frac{\partial M_{xy}}{\partial x} + \frac{\partial M_y}{\partial y} - Q_y = 0 \tag{b}$$

Higher-order terms, such as the moment of p and the moment owing to the change in Q_y, have been neglected. The equilibrium of moments about the y axis yields an expression similar to Eq. (b):

$$\frac{\partial M_{xy}}{\partial y} + \frac{\partial M_x}{\partial x} - Q_x = 0 \tag{c}$$

Equations (b) and (c), when combined with Eq. (13.7), lead to

$$Q_x = -D \frac{\partial}{\partial x}\left(\frac{\partial^2 w}{\partial x^2} + \frac{\partial^2 w}{\partial y^2}\right), \qquad Q_y = -D \frac{\partial}{\partial y}\left(\frac{\partial^2 w}{\partial x^2} + \frac{\partial^2 w}{\partial y^2}\right) \tag{13.11}$$

Finally, substituting the above into Eq. (a) results in the basic differential equation of plate theory (Lagrange, 1811):

$$\frac{\partial^4 w}{\partial x^4} + 2\frac{\partial^4 w}{\partial x^2 \partial y^2} + \frac{\partial^4 w}{\partial y^4} = \frac{p}{D} \tag{13.12a}$$

or in concise form,

$$\nabla^4 w = \frac{p}{D} \tag{13.12b}$$

The bending of plates subject to a lateral loading p per unit area thus reduces to a single differential equation. Determination of $w(x, y)$ relies upon the integration of Eq. (13.12) with the constants of integration dependent upon the identification of appropriate boundary conditions. The shearing stresses τ_{xz} and τ_{yz} can readily be determined by applying Eqs. (13.11) and (13.10) once $w(x, y)$ is known. These stresses display a parabolic variation over the thickness of the plate. The maximum shearing stress, as in the case of a beam of rectangular section, occurs at $z = 0$:

$$\tau_{xz,\max} = \frac{3}{2}\frac{Q_x}{t}, \qquad \tau_{yz,\max} = \frac{3}{2}\frac{Q_y}{t} \tag{13.13}$$

The key to evaluating all the stresses, employing Eqs. (13.5) or (13.9) and (13.13), is thus the solution of Eq. (13.12) for $w(x, y)$. As already indicated, τ_{xz} and τ_{yz} are regarded as small compared with the remaining plane stresses.

13.4   Boundary Conditions

Solution of the plate equation requires that two boundary conditions be satisfied at each edge. These may relate to deflection and slope, or to forces and moments, or to some combination. The principal feature distinguishing the boundary conditions applied to plates from those applied to beams relates to the existence along the plate edge of twisting moment resultants. These moments, as demonstrated below, may be replaced by an equivalent vertical force, which when added to the vertical shearing force produces an *effective* vertical force.

Consider two successive elemental lengths dy on edge $x=a$ of the plate shown in Fig. 13.4a. On the right-hand element, a twisting moment $M_{xy}\,dy$ acts, while the left-hand element is subject to a moment $[M_{xy}+(\partial M_{xy}/\partial y)\,dy]\,dy$. In Fig. 13.4b, we observe that these twisting moments have been replaced by equivalent force couples which produce only *local* differences in the distribution of stress on the edge $x=a$. The stress distribution elsewhere in the plate is unaffected by this substitution. Acting at the left-hand edge of the right-hand element is an upward directed force M_{xy}. Adjacent to this force is a downward directed force $M_{xy}+(\partial M_{xy}/\partial y)\,dy$ acting at the right-hand edge of the left-hand element. The difference between these forces (expressed per unit length), $\partial M_{xy}/\partial y$, may be combined with the transverse shearing force Q_x to produce an effective transverse edge force per unit length, V_x, known as *Kirchhoff's force* (Fig. 13.4c):

$$V_x = Q_x + \frac{\partial M_{xy}}{\partial y}$$

Substitution of Eqs. (13.7) and (13.11) into the above leads to

$$V_x = -D\left[\frac{\partial^3 w}{\partial x^3} + (2-\nu)\frac{\partial^3 w}{\partial x\,\partial y^2}\right] \qquad (13.14)$$

We are now in a position to formulate a variety of commonly encountered situations. Consider first the conditions which apply along the *clamped* edge $x=a$ of the rectangular plate with edges parallel to the x and y axes (Fig. 13.5a). As both the deflection and slope are zero,

$$w=0, \qquad \frac{\partial w}{\partial x}=0 \qquad (x=a) \qquad (13.15)$$

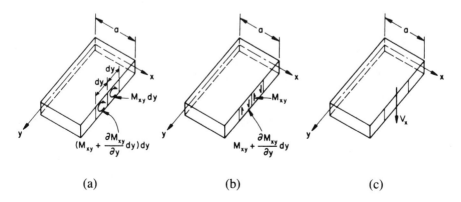

(a)   (b)   (c)

Figure 13.4

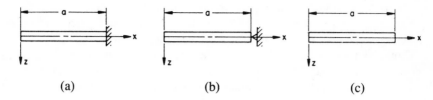

Figure 13.5

For the *simply supported* edge (Fig. 13.5b), the deflection and bending moment are both zero:

$$w=0, \qquad M_x = -D\left(\frac{\partial^2 w}{\partial x^2} + \nu\frac{\partial^2 w}{\partial y^2}\right)=0 \qquad (x=a) \qquad (13.16a)$$

Because the first of these equations implies that along edge $x=a$, $\partial w/\partial y=0$ and $\partial^2 w/\partial y^2=0$, the conditions expressed by Eq. (13.16a) may be restated in the following equivalent form:

$$w=0, \qquad \frac{\partial^2 w}{\partial x^2}=0 \qquad (x=a) \qquad (13.16b)$$

For the case of the *free edge* (Fig. 13.5c), the moment and vertical edge force are zero:

$$\frac{\partial^2 w}{\partial x^2} + \nu\frac{\partial^2 w}{\partial y^2}=0, \qquad \frac{\partial^3 w}{\partial x^3} + (2-\nu)\frac{\partial^3 w}{\partial x\,\partial y^2}=0 \qquad (x=a) \quad (13.17)$$

Example 13.1. Derive the equation describing the deflection of a long (infinite) plate, simply supported at edges $y=0$ and $y=b$ (Fig. 13.6). The plate is subjected to nonuniform loading

$$p(y)=p_0\sin\left(\frac{\pi y}{b}\right) \tag{a}$$

so that it deforms into a cylindrical surface with its generating line parallel to the x axis. The constant p_0 thus represents the load intensity along the line passing through $y=b/2$, parallel to x.

SOLUTION. Because, for this situation $\partial w/\partial x=0$ and $\partial^2 w/\partial x\,\partial y=0$, Eq. (13.7) reduces to

$$M_x = -\nu D\frac{d^2 w}{dy^2}, \qquad M_y = -D\frac{d^2 w}{dy^2} \tag{b}$$

while Eq. (13.12) becomes

$$\frac{d^4 w}{dy^4} = \frac{p}{D} \tag{c}$$

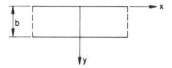

Figure 13.6

The latter expression is the same as the *wide beam equation*, and we conclude that the solution proceeds as in the case of a beam. A wide beam shall, in the context of this chapter, mean a rectangular plate supported on one edge or on two opposite edges in such a way that these edges are free to approach one another as deflection occurs.

Substituting Eq. (a) into Eq. (c), integrating, and satisfying the boundary conditions at $y=0$ and $y=b$, we obtain

$$w = \left(\frac{b}{\pi}\right)^4 \frac{p_0}{D} \sin\left(\frac{\pi y}{b}\right) \tag{d}$$

The stresses are now readily determined through application of Eq. (13.5) or (13.9) and Eq. (13.13).

13.5 Simply Supported Rectangular Plates

In general, the solution of the plate problem for a geometry as in Fig. 13.7, with simple supports along all edges, may be obtained by the application of Fourier series for load and deflection*:

$$p(x, y) = \sum_{m=1}^{\infty} \sum_{n=1}^{\infty} p_{mn} \sin\frac{m\pi x}{a} \sin\frac{n\pi y}{b} \tag{13.18a}$$

$$w(x, y) = \sum_{m=1}^{\infty} \sum_{n=1}^{\infty} a_{mn} \sin\frac{m\pi x}{a} \sin\frac{n\pi y}{b} \tag{13.18b}$$

Here p_{mn} and a_{mn} represent coefficients to be determined. The problem at hand is described by

$$\frac{\partial^4 w}{\partial x^4} + 2\frac{\partial^4 w}{\partial x^2 \partial y^2} + \frac{\partial^4 w}{\partial y^4} = \frac{p(x, y)}{D} \tag{a}$$

---

*This approach was introduced by Navier in 1820. For details, see A. C. Ugural, *Stresses in Plates and Shells*, New York: McGraw-Hill, 1981, Chapter 3.

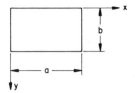

Figure 13.7

and

$$w=0, \qquad \frac{\partial^2 w}{\partial x^2}=0 \qquad (x=0, x=a)$$

$$w=0, \qquad \frac{\partial^2 w}{\partial y^2}=0 \qquad (y=0, y=b)$$

(b)

The boundary conditions given above are satisfied by Eq. (13.18b), and the coefficients a_{mn} must be such as to satisfy Eq. (a). The solution corresponding to the loading $p(x, y)$ thus requires a determination of p_{mn} and a_{mn}. We proceed by dealing first with a general load configuration, subsequently treating specific loadings.

To determine the coefficients p_{mn}, each side of Eq. (13.18a) is multiplied by

$$\sin\frac{m'\pi x}{a}\sin\frac{n'\pi y}{b}\,dx\,dy$$

Integrating between the limits 0, a and 0, b yields

$$\int_0^b\int_0^a p(x, y)\sin\frac{m'\pi x}{a}\sin\frac{n'\pi y}{b}\,dx\,dy$$

$$= \sum_{m=1}^{\infty}\sum_{n=1}^{\infty} p_{mn}\int_0^b\int_0^a \sin\frac{m\pi x}{a}\sin\frac{m'\pi x}{a}\sin\frac{n\pi y}{b}\sin\frac{n'\pi y}{b}\,dx\,dy$$

Applying the orthogonality relation (10.17) and integrating the right side of the above equation, we obtain

$$p_{mn}= \frac{4}{ab}\int_0^b\int_0^a p(x, y)\sin\frac{m\pi x}{a}\sin\frac{n\pi y}{b}\,dx\,dy \qquad (13.19)$$

Evaluation of a_{mn} in Eq. (13.18b) requires substitution of Eqs. (13.18a) and (13.18b) into Eq. (a), with the result

$$\sum_{m=1}^{\infty}\sum_{n=1}^{\infty}\left\{a_{mn}\left[\left(\frac{m\pi}{a}\right)^4+2\left(\frac{m\pi}{a}\right)^2\left(\frac{n\pi}{b}\right)^2+\left(\frac{n\pi}{b}\right)^4\right]-\frac{p_{mn}}{D}\right\}$$

$$\times\sin\frac{m\pi x}{a}\sin\frac{n\pi y}{b}=0$$

This expression must apply for all x and y; we conclude therefore that

$$a_{mn}\pi^4\left(\frac{m^2}{a^2}+\frac{n^2}{b^2}\right)^2-\frac{p_{mn}}{D}=0 \qquad (c)$$

Solving for a_{mn} and substituting into Eq. (13.18b), the equation of the deflection surface of a thin plate is

$$w=\frac{1}{\pi^4 D}\sum_{m=1}^{\infty}\sum_{n=1}^{\infty}\frac{p_{mn}}{\left[(m/a)^2+(n/b)^2\right]^2}\sin\frac{m\pi x}{a}\sin\frac{n\pi y}{b} \qquad (13.20)$$

in which p_{mn} is given by Eq. (13.19).

Example 13.2 (a) Determine the deflections and moments in a simply supported rectangular plate of thickness t (Fig. 13.7). The plate is subjected to a uniformly distributed load p_0. (b) Setting $a=b$, obtain the deflections, moments, and stresses in the plate.

SOLUTION

(a) For this case, $p(x,y)=p_0$, and Eq. (13.19) is thus

$$p_{mn}=\frac{4p_0}{ab}\int_0^b\int_0^a\sin\frac{m\pi x}{a}\sin\frac{n\pi y}{b}\,dx\,dy=\frac{16p_0}{\pi^2 mn}$$

It is seen that because $p_{mn}=0$ for even values of m and n, they can be taken as odd integers. Substituting p_{mn} into Eq. (13.20) results in

$$w=\frac{16p_0}{\pi^6 D}\sum_m^{\infty}\sum_n^{\infty}\frac{\sin(m\pi x/a)\sin(n\pi y/b)}{mn\left[(m/a)^2+(n/b)^2\right]^2} \qquad (m,n=1,3,5,\ldots) \qquad (d)$$

On physical grounds we know that the uniformly loaded plate must deflect into a symmetrical shape. Such a configuration results when m and n are odd. The maximum deflection occurs at $x=a/2$, $y=b/2$. From Eq. (d), we thus have

$$w_{max}=\frac{16p_0}{\pi^6 D}\sum_m^{\infty}\sum_n^{\infty}\frac{(-1)^{(m+n)/2-1}}{mn\left[(m/a)^2+(n/b)^2\right]^2} \qquad (m,n=1,3,5,\ldots) \qquad (e)$$

By substituting Eq. (d) into Eq. (13.7), the bending moments M_x, M_y are obtained:

$$M_x=\frac{16p_0}{\pi^4}\sum_m^{\infty}\sum_n^{\infty}\frac{(m/a)^2+\nu(n/b)^2}{mn\left[(m/a)^2+(n/b)^2\right]^2}\sin\frac{m\pi x}{a}\sin\frac{n\pi y}{b} \qquad (f)$$

$$M_y=\frac{16p_0}{\pi^4}\sum_m^{\infty}\sum_n^{\infty}\frac{\nu(m/a)^2+(n/b)^2}{mn\left[(m/a)^2+(n/b)^2\right]^2}\sin\frac{m\pi x}{a}\sin\frac{n\pi y}{b}$$

(b) For the case of a square plate (setting $a=b$), substituting $\nu=0.3$, the first term of Eq. (e) gives

$$w_{max}=0.0454p_0\frac{a^4}{Et^3}$$

The rapid convergence of Eq. (e) is demonstrated by noting that retaining the first four terms gives the results $w_{max} = 0.0443 p_0(a^4/Et^3)$.

The bending moments occurring at the center of the plate are found from Eq. (f). Retaining only the first term, the result is

$$M_{x,max} = M_{y,max} = 0.0534 p_0 a^2$$

while the first four terms yield

$$M_{x,max} = M_{y,max} = 0.0472 p_0 a^2$$

It is observed from a comparison of the above values that the series given by Eq. (f) does not converge as rapidly as that of Eq. (e).

13.6 Axisymmetrically Loaded Circular Plates

The deflection w of a circular plate will manifest dependence upon radial position r only, if the applied load and conditions of end restraint are independent of the angle θ. In other words, symmetry in plate deflection follows from symmetry in applied load. For this case, only radial and tangential moments, M_r and M_θ, per unit length, and force Q_r, per unit length, act on the circular plate element shown in Fig. 13.8. To derive the fundamental equations of a circular plate, one need only transform the appropriate formulations of previous sections from Cartesian to polar coordinates.

Through the application of the coordinate transformation relationships of Sec. 3.6, the bending moments and vertical shear forces are found from Eqs. (13.7) and (13.11) to be $Q_\theta = 0$, $M_{r\theta} = 0$, and

$$M_r = -D\left(\frac{d^2w}{dr^2} + \frac{\nu}{r}\frac{dw}{dr}\right) \tag{13.21a}$$

$$M_\theta = -D\left(\frac{1}{r}\frac{dw}{dr} + \nu\frac{d^2w}{dr^2}\right) \tag{13.21b}$$

$$Q_r = -D\frac{d}{dr}\left(\frac{d^2w}{dr^2} + \frac{1}{r}\frac{dw}{dr}\right) \tag{13.21c}$$

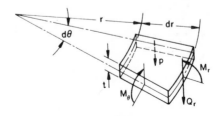

Figure 13.8

The differential equation describing the surface deflection is obtained from Eq. (13.12) in a similar fashion:

$$\nabla^4 w = \left(\frac{d^2}{dr^2} + \frac{1}{r}\frac{d}{dr} \right)\left(\frac{d^2 w}{dr^2} + \frac{1}{r}\frac{dw}{dr} \right) = \frac{p}{D} \qquad (13.22a)$$

where p, as before, represents the load acting per unit of surface area, and D is the plate rigidity. By introducing the identity

$$\frac{d^2 w}{dr^2} + \frac{1}{r}\frac{dw}{dr} = \frac{1}{r}\frac{d}{dr}\left(r\frac{dw}{dr} \right) \qquad (a)$$

Eq. (13.22a) assumes the form

$$\frac{1}{r}\frac{d}{dr}\left\{ r\frac{d}{dr}\left[\frac{1}{r}\frac{d}{dr}\left(r\frac{dw}{dr} \right) \right] \right\} = \frac{p}{D} \qquad (13.22b)$$

For applied loads varying with radius, $p(r)$, the above representation is preferred.

The *boundary conditions* at the edge of the plate of radius a may readily be written by referring to Eqs. (13.15) to (13.17) and (13.21):

Clamped edge:

$$w = 0, \qquad \frac{\partial w}{\partial r} = 0 \qquad (b)$$

Simply supported edge:

$$w = 0, \qquad M_r = 0 \qquad (c)$$

Free edge:

$$M_r = 0, \qquad Q_r = 0 \qquad (d)$$

Equation (13.22), together with the boundary conditions, is sufficient to solve the axisymmetrically loaded circular plate problem.

Example 13.3. Determine the stress and deflection for a built-in circular plate of radius a subjected to uniformly distributed loading p_0.

SOLUTION. The origin of coordinates is located at the center of the plate. The displacement w is obtained by successive integration of Eq. (13.22b):

$$w = \int \frac{1}{r} \int r \int \frac{1}{r} \int \frac{rp}{D}\,dr\,dr\,dr\,dr$$

or

$$\frac{D}{p_0} w = \int \frac{1}{r} \int r \int \left[\frac{1}{r}\left(\frac{r^2}{2} + c_1 \right) \right] dr\,dr\,dr$$

$$= \frac{r^4}{64} + \frac{c_1 r^2}{4}(\ln r - 1) + \frac{c_2 r^2}{4} + c_3 \ln r + c_4 \qquad (e)$$

where the c's are constants of integration.

The boundary conditions are

$$w=0, \qquad \frac{dw}{dr}=0 \qquad (r=a) \qquad \text{(f)}$$

The terms involving logarithms in Eq. (e) lead to an infinite displacement at the center of the plate ($r=0$) for all values of c_1 and c_3 except zero; therefore $c_1=c_3=0$. Satisfying the boundary conditions (f), we find that $c_2=-a^2/8$ and $c_4=a^4/64$. The deflection is then

$$w=\frac{p_0}{64D}(a^2-r^2)^2 \qquad \text{(g)}$$

The maximum deflection occurs at the center of the plate:

$$w_{max}=\frac{p_0 a^4}{64D}$$

Substituting the deflection given by Eq. (g) into Eq. (13.21), we have

$$M_r=\frac{p_0}{16}\left[(1+\nu)a^2-(3+\nu)r^2\right]$$
$$\text{(h)}$$
$$M_\theta=\frac{p_0}{16}\left[(1+\nu)a^2-(1+3\nu)r^2\right]$$

The extreme values of the moments are found at the center and edge. At the center,

$$M_r=M_\theta=(1+\nu)\frac{p_0 a^2}{16} \qquad (r=0)$$

At the edges, Eq. (h) yields

$$M_r=-\frac{p_0 a^2}{8}, \qquad M_\theta=-\frac{\nu p_0 a^2}{8} \qquad (r=a)$$

Examining the above results, it is clear that the maximum stress occurs at the edge:

$$\sigma_{r,max}=-\frac{6M_r}{t^2}=\frac{3}{4}\frac{p_0 a^2}{t^2}$$

A similar procedure may be applied to symmetrically loaded circular plates subject to different end conditions.

13.7 The Finite Element Solution

In this section is presented the finite element method (Chapter 7) for computation of deflection, strain, and stress in a thin plate. The plate, in general, may have any irregular geometry and loading. The derivations are based on the assumptions of small deformation theory, described in Sec. 13.1. A triangular plate element *ijm* coinciding with the *xy* plane will be employed as the finite element model (Fig. 13.9). Each nodal displacement of the element possesses three components: a displacement in the *z* direction, *w*; a rotation about the *x* axis, θ_x, and a rotation about the *y* axis, θ_y. Positive directions of the rotations are determined by the right-hand rule, as illustrated in the figure. It is clear that rotations θ_x and θ_y represent the slopes of *w*: $\partial w/\partial y$ and $\partial w/\partial x$, respectively.

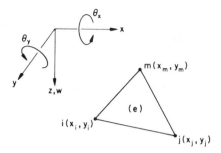

Figure 13.9

Strain, Stress, and Elasticity Matrices

Referring to Eqs. (13.3), we define, for the finite element analysis, a generalized "strain"-displacement matrix as follows:

$$\{\varepsilon\}_e = \left\{ -\frac{\partial^2 w}{\partial x^2},\ -\frac{\partial^2 w}{\partial y^2},\ -2\frac{\partial^2 w}{\partial x\,\partial y} \right\} \tag{13.23}$$

The moment generalized "strain" relationship, from Eq. (13.7), is given in the matrix form

$$\{M\}_e = [D]\{\varepsilon\}_e \tag{13.24}$$

where

$$[D] = \frac{Et^3}{12(1-\nu^2)} \begin{bmatrix} 1 & \nu & 0 \\ \nu & 1 & 0 \\ 0 & 0 & (1-\nu)/2 \end{bmatrix} \tag{13.25}$$

The stresses $\{\sigma\}_e$ and the moments $\{M\}_e$ are related by Eq. (13.9).

Displacement Function

The nodal displacement can be defined, for node i, as follows:

$$\{\delta_i\} = \{w_i,\, \theta_{xi},\, \theta_{yi}\} = \left\{ w_i,\, \left(\frac{\partial w}{\partial y}\right)_i,\, \left(\frac{\partial w}{\partial x}\right)_i \right\} \tag{a}$$

The element nodal displacements are expressed in terms of submatrices δ_i, δ_j, and δ_m:

$$\{\delta\}_e = \left\{ \begin{array}{c} \delta_i \\ \hline \delta_j \\ \hline \delta_m \end{array} \right\} = \{w_i,\quad \theta_{xi},\quad \theta_{yi},\quad w_j,\quad \theta_{xj},\quad \theta_{yj},\quad w_m,\quad \theta_{xm},\quad \theta_{ym}\} \tag{b}$$

The displacement function $\{f\}=w$ is assumed to be a modified third-order polynomial of the form

$$w=a_1+a_2x+a_3y+a_4x^2+a_5xy+a_6y^2$$
$$+a_7x^3+a_8(x^2y+xy^2)+a_9y^3 \tag{c}$$

Note that the number of terms in the above is the same as the number of nodal displacements of the element. This function satisfies displacement compatibility at interfaces between elements but does not satisfy the compatibility of slopes at these boundaries. Solutions based upon Eq. (c) do, however, yield results of acceptable accuracy.

The constants a_1 through a_9 can be evaluated by writing the nine equations involving the values of w and θ at the nodes:

$$\begin{Bmatrix} w_i \\ \theta_{xi} \\ \theta_{yi} \\ w_j \\ \theta_{xj} \\ \theta_{yj} \\ w_m \\ \theta_{xm} \\ \theta_{ym} \end{Bmatrix}_e = \begin{bmatrix} 1 & x_i & y_i & x_i^2 & x_iy_i & y_i^2 & x_i^3 & x_i^2y_i+x_iy_i^2 & y_i^3 \\ 0 & 0 & 1 & 0 & x_i & 2y_i & 0 & x_i^2+2x_iy_i & 3y_i^2 \\ 0 & 1 & 0 & 2x_i & y_i & 0 & 3x_i^3 & 2x_iy_i+y_i^2 & 0 \\ 1 & x_j & y_j & x_j^2 & x_jy_j & y_j^2 & x_j^3 & x_j^2y_j+x_jy_j^2 & y_j^3 \\ 0 & 0 & 1 & 0 & x_j & 2y_j & 0 & x_j^2+2x_jy_j & 3y_j^2 \\ 0 & 1 & 0 & 2x_j & y_j & 0 & 3x_j^2 & 2x_jy_j+y_j^2 & 0 \\ 1 & x_m & y_m & x_m^2 & x_my_m & y_m^2 & x_m^3 & x_m^2y_m+x_my_m^2 & y_m^3 \\ 0 & 0 & 1 & 0 & x_m & 2y_m & 0 & x_m^2+2x_my_m & 3y_m^2 \\ 0 & 1 & 0 & 2x_m & y_m & 0 & 3x_m^2 & 2x_my_m+y_m^2 & 0 \end{bmatrix} \begin{Bmatrix} a_1 \\ a_2 \\ a_3 \\ a_4 \\ a_5 \\ a_6 \\ a_7 \\ a_8 \\ a_9 \end{Bmatrix}$$

$$(13.26a)$$

or

$$\{\delta\}_e=[C]\{a\} \tag{13.26b}$$

Inverting,

$$\{a\}=[C]^{-1}\{\delta\}_e \tag{13.27}$$

It is observed in Eq. (13.26) that the matrix $[C]$ is dependent upon the coordinate dimensions of the nodal points.

Note that the displacement function may now be written in the usual form of Eq. (7.33) as

$$\{f\}_e=w=[N]\{\delta\}_e=[P][C]^{-1}\{\delta\}_e \tag{13.28}$$

in which

$$[P]=\begin{bmatrix} 1, & x, & y, & x^2, & xy, & y^2, & x^3, & (x^2y+xy^2), & y^3 \end{bmatrix} \tag{13.29}$$

Introducing Eq. (c) into Eq. (13.23), we have

$$\{\varepsilon\}_e=\begin{bmatrix} 0 & 0 & 0 & -2 & 0 & 0 & -6x & -2y & 0 \\ 0 & 0 & 0 & 0 & 0 & -2 & 0 & -2x & -6y \\ 0 & 0 & 0 & 0 & -2 & 0 & 0 & -4(x+y) & 0 \end{bmatrix}$$
$$\times\{a_1, a_2,\ldots, a_9\}$$

$$(13.30a)$$

or

$$\{\varepsilon\}_e = [H]\{a\} \tag{13.30b}$$

Upon substituting the values of the constants $\{a\}$ from Eq. (13.27) into the above, we can obtain the generalized "strain"-displacement matrix in the following common form:

$$\{\varepsilon\}_e = [B]\{\delta\}_e = [H][C]^{-1}\{\delta\}_e$$

Thus,

$$[B] = [H][C]^{-1} \tag{13.31}$$

The Stiffness Matrix

The element stiffness matrix given by Eq. (7.38), treating the thickness t as a constant within the element and introducing $[B]$ from Eq. (13.31), becomes

$$[k]_e = [[C]^{-1}]^{\mathrm{T}}\left(\int\int [H]^{\mathrm{T}}[D][H]\, dx\, dy\right)[C]^{-1} \tag{13.32}$$

where the matrices $[H]$, $[D]$, and $[C]^{-1}$ are defined by Eqs. (13.30), (13.25), and (13.26), respectively. After expansion of the expression under the integral sign, the integrations can be carried out to obtain the element stiffness matrix.

External Nodal Forces

As in two-dimensional and axisymmetrical problems, the nodal forces due to the distributed *surface* loading may also be obtained through the use of Eq. (7.39) or by physical intuition.

The standard finite element method procedure described in Sec. 7.10 may now be followed to obtain the unknown displacement, strain, and stress in any element of the plate.

PART 2—Thin Shells

13.8 Definitions

Structural elements resembling curved plates are referred to as shells. Included among the more familiar examples of shells are soap bubbles, incandescent lamps, aircraft fuselages, pressure vessels, and a variety of metal, glass, and plastic containers. As was the case for plates, we shall limit our treatment to isotropic, homogeneous, elastic shells having a constant thickness which is small relative to the remaining dimensions. The surface bisecting the shell thickness is referred to as the *midsurface*. To specify the geometry of a shell, one need only know the configuration of the midsurface and the thickness of the shell at each point. According to the criterion often applied to define a thin shell (for purposes of technical calculations), the ratio of thickness t to radius of curvature r should be equal to or less than $\frac{1}{20}$.

The stress analysis of shells normally embraces two distinct theories. There is the *membrane theory*, limited to moment-free membranes, which often applies to a rather large proportion of the entire shell. The *bending theory* or *general theory* includes the influences of bending and thus enables one to treat discontinuities in the field of stress occurring in a limited region in the vicinity of a load application or a structural discontinuity. This method generally involves a membrane solution, corrected in those areas in which discontinuity effects are pronounced. The principal objective is thus not the improvement of the membrane solution, but rather the analysis of stresses associated with edge loading, which cannot be accomplished by the membrane theory alone.

The following assumptions are generally made in the *small deflection analysis* of thin shells:

1. The ratio of the shell thickness to the radius of curvature of the midsurface is small compared with unity.
2. Displacements are very small compared with the shell thickness.
3. Straight sections of an element, which are perpendicular to the midsurface, remain perpendicular and straight to the *deformed* midsurface subsequent to bending. The implication of this assumption is that the strains γ_{xz} and γ_{yz} are negligible. Normal strain, ε_z, due to transverse loading may also be omitted.
4. The z-directed stress σ_z is negligible.

13.9 Simple Membrane Action

As testimony to the fact that the load-carrying mechanism of a shell differs from that of other elements, we have only to note the extraordinary capacity of an eggshell to withstand normal forces, despite its thinness and fragility. This contrasts markedly with a similar material in a plate configuration subjected to lateral loading.

To understand the phenomenon, consider a portion of a spherical shell of radius r and thickness t, subjected to a uniform pressure p (Fig. 13.10). Denoting by N the normal force per unit length required to maintain the shell in a state of equilibrium, static equilibrium of vertical forces is expressed by

$$2\pi r_0 N \sin\phi = p\pi r_0^2$$

or

$$N = \frac{pr_0}{2\sin\phi} = \frac{pr}{2}$$

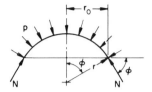

Figure 13.10

This result is valid anywhere in the shell, as N is observed not to vary with ϕ. Note that, in contrast with the case of plates, it is the midsurface which sustains the applied load.

Once again referring to the simple shell shown in Fig. 13.10, we demonstrate that the bending stresses play an insignificant role in the load-carrying mechanism. On the basis of the symmetry of the shell and the loading, the stresses (equal at any point) are given by

$$\sigma_n = -\frac{N}{t} = -\frac{pr}{2t} \tag{a}$$

Here σ_n represents the compressive, in-plane stress. The stress normal to the midsurface is negligible, and thus the in-plane strain involves only σ_n:

$$\varepsilon_n = \frac{\sigma_n}{E} - \nu\frac{\sigma_n}{E} = -(1-\nu)\frac{pr}{2tE} \tag{b}$$

The reduced circumference associated with this strain is

$$2\pi r' = 2\pi(r + r\varepsilon_n)$$

so that

$$r' = r(1 + \varepsilon_n)$$

The change in curvature is therefore

$$\Delta\left(\frac{1}{r}\right) = \frac{1}{r'} - \frac{1}{r} = \frac{1}{r}\left(\frac{1}{1+\varepsilon_n} - 1\right)$$

$$= -\frac{\varepsilon_n}{r}\left(\frac{1}{1+\varepsilon_n}\right) = -\frac{\varepsilon_n}{r}\left(1 - \varepsilon_n + \varepsilon_n^2 - \cdots\right)$$

Dropping higher order terms because of their negligible magnitude, and substituting Eq. (b), the above expression becomes

$$\Delta\left(\frac{1}{r}\right) \approx -\frac{\varepsilon_n}{r} = \frac{(1-\nu)p}{2tE} \tag{c}$$

The bending moment in the shell is determined from the plate equations. Noting that $1/r_x$ and $1/r_y$ in Eq. (13.7) refer to the change in plate curvatures between the undeformed and deformed conditions, we see that for the spherical shell under consideration $\Delta(1/r) = 1/r_x = 1/r_y$. Therefore, Eq. (13.7) yields

$$M_b = -D\left(\frac{1}{r_x} + \frac{\nu}{r_y}\right) = -D(1-\nu^2)\frac{p}{2tE} = -\frac{pt^2}{24} \qquad \text{(d)}$$

and the maximum corresponding stress is

$$\sigma_b = \frac{6M_b}{t^2} = -\frac{p}{4} \qquad \text{(e)}$$

Comparing σ_b and σ_n [Eqs. (e) and (a)], we have

$$\frac{\sigma_n}{\sigma_b} = \frac{2r}{t} \qquad \text{(f)}$$

demonstrating that the in-plane or direct stress is very much larger than the bending stress, inasmuch as $t/2r \ll 1$. It may be concluded, therefore, that the applied load is resisted primarily by the in-plane stressing of the shell.

While the preceding discussion relates to the simplest shell configuration, the conclusions drawn with respect to the fundamental mechanism apply to any shape and loading at locations away from the boundaries or points of concentrated load application. If there are asymmetries in load or geometry, shearing stresses will exist in addition to the normal and bending stresses.

In the following sections we discuss the membrane theory of two common structures: the shell of revolution and cylindrical shells.

13.10   Symmetrically Loaded Shells of Revolution

A surface of revolution, such as in Fig. 13.11a is formed by the rotation of a so-called *meridian curve* (eo') about the OO axis. As shown, a point on the shell is located by coordinates θ, ϕ, r_0. This figure indicates that the elemental surface $abcd$ is defined by two meridian and two parallel circles. The planes containing the principal radii of curvature at any point on the surface of the shell are the meridian plane and a plane perpendicular to it at the point in question. The meridian plane will thus contain r_θ, which is related to side ab. The other principal radius of curvature, r_ϕ, is found in the perpendicular plane, and is therefore related to side bd. Thus length $ab = (r_\theta \sin\theta)\, d\theta = r_0\, d\theta$, and length $bd = r_\phi\, d\phi$.

The condition of symmetry prescribes that no shearing forces act on the element and that the normal forces N_θ and N_ϕ per unit length display no variation with θ (Fig. 13.11b). The externally applied load is represented by the perpendicular components p_y and p_z. We turn now to a derivation of the equations governing the force equilibrium of the element.

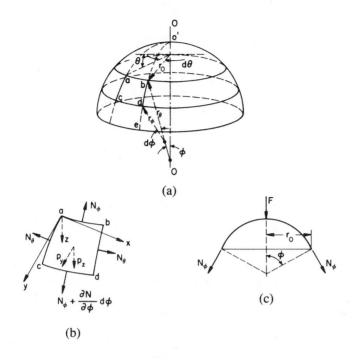

(a)

(b)

(c)

Figure 13.11

To describe equilibrium in the z direction, it is necessary to consider the z components of the external loading as well as of the forces acting on each edge of the element. The external z-directed load acting on an element of area $(r_0\,d\theta)(r_\phi\,d\phi)$ is

$$p_z r_0 r_\phi\,d\theta\,d\phi \tag{a}$$

The force acting on the top edge of the element is $N_\phi r_0\,d\theta$. Neglecting terms of higher order, the force acting on the bottom edge is also $N_\phi r_0\,d\theta$. The z component at each edge is then $N_\phi r_0\,d\theta\sin(d\phi/2)$, which may be approximated $N\phi r_0\,d\theta\,d\phi/2$, leading to a resultant for both edges of

$$N_\phi r_0\,d\theta\,d\phi \tag{b}$$

The force on each side of the element is $N_\theta r_\phi\,d\phi$. The radial resultant for both such forces is $(N_\theta r_\phi\,d\phi)\,d\theta$, having a z-directed component

$$N_\theta r_\phi\,d\phi\,d\theta\sin\phi \tag{c}$$

Adding the z forces, equating to zero, and canceling $d\theta\,d\phi$, we obtain

$$N_\phi r_0 + N_\theta r_\phi\sin\phi + p_z r_0 r_\phi = 0$$

Dividing by $r_0 r_\phi$ and replacing r_0 by $r_\theta\sin\phi$, the above expression is

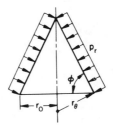

Figure 13.12

converted to the following form:

$$\frac{N_\phi}{r_\phi} + \frac{N_\theta}{r_\theta} = -p_z \tag{13.33}$$

Similarly, an equation for the y equilibrium of the element of the shell may also be derived. But instead of solving the z and y equilibrium equations simultaneously, it is more convenient to obtain N_ϕ from the equilibrium of the portion of the shell corresponding to the angle ϕ (Fig. 13.11c):

$$2\pi r_0 N_\phi \sin\phi + F = 0 \tag{13.34}$$

and then to calculate N_θ from Eq. (13.33). Here F represents the resultant of all external loading acting on the portion of the shell.

To treat the case of a *conical shell*, Fig. 13.12, one need only set $r_\phi = \infty$ in Eq. (13.33). This, together with Eq. (13.34), provides the following pair of equations for determining the stress resultants, under a distributed load $p_z = p_r$:

$$N_\theta = -p_r r_\theta = -\frac{p_r r_0}{\sin\phi} \tag{13.35}$$

$$2\pi r_0 N_\phi \sin\phi + F = 0 \tag{13.36}$$

For the axisymmetrical shells considered, owing to their freedom of motion in the z direction, strains are produced such as to assure consistency with the field of stress. These strains are compatible with one another. It is clear that when a shell is subjected to concentrated surface loadings or is constrained at its boundaries, membrane theory cannot satisfy the conditions on deformation everywhere. In such cases, however, the departure from membrane behavior is limited to a narrow zone in the vicinity of the boundary or the loading. Membrane theory remains valid for the major portion of the shell, but the complete solution can be obtained only through application of bending theory.

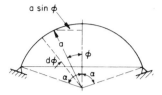

Figure 13.13

Example 13.4. Derive expressions for the stress resultants in a spherical dome of radius a and thickness t, loaded only by its own weight, p per unit area (Fig. 13.13).

SOLUTION. The weight of that portion of the dome intercepted by ϕ is

$$F=\int_0^\phi p(2\pi a\sin\phi\cdot a\cdot d\phi)=2\pi a^2 p(1-\cos\phi)$$

In addition,

$$p_z=ap\cos\phi$$

Substituting into Eqs. (13.33) and (13.34) for p_z and F, we obtain

$$N_\phi=\frac{ap(1-\cos\phi)}{\sin^2\phi}=-\frac{ap}{1+\cos\phi}$$

$$N_\theta=-ap\left(a\cos\phi-\frac{1}{1+\cos\phi}\right)$$

where the negative signs indicate compression. It is clear that N_ϕ is always compressive. The sign of N_θ, on the other hand, depends upon ϕ. From the second expression above, when $N_\theta=0$, $\phi=51°50'$. For ϕ smaller than this value, N_θ is compressive. For $\phi>51°50'$, N_θ is tensile.

13.11   Cylindrical Shells

A cylindrical shell is generated as a straight line, the *generator*, moves parallel to itself along a closed path. In Fig. 13.14 is shown an element isolated from a cylindrical shell of arbitrary cross section. The element is located by coordinates x (axial) and θ, on the cylindrical surface.

The forces acting on the sides of the element are depicted in the figure. The x and θ components of the externally applied forces per unit area are denoted p_x and p_θ, and are shown to act in the directions of increasing x and θ (or y). In addition, a radial (or normal) component of the external loading p_r acts in the positive z direction. The following expressions describe the requirements for equilibrium in the x, θ, and r directions:

$$\frac{\partial N_x}{\partial x}dx(r\,d\theta)+\frac{\partial N_{\theta x}}{\partial\theta}d\theta(dx)+p_x(dx)r\,d\theta=0$$

$$\frac{\partial N_\theta}{\partial\theta}d\theta(dx)+\frac{\partial N_{x\theta}}{\partial x}dx(r\,d\theta)+p_\theta(dx)r\,d\theta=0$$

$$N_\theta dx(d\theta)+p_r(dx)r\,d\theta=0$$

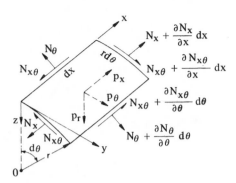

Figure 13.14

Canceling the differential quantities, we obtain the equations of a cylindrical shell:

$$N_\theta = -p_r r$$

$$\frac{\partial N_{x\theta}}{\partial x} + \frac{1}{r}\frac{\partial N_\theta}{\partial \theta} = -p_\theta \qquad (13.37)$$

$$\frac{\partial N_x}{\partial x} + \frac{1}{r}\frac{\partial N_{x\theta}}{\partial \theta} = -p_x$$

Given the external loading, N_θ is readily determined from the first equation given above. Following this, by integrating the second and the third equations, $N_{x\theta}$ and N_x are found:

$$N_\theta = -p_r r$$

$$N_{x\theta} = -\int\left(p_\theta + \frac{1}{r}\frac{\partial N_\theta}{\partial \theta}\right)dx + f_1(\theta) \qquad (13.38)$$

$$N_x = -\int\left(p_x + \frac{1}{r}\frac{\partial N_{x\theta}}{\partial \theta}\right)dx + f_2(\theta)$$

Here $f_1(\theta)$ and $f_2(\theta)$ represent arbitrary functions of integration to be evaluated on the basis of the boundary conditions. They arise, of course, as a result of the integration of partial derivatives.

Example 13.5. Determine the stress resultants in a circular, simply supported tube of thickness t filled to capacity with a liquid of specific weight γ (Fig. 13.15a).

SOLUTION. The pressure at any point in the tube equals the weight of a column of unit cross-sectional area of the liquid at that point. At the arbitrary level mn (Fig. 13.15b), the outward pressure is $-\gamma a(1-\cos\theta)$, where the pressure is positive radially inward—hence the minus sign. Then

$$p_r = -\gamma a(1-\cos\theta), \qquad p_\theta = p_x = 0 \qquad \text{(a)}$$

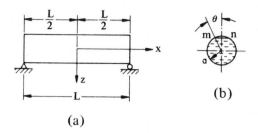

(a)

(b)

Figure 13.15

Substituting the foregoing into Eqs. (13.38), we obtain

$$N_\theta = \gamma a^2 (1 - \cos \theta)$$

$$N_{x\theta} = -\int \gamma a \sin \theta \, dx + f_1(\theta) = -\gamma a x \sin \theta + f_1(\theta)$$

$$N_x = \int \gamma x \cos \theta \, dx - \frac{1}{a} \int \frac{df_1}{d\theta} \, dx + f_2(\theta)$$

$$= \frac{\gamma x^2}{2} \cos \theta - \frac{x}{a} \frac{df_1}{d\theta} + f_2(\theta)$$

(b)

The boundary conditions are

$$N_x = 0 \qquad (x = L/2, x = -L/2) \tag{c}$$

The introduction of Eq. (b) into Eq. (c) leads to

$$0 = \frac{\gamma L^2}{8} \cos \theta - \frac{L}{2a} \frac{df_1}{d\theta} + f_2(\theta)$$

$$0 = \frac{\gamma L^2}{8} \cos \theta + \frac{L}{2a} \frac{df_1}{d\theta} + f_2(\theta)$$

Addition and subtraction of the above give, respectively,

$$f_2(\theta) = -\frac{\gamma L^2}{8} \cos \theta$$

$$\frac{df_1}{d\theta} = 0 \qquad \text{or} \qquad f_1(\theta) = 0 + c$$

(d)

We observe from the second equation of (b) that c in the second equation of (d) represents the value of the uniform shear load $N_{x\theta}$ at $x = 0$. This load is zero because the tube is subjected to no torque; thus $c = 0$. Then, Eq. (b) together with Eq. (d) provides the solution

$$N_\theta = \gamma a^2 (1 - \cos \theta)$$

$$N_{x\theta} = -\gamma a x \sin \theta$$

$$N_x = -\tfrac{1}{8}\gamma (L^2 - 4x^2) \cos \theta$$

(e)

The stresses are determined upon dividing the above stress resultants by the shell thickness. It is observed that the shear $N_{x\theta}$ and the normal force N_x exhibit the same spanwise distribution as the shear force and the bending moment of a beam. Their values, as may readily be verified, are identical with those obtained by application of the beam formulas, Eqs. (5.39) and (5.38), respectively.

It has already been noted that membrane theory cannot, in all cases, provide solutions compatible with the actual conditions of deformation. This theory also fails to predict the state of stress in certain areas of the shell. To overcome these shortcomings, bending theory is applied in the case of cylindrical shells, taking into account the stress resultants such as the types shown in Fig. 13.2 and N_x, N_θ, and $N_{x\theta}$.

Chapter 13—Problems

Secs. 13.1 to 13.7

13.1. A thin rectangular plate is subjected to uniformly distributed bending moments M_a and M_b, applied along edges a and b, respectively. Derive the equations governing the surface deflection for two cases: (a) $M_a \neq M_b$ and (b) $M_a = -M_b$.

13.2. The simply supported rectangular plate shown in Fig. 13.7 is subjected to a distributed load p given by

$$p = \frac{36P(a-x)(b-y)}{a^3 b^3}$$

Derive an expression for the deflection of the plate in terms of the constants P, a, b, and D.

13.3. A simply supported circular plate of radius a and thickness t is deformed by a moment M_0, uniformly distributed along the edge. Derive an expression for the deflection w as well as for the maximum radial and tangential stresses.

13.4. Given a simply supported circular plate containing a circular hole, supported at its outer edge and subjected to uniformly distributed inner edge moments M (Fig. P13.4), derive an expression for the plate deflection.

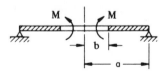

Figure P13.4

13.5. A circular plate of radius a is simply supported at its edge and subjected to uniform loading p. Determine the center point deflection.

Secs. 13.8 to 13.11

13.6. Determine the membrane stress resultants in a steel spherical tank filled with gas of specific weight γ, and supported along circle bb on a cylindrical pipe (Fig. P13.6). Is the deformation due to membrane stresses compatible with the continuity of the structure at the support bb?

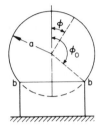

Figure P13.6

13.7. For the toroidal shell of Fig. P13.7, subjected to internal pressure p, determine the membrane forces N_ϕ and N_θ.

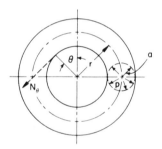

Figure P13.7

13.8. Show that the tangential (circumferential) and longitudinal stresses in a simply supported conical tank filled with liquid of specific weight γ (Fig. P13.8) are given by

$$\sigma_\theta = \frac{\gamma(a-y)y}{t}\frac{\tan\alpha}{\cos\alpha}, \qquad \sigma_\phi = \frac{(a-2y/3)y}{2t}\frac{\tan\alpha}{\cos\alpha} \qquad \text{(P13.8)}$$

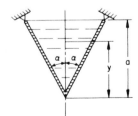

Figure P13.8

13.9. A simply supported circular cylindrical shell of radius a and length L carries its own weight p per unit area (i.e., $p_x = 0$, $p_\theta = -p \cos \theta$, $p_r = p \sin \theta$). Determine the membrane forces. The angle θ is measured from horizontal axis.

13.10. Redo Example 13.5 for the case in which the ends of the cylinder are fixed.

13.11. An edge supported conical shell carries its own weight p per unit area and is subjected to an external pressure p_r (Fig. P13.8). Determine the membrane forces and the maximum stresses in the shell.

Appendix A

Indicial Notation

A particular class of tensor, a *vector*, requires only a single subscript to describe each of its components. Often the components of a tensor require more than a single subscript for definition. For example, second *order* or second *rank* tensors, such as those of stress or inertia, require double subscripting: τ_{ij}, I_{ij}. Quantities such as temperature and mass are scalars, classified as tensors of zero rank.

Tensor or *indicial* notation, here briefly explored, offers the advantage of succinct representation of lengthy equations through the minimization of symbols. In addition, physical laws expressed in tensor form are independent of the choice of coordinate system, and therefore similarities in seemingly different physical systems are often made more apparent.

Two simple conventions enable us to write most of the equations developed in this text in indicial notation. These conventions, relative to range and summation, are as follows:

Range convention: When a lower case alphabetic subscript is *unrepeated*, it takes on all values indicated.

Summation convention: When a lower case alphabetic subscript is *repeated* in a term, then summation over the range of that subscript is indicated, making unnecessary the use of the summation symbol.

For example, on the basis of these conventions, the equations of equilibrium (1.5) may now be written

$$\frac{\partial \tau_{ij}}{\partial x_j} + F_i = 0 \qquad (i, j = x, y, z) \qquad \text{(A.1a)}$$

where $x_x = x$, $x_y = y$, and $x_z = z$. The repeated subscript is j, indicating summation. The unrepeated subscript is i. Here i is termed the *free* index, and j, the *dummy* index.

If in the foregoing expression the symbol $\partial/\partial x$ is replaced by a comma, we have

$$\tau_{ij,j} + F_i = 0 \qquad \text{(A.1b)}$$

where the subscript after the comma denotes the coordinate with respect to which differentiation is performed. If no body forces exist, Eq. (A.1b) reduces to $\tau_{ij,j}=0$, indicating that the *sum of the three stress derivatives is zero*.

Similarly, the strain-displacement relations are expressed more concisely by using commas. Thus, Eq. (2.4) may be stated as follows:

$$\varepsilon_{ij}=\tfrac{1}{2}(u_{i,j}+u_{j,i}) \tag{A.2}$$

The equations of transformation of the components of a stress tensor, in indicial notation, are represented by

$$\tau'_{rs}=l_{ir}l_{js}\tau_{ij} \tag{A.3a}$$

Alternatively,

$$\tau_{rs}=l_{ri}l_{sj}\tau'_{ij} \tag{A.3b}$$

The repeated subscripts i and j imply the double summation in Eq. (A.3a), which, upon expansion, yields

$$\tau'_{rs}=l_{xr}l_{xs}\tau_{xx}+l_{xr}l_{ys}\tau_{xy}+l_{xr}l_{zs}\tau_{xz}$$
$$+l_{yr}l_{xs}\tau_{xy}+l_{yr}l_{ys}\tau_{yy}+l_{ys}l_{zr}\tau_{yz}$$
$$+l_{zr}l_{xs}\tau_{xz}+l_{zr}l_{ys}\tau_{yz}+l_{zr}l_{zs}\tau_{zz}$$

By assigning $r, s=x, y, z$ and noting that $\tau_{rs}=\tau_{sr}$, the foregoing leads to the six expressions of Eq. (1.17).

The transformation relating coordinates x, y, z to x', y', z' is applicable to the components of the strain in a manner analogous to that of the stress:

$$\varepsilon'_{rs}=l_{ir}l_{js}\varepsilon_{ij} \tag{A.4a}$$

Conversely,

$$\varepsilon_{rs}=l_{ri}l_{sj}\varepsilon'_{ij} \tag{A.4b}$$

These equations represent the law of transformation for a strain tensor of rank 2.

Appendix B

A Practical Approach to the Stress Cubic Equation and Direction Cosines

B.1 Principal Stresses

There are many methods in common usage for solving a cubic equation. A simple approach for dealing with Eq. (1.20) is to find one root, say σ_1, by plotting it (σ as abscissa) or by trial and error. The cubic equation is then factored by dividing by $(\sigma_p - \sigma_1)$ to arrive at a quadratic equation. The remaining roots can be obtained by applying the familiar general solution of a quadratic equation. This process requires considerable time and algebraic work, however.

What follows is a practical approach for determining the roots of stress cubic equation (1.20):

$$\sigma_p^3 - I_1\sigma_p^2 + I_2\sigma_p - I_3 = 0 \tag{a}$$

where

$$I_1 = \sigma_x + \sigma_y + \sigma_z$$
$$I_2 = \sigma_x\sigma_y + \sigma_x\sigma_z + \sigma_y\sigma_z - \tau_{xy}^2 - \tau_{yz}^2 - \tau_{xz}^2 \tag{B.1}$$
$$I_3 = \sigma_x\sigma_y\sigma_z + 2\tau_{xy}\tau_{yz}\tau_{xz} - \sigma_x\tau_{yz}^2 - \sigma_y\tau_{xz}^2 - \sigma_z\tau_{xy}^2$$

According to the method, expressions that provide direct means for solving both two-dimensional and three-dimensional stress problems are*

$$\sigma_a = 2S[\cos(\alpha/3)] + \tfrac{1}{3}I_1$$
$$\sigma_b = 2S\{\cos[(\alpha/3) + 120°]\} + \tfrac{1}{3}I_1 \tag{B.2}$$
$$\sigma_c = 2S\{\cos[(\alpha/3) + 240°]\} + \tfrac{1}{3}I_1$$

---

*See E. E. Messal, Finding true maximum shear stress, *Machine Design* pp. 166–169 (December 7, 1978).

Here the constants are given by

$$S = \left(\tfrac{1}{3}R\right)^{1/2}$$

$$\alpha = \cos^{-1}\left(-\frac{Q}{2T}\right)$$

$$R = \tfrac{1}{3}I_1^2 - I_2 \qquad\qquad\qquad (B.3)$$

$$Q = \tfrac{1}{3}I_1 I_2 - I_3 - \tfrac{2}{27}I_1^3$$

$$T = \left(\tfrac{1}{27}R^3\right)^{1/2}$$

and invariants I_1, I_2, and I_3 are represented in terms of the given stress components by Eqs. (B.1).

The principal stresses found from Eqs. (B.2) are *redesignated* using numerical subscripts so that $\sigma_1 > \sigma_2 > \sigma_3$. The above procedure is well adapted to a pocket calculator or digital computer.

B.2   Direction Cosines

The values of the direction cosines of a principal stress are determined through the use of Eqs. (1.18) and (1.13), as already discussed in Sec. 1.9. That is, substitution of a principal stress, say σ_1, into Eqs. (1.18) results in *two independent* equations in three unknown direction cosines. From these expressions together with $l_1^2 + m_1^2 + n_1^2 = 1$, we obtain l_1, m_1, and n_1.

However, instead of solving one second-order and two linear equations simultaneously, the following simpler approach is preferred. Expressions (1.18) are expressed in matrix form as follows:

$$
\begin{bmatrix}
(\sigma_x - \sigma_i) & \tau_{xy} & \tau_{xz} \\
\tau_{xy} & (\sigma_y - \sigma_i) & \tau_{yz} \\
\tau_{xz} & \tau_{yz} & (\sigma_z - \sigma_i)
\end{bmatrix}
\begin{Bmatrix}
l_i \\
m_i \\
n_i
\end{Bmatrix} = 0
\qquad (i = 1, 2, 3)
$$

The cofactors of the determinant of the above matrix on the elements of the first row are

$$
a_i = \begin{vmatrix}
(\sigma_y - \sigma_i) & \tau_{yz} \\
\tau_{yz} & (\sigma_z - \sigma_i)
\end{vmatrix}
$$

$$
b_i = -\begin{vmatrix}
\tau_{xy} & \tau_{yz} \\
\tau_{xz} & (\sigma_z - \sigma_i)
\end{vmatrix}
\qquad\qquad (B.4)
$$

$$
c_i = \begin{vmatrix}
\tau_{xy} & (\sigma_y - \sigma_i) \\
\tau_{xz} & \tau_{yz}
\end{vmatrix}
$$

Upon introduction of the notation

$$k_i = \frac{1}{\left(a_i^2 + b_i^2 + c_i^2\right)^{1/2}} \tag{B.5}$$

the direction cosines are then expressed

$$l_i = a_i k_i, \qquad m_i = b_i k_i, \qquad n_i = c_i k_i \tag{B.6}$$

It is clear that Eqs. (B.6) lead to $l_i^2 + m_i^2 + n_i^2 = 1$.

Application of Eqs. (B.2) and (B.6) to the sample problem described in Example 1.3 provides some algebraic exercise. Substitution of the given data into Eqs. (B.1) results in

$$I_1 = -22.7 \text{ MPa}, \qquad I_2 = -170.8125 \text{ MPa}, \qquad I_3 = 2647.5215 \text{ MPa}$$

We then have

$$R = \tfrac{1}{3}I_1^2 - I_2 = 342.5758$$

$$T = \left(\tfrac{1}{27}R^3\right)^{1/2} = 1220.2623$$

$$Q = \tfrac{1}{3}I_1 I_2 - I_3 - \tfrac{2}{27}I_1^3 = -488.5896$$

$$\alpha = \cos^{-1}(-Q/2T) = 78.4514°$$

Hence, Eqs. (B.2) give

$$\sigma_a = 11.618 \text{ MPa}, \qquad \sigma_b = -25.316 \text{ MPa}, \qquad \sigma_c = -9.002 \text{ MPa}$$

Reordering and redesignating the above values,

$$\sigma_1 = 11.618 \text{ MPa}, \qquad \sigma_2 = -9.002 \text{ MPa}, \qquad \sigma_3 = -25.316 \text{ MPa}$$

from which it follows that

$$a_1 = \begin{vmatrix} (4.6 - 11.618) & 11.8 \\ 11.8 & (-8.3 - 11.618) \end{vmatrix} = 0.5445$$

$$b_1 = -\begin{vmatrix} -4.7 & 11.8 \\ 6.45 & (-8.3 - 11.618) \end{vmatrix} = -17.5046$$

$$c_1 = \begin{vmatrix} -4.7 & (4.6 - 11.618) \\ 6.45 & 11.8 \end{vmatrix} = -10.1939$$

and

$$k_1 = \frac{1}{\left(a_1^2 + b_1^2 + c_1^2\right)^{1/2}} = 0.0493$$

Thus, Eqs. (B.6) yield

$$l_1 = 0.0266, \qquad m_1 = -0.8638, \qquad n_1 = -0.5031$$

As a check, $l_1^2 + m_1^2 + n_1^2 = 0.9999 \approx 1$. Repeating the same procedure for σ_2 and σ_3, we obtain the values of direction cosines given in Example 1.3.

Answers to Selected Problems

Chapter 1

1.2. $F_x = F_y = F_z = 0$

1.3. (a) $\sigma_1 = 121$ MPa, $\sigma_2 = -71$ MPa, $\tau_{max} = 96$ MPa
$\sigma_1 = 200$ MPa, $\sigma_2 = -50$ MPa, $\tau_{max} = 125$ MPa
(b) $\theta_p' = -19.3°$, $\theta_s' = 25.7°$, $\theta_p = 26.55°$, $\theta_s' = 71.55°$

1.5. (a) $\sigma = -0.237\tau_0$, $\tau = 0.347\tau_0$
(b) $\sigma_1 = 3.732\tau_0$, $\sigma_2 = -0.268\tau_0$, $\theta_p' = 15°$

1.7. $\sigma_1 = -\sigma_2 = 51.96$ MPa, $\theta_p' = -30°$

1.9. $P = 32.5$ kN, $\theta = 26.56°$

1.10. (a) $\sigma_x = 186$ MPa, (b) $\sigma_1 = 188$ MPa, $\tau_{max} = 101$ MPa, $\theta_p' = 5.68°$

1.12. $P = 1069$ kN, $p = 0.467$ MPa

1.14. $\sigma_1 = 24.746$ MPa, $\sigma_2 = 8.479$ MPa, $\sigma_3 = 2.773$ MPa, $l_1 = 0.647$
$m_1 = 0.396$, $n_1 = 0.652$

1.15. (a) $\sigma_{x'} = 10$, $\sigma_{y'} = 5$, $\sigma_{z'} = -1$, $\tau_{y'z'} = 3\sqrt{3}$, $\tau_{x'y'} = \tau_{x'z'} = 0$
(b) $\sigma_{x''} = 7.2$, $\sigma_{y''} = -1.2$, $\sigma_{z''} = 8$, $\tau_{x''y''} = 5.6$, $\tau_{x''z''} = \tau_{y''z''} = 0$

1.17. (a) $\sigma_1 = 12.049$ MPa, $\sigma_2 = -1.521$ MPa, $\sigma_3 = -4.528$ MPa, $l_1 = 0.618$, $m_1 = 0.533$, $n_1 = 0.577$
(b) $\sigma_1 = 19.237$ MPa, $\sigma_2 = 13.704$ MPa, $\sigma_3 = 4.648$ MPa, $l_1 = 0.334$, $m_1 = -0.386$, $n_1 = 0.860$

1.19. (a) $\tau_{13} = 8.288$ MPa, $\tau_{12} = 6.785$ MPa, $\tau_{23} = 1.503$ MPa
(b) $\tau_{13} = 7.294$ MPa, $\tau_{12} = 2.766$ MPa, $\tau_{23} = 4.528$ MPa

1.22. $\sigma_{oct} = 12$ MPa, $\tau_{oct} = 9.31$ MPa

1.24. (a) $\tau_{max} = 21$ MPa, (b) $\sigma_{oct} = 35$ MPa, $\tau_{oct} = 17.15$ MPa

1.25. $\sigma = -12.4$ MPa, $\tau = 26.2$ MPa, $T_x = 16.81$ MPa, $T_y = -3.88$ MPa, $T_z = -23.30$ MPa

Chapter 2

2.2. (a) $\varepsilon_x = 0.06667$, $\varepsilon_y = 0.075$, $\gamma_{xy} = -0.125$
(b) $\varepsilon_1 = 133.47 \times 10^{-3}$, $\varepsilon_2 = 8.195 \times 10^{-3}$, $\theta_p' = 133.09°$

2.4. (a) $\gamma_{max} = 0.002$, $\theta_s = 45°$
(b) $\varepsilon_x = 350 \times 10^{-5}$, $\varepsilon_y = 250 \times 10^{-5}$, $\gamma_{xy} = -173 \times 10^{-5}$

2.5. $\varepsilon_1 = -5.92 \times 10^{-5}$, $\varepsilon_2 = -114.08 \times 10^{-5}$, $\theta_p' = 61.85°$

2.6. $\varepsilon_1 = 114.08 \times 10^{-5}$, $\varepsilon_2 = 5.92 \times 10^{-5}$, $\theta_p' = -61.85°$

2.8. (a) $I_1 = -3 \times 10^{-6}$, $I_2 = -44 \times 10^{-12}$, $I_3 = 58 \times 10^{-18}$
(b) $\varepsilon_{x'} = 3.848 \times 10^{-6}$
(c) $\varepsilon_1 = 6.01 \times 10^{-6}$, $\varepsilon_2 = -1.26 \times 10^{-6}$, $\varepsilon_3 = -7.72 \times 10^{-6}$
(d) $\gamma_{max} = 13.73 \times 10^{-6}$

2.9. (a) $I_1 = 10 \times 10^{-6}$, $I_2 = 19 \times 10^{-12}$, $I_3 = -22 \times 10^{-18}$
(b) $\varepsilon_{x'} = 3.866 \times 10^{-6}$
(c) $\varepsilon_1 = 6.18 \times 10^{-6}$, $\varepsilon_2 = 4.08 \times 10^{-6}$, $\varepsilon_3 = -0.798 \times 10^{-6}$
(d) $\gamma_{max} = 6.978 \times 10^{-6}$

2.11. $\Delta_{BD} = 0.282/E$ m

2.13. $\sigma_x = 720$ MPa, $\sigma_y = 990$ MPa, $\sigma_z = 400$ MPa, $\tau_{xy} = 160$ MPa, $\tau_{yz} = 640$ MPa, $\tau_{xz} = 0$

2.15. $\sigma_3 : \sigma_2 : \sigma_1 = 1 : 1.086 : 1.171$, $\sigma_1 = 139,947$ MPa, $\sigma_2 = 129.757$ MPa, $\sigma_3 = 119.513$ MPa

2.18. $\varepsilon_x = \gamma(a-x)/E$, $\varepsilon_y = -\nu\varepsilon_x$, $\sigma_x = \gamma(a-x)$, $\gamma_{xy} = \tau_{xy} = \sigma_y = 0$.
Yes.

2.20 $U_v = \dfrac{1}{12} \dfrac{N^2 L}{\pi r^2 E}$, $U_d = \dfrac{5}{2} \dfrac{M_t^2 L}{\pi r^4 E} + \dfrac{5}{12} \dfrac{N^2 L}{\pi r^2 E}$

Chapter 3

3.1 (b) $\phi = \dfrac{c_1 x^3 y^3}{6} - \dfrac{c_1 x^5 y}{10} + (c_5 y^3 + c_6 y^2 + c_7 y + c_8)x$
$+ c_9 y^3 + c_{10} y^2 + c_{11} y + c_{12}$

3.4. All conditions, except on edge $x = L$, are satisfied by the ϕ.

3.5. $\sigma_x = \dfrac{p}{a^2}(x^2 - 2y^2)$, $\sigma_y = \dfrac{py^2}{a^2}$, $\tau_{xy} = -\dfrac{2pxy}{a^2}$

3.10. (a) $(\sigma_x)_{\text{elast.}} = P/0.512h$, $(\sigma_x)_{\text{elem.}} = P/0.536h$
(b) $(\sigma_x)_{\text{elast.}} = P/1.483h$, $(\sigma_x)_{\text{elem.}} = P/3.464h$

3.12. (a) $(\sigma_x)_{\text{elast.}} = 19.4F/h$, $(\sigma_x)_{\text{elem.}} = 20.9F/h$
$(\tau_{xy})_{\text{elast.}} = 5.2F/h$, $(\tau_{xy})_{\text{elem.}} = 2.8F/h$
(b) $(\sigma_x)_{\text{elast.}} = 0.176F/h$, $(\sigma_x)_{\text{elem.}} = 0.5F/h$
$(\tau_{xy})_{\text{elast.}} = 0.0305F/h$, $(\tau_{xy})_{\text{elem.}} = 0.433F/h$

3.13. $k = 3.4$

3.14. $\sigma_c = 1038.46$ MPa

3.17. $\sigma_x = \sigma_y = E\alpha T_1/(v-1)$, $\varepsilon_z = 2v\alpha T_1/(1-v) + \alpha T_1$

Chapter 4

4.2. (a) $\sigma_{yp} = 116.716$ MPa, (b) $\sigma_{yp} = 152.64$ MPa, (c) $\sigma_{yp} = 134.9$ MPa

4.4. $t = 8.37$ mm

4.5. (a) $d = 27.95$ mm, (b) $d = 36.8$ mm

4.6. (a) $N = 17.3$, (b) $N = 2.87$

4.8. (a) $p = 5.6$ MPa, (b) $p = 6.466$ MPa, (c) $p = 5.6$ MPa

4.9. (a) No; (b) yes

4.10. (b) $\sigma_1 = 75$ MPa, $\sigma_2 = -300$ MPa

4.14. $p = 66.67$ MPa

4.19. $\tau_{\text{max}} = 274.26$ MPa, $\theta_{\text{max}} = 4.76°$

Chapter 5

5.2. (a) $\sigma_x = \dfrac{p}{tL}\left[\dfrac{3xy}{5h} + \dfrac{2x^3y}{h^3} - \dfrac{4xy^3}{h^3}\right]$

$\sigma_y = \dfrac{-p}{tL}\left[-\dfrac{x}{2} + \dfrac{3xy}{2h} - \dfrac{2xy^3}{h^3}\right]$

$\tau_{xy} = \dfrac{p}{tL}\left[\dfrac{h}{80} + \dfrac{3}{4}\dfrac{x^2}{h} - \dfrac{3y^2}{10h} - \dfrac{3x^2y^2}{h^3} + \dfrac{y^4}{h^3}\right]$

(b) $\sigma_x = px^3/Lth^2$

(c) $(\sigma_x)_{elast.} = 0.998(\sigma_x)_{elem.}$

5.4. $e = 4R/\pi$

5.7. $p = 3.88$ kN/m

5.8. $P = 9320$ N

5.9. $R = -13pL/32$

5.16. (a) $P = 18.967$ kN, (b) $\sigma_{\theta 2} = -80.498$ MPa

5.17. $e_p = 1459.827 \, P/E$ m

5.19. (a) $\sigma_\theta = -152.865P$, (b) $e_p = 215.65P/E$ m

Chapter 6

6.1. (a) $\tau_e > \tau_c$; (b) $M_e > M_c$

6.2. $k = G\theta/2a^2(b-1)$

6.5. $\theta_A = aM_t/2r^4G$, $\theta_B = 2\theta_A$

6.6. $\tau_{max} = 15\sqrt{3} \, M_t/2h^3$, $\tau_{min} = 0$, $\theta = 15\sqrt{3} \, M_t/Gh^4$

6.9. $\tau_{max} = 76.8$ MPa, $\theta = 0.192$ rad/m

6.11 (a) $C = 2.1 \times 10^{-7}G$, $\tau_{max} = 112,860M_t$
(b) $C = 8 \times 10^{-7}G$, $\tau_{max} = 3,252,032 \, M_t$
(c) $C = 2.7 \times 10^{-7}G$, $\tau_{max} = 88,643M_t$

6.12. $\theta = 0.1617$ rad/m

6.15. $\tau_{max} = 5.279$ MPa, $\theta = 0.13062$ rad/m

6.16. $\tau_1 = \tau_5 = \tau_{max} = 51.77$ MPa, $\theta = 0.01914$ rad/m

Chapter 7

7.1. $v_{max} = 3320 PL^3/E$

7.2. $v_{max} = v_3 = 0.0065 PL^4/EI$

7.5. $\tau_B = 0.0107 G\theta$

7.7. $v(L) = 7PL^3/36EI$

7.9. $M_B = M_D = 0.02 PL$, $M_C = 0.08 PL$

7.11. $M_A = -8.9$ kN·m, $M_B = 12.3$ kN·m, $M_C = 2.7$ kN·m

7.14. $\{\sigma_x, \sigma_y, \tau_{xy}\}_a = \{66.46, 6.65, -92.12\}$ MPa

Chapter 8

8.2. (a) $r_x = \left[\dfrac{2n^2 a^2 \sigma_\theta + \Delta\sigma_\theta(n^2+1)n^2 a^2}{\Delta\sigma_\theta(n^2+1) + 2\sigma_\theta n^2} \right]^{1/2}$, (b) $r_x = 27.12$ mm

8.4. (a) $p_i = 1.6 p_0$, (b) $p_i = 1.16 p_0$

8.6. (a) $t = 0.825 d_i$, (b) $\Delta d = 0.0074$ mm

8.8. (a) $\sigma_z = 2\nu \dfrac{a^2 p_i - b^2 p_0}{b^2 - a^2}$, (b) $\varepsilon_z = -\dfrac{2\nu}{E} \dfrac{a^2 p_i - b^2 p_0}{b^2 - a^2}$

8.10. $M_t = 5022.356$ N·m

8.12. $\Delta d_s = 0.23\delta_0$ m

8.14. $\sigma_{\theta, max} = 1.95 E_b E_s (T_2 - T_1)/(E_s + 3E_b)10^5$

8.15. $M_t = 1017.36$ N·m

8.19. (a) $\sigma_{\theta, max} = 554.58$ MPa

Chapter 9

9.2. $v = \dfrac{P_1}{k\left[1 + 4(\pi/\beta L)^4\right]} \sin\dfrac{2\pi x}{L}$

9.4. $v_{max} = 3.375$ mm, $\sigma_{max} = 103.023$ MPa

9.6. $v_A = \dfrac{P}{4\beta k L}\left[f_3(\beta a) - f_3(\beta b) - 2\beta L f_4(\beta b) + 4\beta a\right]$

9.8. $v = -M_L f_2(\beta x)/2\beta^2 EI$

9.10. (a) $v = 10$ mm; (b) $v_L = 15$ mm, $v_R = 5$ mm

9.11. $v_C = 186 \times 10^{-6}$ m

9.12. $v_C = 2.81 \times 10^{-8}$ m, $\theta_E = 5 \times 10^{-8}$ rad

Chapter 10

10.3. $v_p = 11 Pc_1 a^4/12E + 7 Pc_2 a^3/3E + Pa^3/3EI_2$

10.4. $e_D = Fab^2/8EI$

10.6. $e = \dfrac{PR^3}{r^4}\left(\dfrac{1}{E} + \dfrac{0.226}{G}\right)$

10.9. $R_f = 4pL/10$, $M_f = pL^2/15$, $R_r = pL/10$

10.10. $R_{AV} = 3(\lambda+1)pL_1/2(3\lambda+4)$, $R_{AH} = \lambda pL_1^2/4(3\lambda+4)L_2$,
$\lambda = E_2 I_2 L_1/E_1 I_1 L_2$

10.12. $N_A = P/2\pi$, $M_A = Pr/4$

10.14. $v = \dfrac{2PL^3}{\pi^4 EI}\displaystyle\sum_{n=1}^{\infty}\dfrac{\sin(n\pi c/L)\sin(n\pi x/L)}{n^2 + (kL^4/\pi^4 EI)}$

10.15. $v = Pc^2(L-c)^2/4EIL$

Chapter 11

11.2. (a) $\sigma_{cr} = 33.91$ MPa, (b) $\sigma_w = 33.534$ MPa

11.4. $L_e = L$

11.6.   (a) $\Delta T = \delta/2\alpha L$, (b) $\Delta T = (\delta/2\alpha L) + (\pi^2 I/4L^2 A\alpha)$

11.8.   Bar BC fails as a column; $P_{cr} = 1296.629$ N

11.10. $\sigma_{cr} = 59.27$ MPa

11.12.   (a) $\sigma_{max} = 93.705$ MPa, $v_{max} = 0.1506$ m
          (b) $\sigma_{max} = 51.019$ MPa, $v_{max} = 0.2275$ m

11.14. $P = 0.89\pi^2 EI/L^2$

11.16. $P_{cr} = 12 EI/L^2$

11.17. $P_{cr} = 1.7\pi^2 EI_1/L^2$

11.18. $P_{cr} = 9EI_1/4L^2$

11.19. $v = \dfrac{4pL^4}{\pi^5 IE} \displaystyle\sum_{n=1,3,5,\ldots}^{\infty} \dfrac{1}{n^3(n^2-b)} \sin\dfrac{n\pi x}{L}$

11.24. $P_{cr} = 16EI_1/L^2$

Chapter 12

12.2. $\alpha = 42.68°$

12.3. $v = \lambda\left[\dfrac{(L-x)^{2/n+2}}{(2/n+1)(2/n+2)} - \dfrac{L^{2/n+2}}{(2/n+1)(2/n+2)} + \dfrac{L^{2/n+1}}{2/n+1}x \right]$

12.5. $\sigma_{max} = 3Mh/4I$

12.7. $M = 11ah^2\sigma_{yp}/54$

12.10. $P = 46.181$ kN

12.12. $P_u = 9M_u/2L$

12.14.   (a) $M_u = 2M_{yp}$, (b) $M_u = 16b(b^3-a^3)M_{yp}/3\pi(b^4-a^4)$

12.19. $t_0 = 6.3$ mm

Chapter 13

13.1. (a) $w = -(M_b - \nu M_a)x^2/2D(1-\nu^2)$
$\qquad - (M_a - \nu M_b)y^2/2D(1-\nu^2)$
(b) $w = M_a(x^2 - y^2)/2D(1-\nu)$

13.2. $\dfrac{144Pa^2b^2}{D\pi^6} \displaystyle\sum_{m=1}^{\infty} \sum_{n=1}^{\infty} \dfrac{1}{mn(b^2m^2 + 4a^2n^2)^2} \sin\dfrac{m\pi x}{a} \sin\dfrac{n\pi y}{b}$

13.3. $w = M_0(a^2 - r^2)/2D(1+\nu), \ \sigma_{r,\max} = \sigma_{\theta,\max} = 6M_0/t^2$

13.5. $\dfrac{p(a^2 - r^2)}{64D}\left(\dfrac{5+\nu}{1+\nu}a^2 - r^2\right)$

13.6. $N_\phi = \dfrac{\gamma a^2}{6}\dfrac{1-\cos\phi}{1+\cos\phi}(1+2\cos\phi),$
$N_\theta = \dfrac{\gamma a^2}{6}\dfrac{1-\cos\phi}{1+\cos\phi}(5+4\cos\phi)$

13.7. $N_\phi = \dfrac{pa}{a\sin\phi + r}\left(\dfrac{a}{2}\sin\phi + r\right), \ N_\theta = \dfrac{pa}{2}$

13.9. $N_\theta = -pa\sin\theta, \ N_x = -\dfrac{1}{a}(L-x)px\sin\theta, \ N_{x\theta} = -p(L-2x).$
$\cos\theta$

13.10. $N_\theta = \gamma a^2(1-\cos\theta), \ N_{x\theta} = -\gamma ax\sin\theta,$
$N_x = \dfrac{1}{2}x^2\cos\theta + \nu\gamma a^2(1-\cos\theta) - \dfrac{1}{24}\gamma L^2\cos\theta$

Index